Online Gaming in Context

There is little question of the social, cultural and economic importance of video games in the world today, with gaming now rivalling the movie and music sectors as a major leisure industry and pastime. The significance of video games within our everyday lives has certainly been increased and shaped by new technologies and gaming patterns, including the rise of home-based games consoles, advances in mobile telephone technology, the rise in more 'sociable' forms of gaming and of course the advent of the Internet.

This book explores the opportunities, challenges and patterns of game play and sociality afforded by the Internet and online gaming. Bringing together a series of original essays from both leading and emerging academics in the field of game studies, many of which employ new empirical work and innovative theoretical approaches to gaming, this book considers key issues crucial to our understanding of online gaming and associated social relations, including: patterns of play, legal and copyright issues, player production, identity construction, gamer communities, communication, patterns of social exclusion and inclusion around gender, race and disability and future directions in online gaming.

Garry Crawford is a Senior Lecturer in Sociology at the University of Salford, UK. His teaching and research primarily focus upon audiences, fan cultures and the everyday uses of media technologies. He is the author of *Consuming Sport* (2004), and the co-author of the second edition of *Introducing Cultural Studies* (2008, with B. Longhurst, G. Smith, G. Bagnall and M. Ogborn) and the *Sage Dictionary of Leisure Studies* (2009, with T. Blackshaw).

Victoria K. Gosling is a Lecturer in Sociology at the University of Salford, UK. Her teaching and research interests encompass gender, urban regeneration, social exclusion, leisure, popular culture and digital games. She is the former editor of the British Sociological Association newsletter *Network*, an editorial board member for the journal *Sociology* and the former post-graduate forum convenor of the BSA.

Ben Light is a Professor of Digital Media in the School of Media, Music and Performance, and a member of the Communication, Cultural and Media Studies Research Centre at the University of Salford, UK. His current research interests centre on analysing the development and use of social media in everyday life.

Routledge Advances in Sociology

1. **Virtual Globalization**
   Virtual spaces/tourist spaces
   Edited by David Holmes

2. **The Criminal Spectre in Law, Literature and Aesthetics**
   Peter Hutchings

3. **Immigrants and National Identity in Europe**
   Anna Triandafyllidou

4. **Constructing Risk and Safety in Technological Practice**
   Edited by Jane Summerton and Boel Berner

5. **Europeanisation, National Identities and Migration**
   Changes in boundary constructions between Western and Eastern Europe
   Willfried Spohn and Anna Triandafyllidou

6. **Language, Identity and Conflict**
   A comparative study of language in ethnic conflict in Europe and Eurasia
   Diarmait Mac Giolla Chríost

7. **Immigrant Life in the U.S.**
   Multi-disciplinary perspectives
   Edited by Donna R. Gabaccia and Colin Wayne Leach

8. **Rave Culture and Religion**
   Edited by Graham St John

9. **Creation and Returns of Social Capital**
   A new research program
   Edited by Henk Flap and Beate Völker

10. **Self-Care**
Embodiment, personal autonomy and the shaping of health consciousness
Christopher Ziguras

11. **Mechanisms of Cooperation**
Werner Raub and Jeroen Weesie

12. **After the Bell**
Educational success, public policy and family background
Edited by Dalton Conley and Karen Albright

13. **Youth Crime and Youth Culture in the Inner City**
Bill Sanders

14. **Emotions and Social Movements**
Edited by Helena Flam and Debra King

15. **Globalization, Uncertainty and Youth in Society**
Edited by Hans-Peter Blossfeld, Erik Klijzing, Melinda Mills and Karin Kurz

16. **Love, Heterosexuality and Society**
Paul Johnson

17. **Agricultural Governance**
Globalization and the new politics of regulation
Edited by Vaughan Higgins and Geoffrey Lawrence

18. **Challenging Hegemonic Masculinity**
Richard Howson

19. **Social Isolation in Modern Society**
Roelof Hortulanus, Anja Machielse and Ludwien Meeuwesen

20. **Weber and the Persistence of Religion**
Social theory, capitalism and the sublime
Joseph W. H. Lough

21. **Globalization, Uncertainty and Late Careers in Society**
Edited by Hans-Peter Blossfeld, Sandra Buchholz and Dirk Hofäcker

22. **Bourdieu's Politics**
Problems and possibilities
Jeremy F. Lane

23. **Media Bias in Reporting Social Research?**
The case of reviewing ethnic inequalities in education
Martyn Hammersley

24. **A General Theory of Emotions and Social Life**
Warren D. TenHouten

25. **Sociology, Religion and Grace**
Arpad Szakolczai

26. **Youth Cultures**
Scenes, subcultures and tribes
Edited by Paul Hodkinson and Wolfgang Deicke

27. **The Obituary as Collective Memory**
Bridget Fowler

28. **Tocqueville's Virus**
Utopia and dystopia in Western social and political thought
Mark Featherstone

29. **Jewish Eating and Identity Through the Ages**
David Kraemer

30. **The Institutionalization of Social Welfare**
A study of medicalizing management
Mikael Holmqvist

31. **The Role of Religion in Modern Societies**
Edited by Detlef Pollack and Daniel V. A. Olson

32. **Sex Research and Sex Therapy**
A sociological analysis of Masters and Johnson
Ross Morrow

33. **A Crisis of Waste?**
Understanding the rubbish society
Martin O'Brien

34. **Globalization and Transformations of Local Socioeconomic Practices**
Edited by Ulrike Schuerkens

35. **The Culture of Welfare Markets**
The international recasting of pension and care systems
Ingo Bode

36. **Cohabitation, Family and Society**
Tiziana Nazio

37. **Latin America and Contemporary Modernity**
A sociological interpretation
José Maurízio Domingues

38. **Exploring the Networked Worlds of Popular Music**
Milieu cultures
Peter Webb

39. **The Cultural Significance of the Child Star**
Jane O'Connor

40. **European Integration as an Elite Process**
The failure of a dream?
Max Haller

41. **Queer Political Performance and Protest**
Benjamin Shepard

42. **Cosmopolitan Spaces**
Europe, globalization, theory
Chris Rumford

43. **Contexts of Social Capital**
Social networks in communities, markets and organizations
Edited by Ray-May Hsung, Nan Lin, and Ronald Breiger

44. **Feminism, Domesticity and Popular Culture**
Edited by Stacy Gillis and Joanne Hollows

45. **Changing Relationships**
Edited by Malcolm Brynin and John Ermisch

46. **Formal and Informal Work**
The hidden work regime in Europe
Edited by Birgit Pfau-Effinger, Lluis Flaquer and Per H. Jensen

47. **Interpreting Human Rights**
Social science perspectives
Edited by Rhiannon Morgan and Bryan S. Turner

48. **Club Cultures**
Boundaries, identities and otherness
Silvia Rief

49. **Eastern European Immigrant Families**
    Mihaela Robila

50. **People and Societies**
    Rom Harré and designing the social sciences
    Luk van Langenhove

51. **Social Theory in Contemporary Asia**
    Ann Brooks

52. **Foundations of Critical Media and Information Studies**
    Christian Fuchs

53. **A Companion to Life Course Studies**
    The social and historical context of the British birth cohort studies
    Michael Wadsworth and John Bynner

54. **Understanding Russianness**
    Risto Alapuro, Arto Mustajoki and Pekka Pesonen

55. **Understanding Religious Ritual**
    Theoretical approaches and innovations
    John Hoffmann

56. **Online Gaming in Context**
    The social and cultural significance of online games
    Garry Crawford, Victoria K. Gosling and Ben Light

Online Gaming in Context

The social and cultural significance of online games

Edited by
Garry Crawford, Victoria K. Gosling and Ben Light

Routledge
Taylor & Francis Group
LONDON AND NEW YORK

First published 2011 by Routledge
2 Park Square, Milton Park, Abingdon, Oxon, OX14 4RN

Simultaneously published in the USA and Canada
by Routledge
711 Third Avenue, New York, NY 10017

Routledge is an imprint of the Taylor & Francis Group

First issued in paperback 2013

British Library Cataloguing in Publication Data
A catalogue record for this book is available from the British Library

Library of Congress Cataloguing in Publication Data
Crawford, Garry.
Online gaming in context : the social and cultural significance of online games/
edited by Garry Crawford, Victoria K. Gosling and Ben Light.
p. cm.—(Routledge advances in sociology)
ISBN 978–0–415–55619–4 (hardback)
1. Internet games—Social aspects. I. Gosling, Victoria K. II. Light, Ben. III. Title.
GV1469.15.C73 2011
306.4'87—dc22
2010050590

ISBN 13: 978–0–415–55619–4 hbk
ISBN 13: 978–0–203–86959–8 ebook
ISBN 13: 978-0-415-71497-6 pbk

Typeset in Baskerville by Swales & Willis Ltd, Exeter, Devon

Contents

Contributors

Douglas Brown is a Lecturer at Brunel University where he examines the role of the suspension of disbelief in video games. His interest centres around narrative and storytelling in modern video games, and his work has been published in several journals, with book chapters forthcoming. He also works for Square-Enix and is credited on many of their recent titles. His website is online at www.geekademic.co.uk

José Carlos Ribeiro has a Ph.D. degree from the Communication and Contemporary Culture Post-Graduation Program of the Federal University of Bahia, where he currently stands as Associate Researcher; parallel to this, he works as Associate Professor in the Department of Psychology of the Federal University of Bahia, focusing his works on the intersection of the Communication and Psychology fields.

Garry Crawford is a Senior Lecturer in Cultural Sociology at the University of Salford (UK). His teaching and research primarily focus upon audiences, sport, fan cultures, video gamers and the everyday uses of media. Garry is the author of *Consuming Sport* (Routledge 2004), and the co-author of the second edition of *Introducing Cultural Studies* (Pearson 2008, with B. Longhurst, G. Smith, G. Bagnall, and M. Ogborn) and the *Sage Dictionary of Leisure Studies* (Sage 2009, with T. Blackshaw). Garry is Director of the University of Salford Digital Cluster (digital.salford.ac.uk), Publications Director of the British Sociological Association (BSA) and review editor for the journal *Cultural Sociology*.

Fern M. Delamere has been a Professor at Concordia University, Montreal, since 2004. She completed her Ph.D. at the University of Waterloo. Fern is a member of Concordia's Technoculture, Art and Games Research Center. Her research is focused on leisure, media cultures and virtual worlds. More specifically, her research explores online leisure contexts and the digitally mediated social networks of marginalized social groups related to disability and gender. Her publications, like her research, are interdisciplinary. She has published extensively within her primary field of leisure studies, but also has other publications in media communications and game studies.

Anders Drachen is a user experience specialist and industry consultant with Game Analytics Technologies, and currently a visiting researcher at the Institute for Informatics at Copenhagen Business School. His research is focused on usability, user experience, user-testing and user-driven innovation. This research spans social, ethnographic and qualitative approaches towards empirical, statistical and computational data analysis. He collaborates with a range of game development companies to improve user-oriented testing in game development.

Astrid Ensslin is a Lecturer in Digital Communication at the School of Creative Studies and Media, Bangor University. Her main research interests are in the areas of digital media (especially digital literature, video games and virtual worlds), semiotics and discourse analysis, and language in the (new) media. She is founding editor of *Journal of Gaming and Virtual Worlds* and Co-Investigator of the Leverhulme Digital Fiction International Network. Her main publications include *The Language of Gaming: Discourse and Ideology* (Palgrave Macmillan 2011), *Creating Second Lives: Community, Identity and Spatiality* (co-edited with Eben Muse, Routledge 2011), *Canonizing Hypertext: Explorations and Constructions* (Continuum 2007), and *Language in the Media: Representations, Identities, Ideologies* (co-edited with Sally Johnson, Continuum 2007).

Thiago Falcão is a Ph.D. Student at the Contemporary Communication and Culture Postgraduate Program of the Federal University of Bahia, where he is both member of the Research Group on Society, Digital Technology and Interaction (GITS) and of the Research Group on Cybercities (GPC). He is currently researching narratives and cognitive processes involved in online gaming. He is also editor of the peer-reviewed journal *Realidade SintÈtica*; the first Portuguese-language based resource to fully dedicate itself to issues regarding research on video game culture.

René Glas is an Assistant Professor in New Media and Digital Culture at the Department of Media and Culture Studies at the Utrecht University. He is a founding member of the Utrecht University's Center for the Study of Digital Games and Play.

Victoria K. Gosling is a Lecturer in Sociology at the University of Salford. Her doctoral research focused on gender, urban regeneration and social exclusion, and she continues to teach and publish in these fields. Additionally, she teaches and researches leisure and popular culture and, in particular, she published on female sport fans and gamers. Victoria Gosling is an editorial board member for the journal *Sociology*, and former editor of the British Sociological Association newsletter *Network*.

Kristine Jørgensen is a Postdoctoral Research Fellow at the Department of Information Science and Media Studies, University of Bergen, Norway, and member of the board of Joingame, the Norwegian network for games research and development since 2009.

Aphra Kerr is a Lecturer at the National University of Ireland Maynooth where she teaches courses on the sociology and culture of media and technology. She has been researching the production, consumption and regulation of digital media on European and nationally funded projects for the past 10 years. Her publications include *The Business and Culture of Digital Games* (Sage 2006) and she has had journal articles published in *Media, Culture and Society*, *New Media and Society*, *The International Journal of Cultural Studies*, *Convergence*, *Fibreculture and Information Communication and Society*. Aphra is a committee member of Women in Games (Europe), an editorial board member of *Popular Communication and Eludamos: Journal for Computer Game Culture*, a founding member of DIGRA and she runs the community website www.gamedevelopers.ie

Ben Light is Professor of Digital Media in the School of Media, Music and Performance at the University of Salford. His research concerns how people interact with socio-technical sets of arrangements on an everyday basis. He began investigating this area in the 1990s, generally in the workplace, and since then has moved to consider arrangements in wider society. His current research agenda centres on analysing the Internet and digital games which includes ethnographies of interactions with *Gaydar*, *SingStar*, *Facebook* and *YouTube*.

Holin Lin is a Professor of the Department of Sociology at National Taiwan University. Her major research interests include computer-mediated communication, information technology and society and gender studies. She has written articles on the social dynamics of in-game communities, game tips sharing behaviours, cash trades of in-game assets, formation of norms and deviance negotiation in MMORPGs, and gendered gaming experience in different physical spaces. Currently, she serves on the editorial board of *Games and Culture*.

Keith Massie is a Ph.D. candidate at the University of Utah and Visiting Assistant Professor at New Mexico State University. His research has been presented at numerous regional, national and international conference venues. He has published in *Convergence: The International Journal of Research into New Media Technologies*, *The Rocky Mountain Communication Review* and *The Journal of Radio Studies*. His research focuses on new media, deconstruction, media ecology, the intersection of technology and culture and race/gender representation in the media.

Esther MacCallum-Stewart is a Lecturer at The University of Chichester and a Research Fellow at SMARTlab, The University of East London. She is also Vice-President of DiGRA (the Digital Games Research Association). Her work investigates how players interpret and understand the narratives within games, and how they form social communities based upon these structures. She has written on roleplaying in games, protests by players, the interpretation of gender in games and gaming narratives. She is currently writing a book about the role of games narratives and players in online gaming.

Frans Mäyrä has studied the relationship between culture and technology from the early nineties. He is Professor and the head of Games Research Laboratory in the University of Tampere where he is currently teaching, researching and heading numerous research projects in the study and development of games, interactive media and digital culture. He has also served as the founding President of Digital Games Research Association, DiGRA. His publications include *An Introduction to Game Studies* (2008), as well as a number of other books and articles on games, technology, science fiction and fantasy.

Torill Elvira Mortensen is an Associate Professor at the IT University of Copenhagen, Denmark and member of the board of DiGRA since 2007.

Christopher A. Paul is an Assistant Professor at Seattle University and his research focuses on the intersections of rhetoric and video games. He is particularly interested in the discourse of massively multiplayer online games and how discursive practices shape the terrain for interaction with games. He has published work in *Games and Culture*, the *Journal of Virtual Worlds Research*, the *International Journal of Role-Playing* and *First Monday*.

Celia Pearce is a game designer, author, researcher, teacher, curator and artist, specialising in multiplayer gaming and virtual worlds, independent, art and alternative game genres, as well as games and gender. She currently is an Assistant Professor of Digital Media in the School of Literature, Communication and Culture at Georgia Tech, where she also directs the *Experimental Game Lab and the Emergent Game Group*. Her game designs include the award-winning virtual reality attraction *Virtual Adventures* and the *Purple Moon Friendship Adventure Cards for Girls*. She is the author of *The Interactive Book* (Macmillan 1997) and *Communities of Play: Emergent Cultures in Multiplayer Games and Virtual Worlds* (2009 MIT Press) as well as numerous papers and book chapters. She has also curated new media, virtual reality and game exhibitions and is currently Festival Chair for IndieCade, an international independent games festival and showcase series. She is a co-founder of the *Ludica* women's game collective. Her work can be found online at http: //cpandfriends.com/

Jeffrey Philpott is a tenured Assistant Professor in the Communication Department at Seattle University, where he has taught since 1992. With a background in rhetorical theory and criticism, most of his research has focused on the epistemology of public discourse, particularly discourse surrounding major public events, or the construction of identity in social protest movements. The paper in this book represents a new foray into the analysis of public discourse in online gaming communities.

Neil Randall teaches digital media and communication design in the Department of English at the University of Waterloo, Canada, where he researches the intersections among rhetorical theory, semiotics and technology, particularly in the field of Games Studies. He has developed seven historical simulation boardgames, published by GMT Games and Compass Games in the

US, and is currently contracted to design two historical simulation games for Compass. He is a long-time contributing editor for PC Magazine (New York), in which he has published numerous columns, features, reviews and technology explanations. For several years he was a primary contributor to the *Compute* line of magazines and their successors, publishing columns, features and reviews on video and computer games.

Luca Rossi is a Research Fellow at the department of Communication Studies at University of Urbino 'Carlo Bo', Italy. His interests are in the emergence of social structures in massively multiplayer online games and games-related cultures. He is author of the chapter 'Identity, Interactivity and Interaction' in *VideoMondi* (2008).

Kate E. Taylor is a Lecturer in Visual Culture at the University of Bangor, Wales. Her main research interests are the visual culture of the Japanese Colonial Empire and the visual representation of prostitution and sex work in Asia since 1900. She has published on a variety of topics such as representing the Nanking massacre, de-Westernising film studies and the figure of the prostitute in 1950s Japanese cinema and is closely involved with a variety of global women's organisations. She is currently working on a monograph study examining Japanese colonial propaganda in Korea, Singapore, Taiwan and Manchuria.

Chuen-Tsai Sun is a Joint Professor of Department of Computer Science and Graduate Institute of Education, National Chiao Tung University, Taiwan. He has been working in the fields of digital games, digital learning, social network analysis and agent-based social simulation. He has published digital-game-related articles on players' altruistic behaviour, negotiation of social norms in game worlds, dynamics of player guilds, onlooker behaviour in gaming arcades and free-to-play online game models, among others.

Acknowledgements

This book developed initially out of the game stream of the 2008 Association of Internet Researchers (AoIR) conference in Copenhagen, where the high quality of papers within the stream inspired the editors to compile a selection of presented work in an edited collection. The book you have before you is the culmination of work that evolved out of papers presented at that conference, as well as a few invited contributions from authors not present in Copenhagen, but who nevertheless, we felt would enhance this volume. Therefore the editors would like to acknowledge and thank the Association of Internet Researchers, and most especially, Brian Loader from the University of York, who was chiefly responsible for the organization of the conference that year.

1 The social and cultural significance of online gaming

Garry Crawford, Victoria K. Gosling and Ben Light

The chapters in this book cover scholarship on a wide range of subjects such as communities, empowerment, consumption, game aesthetics and forms, identities, language, productivity and patterns of exclusion, by authors from a number of fields including communication studies, English, media studies, game studies, journalism, informatics, religious studies, sociology and visual culture. The uniting theme is, of course, online games and their game play.

Online games, i.e. games (and most typically what we would term 'video games') played on or over an Internet connection, have, in a relatively short period, become a significant cultural phenomenon. The most obvious, and frequently discussed, examples of this being the large number of massively multiplayer online role playing games (frequently abbreviated to MMORPGs or MMOs), which attract player communities larger than the population of many small- to medium-sized countries. The most popular game currently is *World of Warcraft*, which (at time of writing) has more than 11 million registered players. However, online gaming is not just about MMORPGs. Beyond these contemporary examples, video gaming and the Internet have a long and mutual (even, at times, symbiotic) relationship, with video games providing a key early use for computer networks, which in turn have played a significant role in video game development and culture, by allowing games to be accessed, distributed, modified and discussed over network connections. But the history of the Internet, and its links to video gaming, is rather complex.

This chapter therefore provides a brief introduction to the study of online gaming. It begins with a short history of the Internet and online gaming, and then (following Burn and Carr 2006) considers three key areas of debate in game research – the ludic, representational and communal aspects of online gaming. The final section of the chapter provides a brief overview of the structure and content of this book.

A brief history of online gaming

As any historian of worth would tell you, *all* history is contested. History is made, and then told by people, and people are inevitably subjective. Therefore, there is neither one history of the Internet nor that of video games, but rather a variety of fragments of information, some more or less contested than others, which are frequently drawn together to tell the 'story'. Inevitably though, there will be differences, depending on who is telling the story. In the case of both the Internet and video games, their history is more complex because of the developmental paths both have followed.

Contrary to popular myths, most technologies are not, one day, simply 'invented' by one or more brilliant scientists. More commonly, technologies advance and develop in small increments, as it would appear was the case with the Internet. It is generally accepted that the origins of the Internet can be found in the loose conglomerate of computer networks that developed around the Advanced Research Projects Agency Network (ARPANET), created by the Defense Advanced Research Projects Agency (DARPA) of the United States (US) Department of Defense in the late 1960s (Gere 2008). This was helped further by the development of a number of Internet protocols for the transmission of data in the 1970s, and most notably 'Ethernet' developed at Harvard University in 1974.

Contrary to its origins, advancement of the Internet was most notably driven by non-military needs and personal, and in particular newsgroups, bulletin-boards and multi-user dungeons or domains (MUDs), which helped establish new and more common applications of this technology. The increasing non-military use of the ARPANET led to a formal division into military and non-military networks, and then to its dissolution in 1989 (Gere 1989). Modems (which had been invented in the 1960s) allowed increasingly large networks to develop around educational and industrial, as well as military, usage. Access to modems, and their ease of use, was significantly assisted by the advent of the world wide web, first devised as an academic tool by Swiss physicist Tim Berners-Lee in the late 1980s, and then later the first web browser, 'Mosaic', developed by the National Center for Supercomputing Applicants at the University of Illinois in 1993 (Gere 2008). By the mid-1990s, the Internet was rapidly spreading out of universities and industries into people's homes.

It appears that, whenever and wherever humanly possible, people will always strive to find ways of playing with, and gaining pleasure from, new technologies. Even the most 'serious' of technologies, such as weapons, are utilized as the basis of sport, leisure and fun – such as target shooting – as it seems was the case for the early days of the Internet. Initially developed for military purposes, and finding immense value in the workplace, it was initially social networking, video gaming and, then later, music, video and pornography, which were crucial in cementing the significant place of the Internet within our culture.

Of particular interest here are MUDs. The origins of MUDs can be found in pen and paper (PnP) role-playing games (RPGs), such as and most notably, *Dungeons and Dragons*, which, though not the first, was (and continues to be) the most popular system within this genre of gaming. The rules of *Dungeons and Dragons* were first published in 1974, devised by Gary Gygax and Dave Arneson, but this was simply a development and greater formalization (and commercialization) of gaming trends and systems that had been growing in popularity through the 1960s, most notably on American university campuses. But PnP RPGs (particularly *Dungeons and Dragons*) have their origins in the much older practice of war gaming and, of course, the practice of acting out roles for fun is timeless (Fine 1983).

Dungeons and Dragons, and similar RPGs such as *Traveller*, *Call of Cthulhu* and *Tunnels and Trolls* were (and still are) normally played with other people in the same room, and hence are often referred to as face-to-face (FTF) or table-top RPGs. But RPGs (even before the publication of *Dungeons and Dragons*), and before them, war games, and other boardgames such as chess, were also played with other players at a distance, most commonly by post, frequently referred to as 'play by mail' (or PBMs), or in the case of chess, 'correspondence chess', which is centuries old. Hence, the rise of the Internet was concurrent with, if not pre-dated by, the increasing popularity of PnP RPGs; a leisure activity that already had a long history of gaming at a distance. Significantly, computer and Internet pioneers were drawn from the same demographic and located within the same locales (most notably universities) as PnP, play by mail and war gamers. Hence, it is of little surprise that one of the first non-military, (and back then) non-commercial, uses of evolving Internet technologies was RPGs.

What is generally seen as the first online RPG, *MUD* (which is where the genre takes its name from), was developed between 1978 and 1980 by, first, Roy Trubshaw, and then, Richard Bartle, at the University of Essex (Fox 2006). *MUD* was a text-based adventure game in which players were presented with written descriptions of an environment, which they responded to by typing in a number of pre-set options – such as 'go west' or 'look'. The first text-based video game (which was single player and offline), called *Adventure*, had been developed a few years earlier in 1975 by Will Crowther, with others such as *Zork*, following quickly. Here, the Internet also played a key role in the development and spread of these games; games such as *Adventure*, *Zork* and those that pre-dated them, such as *Spacewar!*, were distributed over fledgling computer networks to others who played with, added to and expanded these games, as well as those taking them as inspirations for developing their own, new, video games.

Taylor (2006: 23) in her discussion of *MUD* describes this game as signalling 'a new turn in which multi-user spaces were to become one of the most innovative developments within internet technologies and certainly a genre that excited many computer users'. *MUD* also paved the way for a closely linked phenomena, that of multi-user 'persistent' (or sometimes called 'virtual') worlds (Taylor 2006), such as *Second Life*.

Probably the first multi-user persistent world was *TinyMUD* developed by Carnegie Mellon student James Aspnes in 1989. *TinyMUD* was different from previous MUDs as it did away with much of the 'game' elements of the genre, instead focusing most notably on socializing (Keegan 1997; Taylor 2006). *TinyMUD* in turn paved the way for 'world-building' activities through MUD object oriented (MOO), developed by Stephen White and released in 1990. *MOO*, though still text-based, allowed users to create new 'environments', such as text described rooms and objects, which they and others could interact with. But a few years prior to this, in 1985, Lucasfilm Games (later to be renamed Lucas Arts in 1990) had developed the first online 'graphical' persistent world called *Habitat*.

The 1990s saw a rapid rise in the number and advancement of both graphical online persistent worlds and games. These included the graphic-based online chat room, *OnLive Traveller*, and increasingly video games, such as *Diablo* (released by Blizzard in 1996) were allowing gamers to play online with others. But it was *Ultima Online* in 1997 that safely secured the popularity of MMOs, adding massively multiplayer and world-building elements to an already very well established and successful series of computer-based fantasy RPGs, the *Ultima* series developed by Richard Garriott for Origin System games.

The previous decade or so has increasingly secured the importance and location of both the Internet and video gaming within our social and cultural lives, as well as strengthening the relationship between these technologies. Video gaming is now a multi-billion dollar industry, and a recent poll by the Entertainment Software Association (2007) suggested that 67 percent of the heads of households in America 'play computer or video games'. Global video games sales are now at levels comparable to box-office cinema receipts and more video games are now sold in the United States and United Kingdom than books (Bryce and Rutter 2006). Similarly, by 2009, almost three-quarters of the entire US population were 'Internet users' (Nielsen 2010). Certainly for those of us in the Western world, it is hard to imagine the domains of work or leisure without the Internet, which, for most of us, has become an integral part of our everyday lives.

The relationship between video games and the Internet is most vividly illustrated by MMORPGs, and this book is testament to that, including several chapters on MMORPGs by Douglas Brown, Esther MacCallum-Stewart, Keith Massie, Christopher Paul, Jeffrey Philpott and Celia Pearce. But beyond the long list of current and previous massively multiplayer games, the Internet has provided a rich history of gaming opportunities, such as allowing the conversion of boardgames into online digital form (such as those discussed by Neil Randall in this book), the exploration of virtual worlds (like *Second Life* considered by Fern M. Delamere in this book), the production, distribution and play of 'serious' games (such as *Wordslinger* as outlined by Kate Taylor in this book) and game modifications (such as Aphra Kerr's discussion of gamer productivity and innovation, and Holin Lin and Chuen-Tsai Sun's consideration of the construction and use of private game servers).

It is also worth noting that online gaming is not restricted to PCs, and even the current generation of video game consoles (such as the Nintendo Wii, Sony Playstation 3 and Microsoft Xbox 360), a form of hardware that was once considered paradigmatically different to the Internet-linked home computer, are making the Internet, online content and play a fundamental part of their gaming experience. Furthermore, in this book, Frans Mäyrä alerts us to the fact that online gaming no longer has to be location-specific, as online gaming has now become mobile, such as by being played on cellular (mobile) telephones.

It is apparent that online gaming forms an important part of (at least some) peoples' everyday lives and identities, and is important and worthy of academic consideration. As Castronova (2005) has argued, online games (or 'synthetic worlds' as he calls them) are important because they may have economic consequences outside of the game. As he writes:

> synthetic worlds . . . are now worthy of study . . . Once one recognizes that a silver piece in Sabert's [his game character] world can have value just like a US Dollar, one must realize that a silver piece is not merely like money, it is money
>
> (Castronova 2005: 47, emphasis in original)

But beyond their economic significance, online games also have a much wider and frequent social and cultural impact and significance, such as providing help for women suffering from domestic violence (Taylor, Chapter 16, this book), a sense of community and identity (Pearce, Chapter 10) or social networks and capital for people with disabilities (Delamere, Chapter 14), as well as so much more.

In considering the social and cultural significance of online games, Burn and Carr (2006) highlight what they see as three key motivations for playing online games, which they categorize as 'ludic', 'representation' and 'communal'. These three categories, we wish to suggest, are not only useful ways of categorizing types of gamer motivation but also highlight key areas of debate that have been frequent, if not dominant, in online gaming discussions and literatures.

Key debates on online gaming

Turning first to 'ludic' game motives and debates, the term ludic refers directly to aspects and definitions of game play. The starting point (or at least the fundamental assumption) of many ludic (or ludology) debates is the argument for considering video games as first and foremost 'games', and the use of which as 'play'.

There is now a very well established and extensive literature on what constitutes a 'game', be it computer-based or otherwise. Mostly, this work draws upon earlier literature, and most notably the writings of John Huizinga (1949), Roger Caillois (1962) and Brian Sutton-Smith (both alone and in his work with E.M. Avedon, i.e. Avedon and Sutton-Smith 1971), which attempts to formally set out the characteristics and boundaries of games and/or play. In relation to video games, these theories are brought together by Jesper Juul (2005) in his formalization of, what he calls, a 'classic game model', in which he defines a classic game as

> a rule-based system with a variable and quantifiable outcome, where different outcomes are assigned different values, the player exerts effort in order to influence the outcome, the player feels emotionally attached to the outcome and the consequences of the activity are negotiable
>
> (Juul 2005: 36).

Utilizing a ludic focused approach, such as Juul's classic game model, Burn and Carr (2006) suggest that this allows us to understand many of the pleasures and motivations for playing online games. For instance, many online games such as what is crucial is that these are understood as *reflections* of 'reality'. There may be similarities and continuities between the representation and what it represents, but these are crucially *not the same thing* (Longhurst *et al.* 2008). For instance, an oil painting or photograph may depict a bowl of fruit, but crucially they are not the bowl of fruit, they are a picture, or representation, of it.

It is this approach (that of 'representation') that has frequently been taken in understanding video games, game play and online presence. That is, when a person plays a video game, talks in an online chatroom or does something similar, what they are utilizing is a representation, or presentation, of themselves. This theorization is closely linked to, and informed by, the categorization of games and online worlds as 'spaces', which are (at least to some degree) demarcated from 'normal' everyday life.

For instance, for many writers, the Internet was, and continues to be, perceived as a place away from our ordinary lives, a space we travel to and explore, as Newman (2004) points out, often understood using the metaphor (drawn from popular fiction, such as most notably the work of William Gibson) of 'cyberspace', or as Turkle (1995) puts it, a 'life on the screen'. Similarly, the concept of space has been significant in the video game debates, with Aarseth (1997) suggesting that this was a universal and unifying theme within this literature. In particular, this debate has manifested itself, most recently, in arguments surrounding the idea of a gaming 'magic circle'.

The origins of 'magic circle' can be found in work of the Dutch historian Johan Huizinga, and in particular his 1944 (first published in English in 1949) book *Homo Ludens* – which roughly translates into English as 'human player'. For Huizinga, the magic circle is one example, within a list of others, of places where play takes place; places that are bound and defined by specific and temporary rules and norms, which do not necessarily apply outside of the circle. For some writers, such as Salen and Zimmerman (2004) and Juul (2005), this has proved a useful tool for framing and understanding patterns of play and gamer interactions. However, others argue that it has become an 'unproductive orthodoxy' (Neiborg and Hermes 2008: 135) within video game research. Most commonly, the argument made against the use of the magic circle, or similar concepts, is how it is used to draw distinctions between play and everyday life. For instance, Crawford (2009) argues that while it is useful to understand that game play involves different rules, norms and patterns (or a different 'frame' to use Erving Goffman's, 1974, terminology), what is problematic is that this is then frequently used as an excuse to ignore all that takes place outside of this perceived 'magic circle'. Hence, in this book, Thiago Falcão and José Carlos Ribeiro offer a re-consideration and re-interpretation of the magic circle, while René Glas, Kristine Jørgensen, Torill Mortensen and Luca Rossi consider the use of Goffman's 'frame analysis' as an alternative to this concept.

The idea of 'separation' from everyday life has similarly been a problem with some of the (particularly, early) Internet and cyberspace literature, which either sought to isolate Internet use and understand it as a different and separate space or world, or failed to locate Internet use within a wider social and cultural context.

Put simply, video gaming and Internet use are not separate domains or worlds, but rather a (frequently, relatively small) part of peoples' everyday lives, social patterns and interactions, and it is important that we understand them as such.

For instance, Burn and Carr (2006) argue that though online video games, such as MMORPGs or persistent worlds, are frequently seen as the gamer taking on a new and different identity, a 'representation' of themselves; this can be better understood as an extension of our identity – and similarities can be drawn here with the work of Cogburn and Silcox (2009) on the 'extended self' highlighted earlier. This is because the choices made in-game and online worlds are *our* choices, just as, the mistakes made are *our* mistakes. Furthermore, they suggest that even the most rudimentary game choices, such as naming our characters/avatars, are undertaken by us, the gamer, and hence reflect our choices, opinions and (at least some aspect of) *our* personalities. It is we who interact with a computer, and we who interact with other people. This interaction maybe 'mediated' but so is all interaction, most notably through language, as well as the complex patterns of the presentation of self we utilize in our everyday lives, irrespective of whether we are online or not (see Goffman 1969).

This then links neatly onto the third key area of motivation and debate within contemporary online game studies highlighted by Burn and Carr (2006), that of 'communal motivation'. This third key motivator is a less contested area than the previous two identified, as most video game scholars would readily accept that interacting with other gamers and the communal feelings that this can bring is, for many, a key pleasure and motivation of playing online video games.

Here, Burn and Carr highlight how most online games allow direct player interaction, most commonly through text or sometimes voice-based communicative tools, depending on the specific game and genre. For instance, most MMORPGs rely mainly on text-based communication. This form of communication probably best suits this genre of game for a number of reasons.

First, many MMORPGs involve (to varying degrees) role-playing fantasy characters, the illusion of which might be easily broken by, for example, the regional accent of the gamer. As Astrid Ensslin illustrates in this book, accents are very important in games for defining characters, their personalities and ideologies. The wrong kind of accent, such as a Blood Elf Paladin in *World of Warcraft* speaking with a Black Country accent, could easily disrupt game immersion.

Second, many MMORPGs involve both more formal and long-term group (often called 'guilds') memberships, as well as more temporary associations, such as 'hunting parties'; communicating with these (sometimes very large) groups is therefore much easier in text-based form, which can easily be sent to all members. Third, MMORPGs can often involve complex planning, organization and strategizing, again all of which are often best undertaken in written form. In contrast, many online FPS, such as the *Call of Duty* series of games, rely largely on verbal communication for three main reasons. First, FPS tend primarily to be played with a gamepad rather than a keyboard, making typing difficult, if not impossible. Second, given the fast-paced nature of these games, the speed of verbal communication (over typing) is more desirable. Finally, the hurried and hassled voices of similarly, pre-existing social networks (and with them, social capital) are also often replicated online, with people playing online with those they already know prior to the game (Taylor 2006). The links between game play and other social and cultural fields are not simply about online social networks, as Burn and Carr (2006) emphasize, the 'communal' nature of play also involves the social markers, characteristics, identities and opinions that we *bring with us* to the game.

The controversy and debate here then begins to surface when one begins to consider, in more detail, just how important and significant interpersonal game interactions are, as well as, the location of gaming and Internet use within a wider social and cultural context. Of course, for us (the editors of this book) as sociologists and a 'social focused' media scholar, the 'social' nature of gaming is extremely important and a key area of our collective interest. But if, as research increasingly would seem to suggest, a key pleasure, motivation and appeal of online gaming is its communal nature, as well as the need to locate gaming within a wider social setting, then we can begin to question the validity of focusing just on the *act* of video gaming. This shifts the study of video games away from play, and even psychology, towards the social. Therefore, we think there is little doubt (as this book clearly demonstrates) that there is a need to compliment (and not necessarily replace) focus on games, play and the online world, with a wider analysis of the social and cultural significance of online gaming.

Organization of this book

This collection of essays is divided into four key parts. Part I includes this first chapter, which introduces the book, the field of study and highlights key contemporary debates that have relevance to the discussions that follow. The next two parts form the substantive part of the book. The first of these, Part II, includes eight chapters that share the common themes of production and play. The third part consists of seven chapters relating to communities and communication. The final part of the book, and our conclusion, specifically considers Perron and Wolf's (2009) seven key challenges for game studies, and reflects upon these in light of the arguments set out in this book, as well as wider debates relating to online gaming. Before moving on to our first contribution, we set out here below a brief overview of the 15 chapters in the two substantive parts of the book.

The first chapter in Part II is provided by Aphra Kerr. In this chapter, Kerr argues that we need to reconsider discourses around the gamer as 'producer' and their relationship to 'formal' product development within the games industry. Drawing from science and technology studies, Kerr's starting point is that the digital games industry displays characteristics of many innovation-based environments and that a strong rhetoric of the 'democratization' of innovation persists. This, she argues, is fuelled by the range of ways game companies and players choose to mediate their relationship. Kerr unpacks such interactions and raises the question of whether we are witnessing the democratization of innovation or in fact the outsourcing of risk and value creation to game players. We thus need to consider the power dynamics in such relationships and finesse these.

She argues that there may not be the symmetry that the democratization thesis suggests.

Next, Esther MacCallum-Stewart examines how multiple readings of a MMORPG text can be articulated by player behaviour with a focus upon how the associated tensions that are provoked are negotiated by players. Rooted in a firm belief in the importance of the agency of the player, MacCallum-Stewart considers two interrelated areas. First, she presents how different social rule structures impinge on understandings of the game text and investigates the potentially conflicting ways in which the game world can be interpreted or lead the player into 'fannish' practices in order to reinforce their views. Second, she examines the mechanisms by which such complex social expectations are enrolled in manipulating players and their worlds in order to facilitate social interaction and to include fannish appropriation as a valid part of game design. In conclusion, she argues that the autonomy of players can have mixed outcomes and that it remains to be seen how far the formal designers of games will consider their input.

In Chapter 4, Holin Lin and Chuen-Tsai Sun provide groundbreaking insight into gamers who play MMORPGs on 'private servers'. Private servers are unauthorized, usually illegal, game servers set up to run alternative versions of commercially available MMORPG games. To date, very little is known about those who set up and play on private servers, their motivations or the nature of their game play. The common view of these gamers and servers is often akin to music or software pirates, and hence, ultimately criminals. Those who play on private servers are often seen as a small minority, who exist on the peripheries of mainstream gaming. However, Lin and Sun drawing on primary data highlight a far more complex picture of game play and the relationship between private and official servers. Lin and Sun highlight a plethora of motivations and reasons for using private game servers. These include allowing 'nostalgic' players to play games that are no longer officially supported (such as was the case with the *Uru* players, highlighted elsewhere in this book by Pearce), or gamers who are dissatisfied with the way official game servers are managed or structured. Private servers also offer greater customization and individualization of the gaming experience, rather than the sometimes 'one-size-fits-all' attitude to how games should be played evident in many official game servers. This discussion therefore demonstrates a much more complex relationship between official and private servers, where this is not necessarily a one-dimensional parasitic draining of gamers away from official servers. Private servers allow gamers an alternative where they can circumnavigate the parts of the official game that they do not enjoy or experiment with new ideas and styles of play, which may help keep them as players on the official servers, rather than drifting away from the game altogether.

Douglas Brown in Chapter 5 offers a deconstruction of the success of *World of Warcraft*, focusing specifically upon the centrality, complexity and incompleteness of its endgame and the approach Blizzard Entertainment has taken to the trajectory of the games development. Brown points to the role of Blizzard in unashamedly appropriating player developments and features of *EverQuest* particularly in this respect. Further he points to the first mover advantage around the player experience and path dependency associated with this strategy that has enabled *World of Warcraft* to remain, at least until now, ahead of the competition. Brown argues that, due to Blizzard Entertainment's approach, competitors' products appear to be under-designed in comparison to the *World of Warcraft* experience. In conclusion, he suggests that radically new approaches will not work in this field, where older is better and repetition is not a bad thing.

Next, Neil Randall explores the importance of playing spaces and simulation techniques in online boardgames. Notwithstanding the rise of other game forms, and most notably video games, boardgames, both online and offline, continue to be an extremely popular leisure pursuit. Here, Randall takes a closer look at how programmes for online boardgames play replicate traditional boardgame components and attempt to simulate the experience of FTF play. Randall draws on social semiotic theories of modality to explore the specific ways in which online boardgames seek to represent the offline boardgames they seek to replicate. It is argued that if boardgames are to be accepted by players, they need to simulate the game play of traditional boardgames in terms of physical space, key components and player communication. This then highlights the importance of gaming location, place and space, and articulates the complex nature of this, as it is mediated both online and offline. Moreover, and contrary to dominant arguments within contemporary games studies, Randall argues that boardgames are crucially quite different to video games, in that there are key differences between the nature of immersion, narrative generation, simulation, styles of play, interactions, norms and significantly, the creation and maintenance of game rules. This chapter therefore not only contributes to our understanding of an under-researched area – online boardgames – but also offers a significant challenge to a key foundation of game studies as a discipline.

Frans Mäyrä's chapter introduces the idea of contextual play, those playful behaviours that are rooted in, or emerge from, social relations and exchanges of information that are used to maintain and expand such networks of relationships. Essentially, Mäyrä offers various cases of game playing within a variety of social media such as *Flickr* and *Facebook*. The case is made for the convergence of playful activity in and among other activity, such as online photograph sharing and social networking sites. Thus, he demonstrates how game play is augmented by other activities afforded by the sites in question. For instance, status updates are made in *Facebook* while games like *Scrabulous* are played simultaneously. Through this, Mäyrä discusses the possibility for gaming, and playful activity more generally, as facilitating social networking. He also points to the further blurring of the boundaries within our lives brought about this and the hybrid nature of mobility with respect to online communication.

In Chapter 7, Thiago Falcão and José Carlos Ribeiro provide a significant contribution to the discussion of the magic circle. The concept of the magic circle originates in the work of Dutch historian Johan Huizinga (1949) but its relevance to video game studies is to be found most notably in its application by Salen and Zimmerman (2004), who use this to highlight the specific rules and social patterns that govern incidents of play. However, the applicability of this term to video game play has been questioned by numerous authors such as Pargman and Jakobsson (2008) and Crawford (2009). Falcão and Ribeiro argue that the magic circle proves a useful concept for understanding video games that are narrative-led, which can immerse the gamer in a (at times) detailed and evolving storyline, but question the validity of applying this concept (in its original form) to other game forms such as MMORPGs. For Falcão and Ribeiro, the magic circle needs to be re-conceived using the metaphor of Lewis Caroll's Alice stepping 'Through the Looking Glass' (1871). What they mean by this is that the magic circle, rather than being a divider of worlds, is rather a point of intersection. What determines the relative separation of the game world from the world outside is the degree to which the gamer is immersed in the gaming environment, and this immersion can be strengthened by game narratives, spatial navigation within games as well as interactions with other players. Hence, virtual worlds are the product of game objects (such as technologies, structures and narratives) but are also shaped by the gamers themselves, while in turn the game world plays a role in shaping gamer identities and interactions.

In a similar vein to Falcão and Ribeiro, Glas *et al.* in Chapter 8 offer a consideration of a key theory in the video game analysis. Unlike Falcão and Ribeiro, Glas *et al.* in this chapter, Jørgensen, Mortensen and Rossi reject the magic circle, for what they see as its too rigid distinction between the game and the outside world, and instead turn to the work of Erving Goffman on frame analysis, as a suitable way of understanding video gaming. Glas, Jørgensen, Mortensen and Rossi argue that frame analysis is particularly useful in understanding game play, as unlike the magic circle which is a largely linear divide, frames can be understood and multilayered, stacked upon each other and at times even conflictual. In this chapter, they consider four frames of play within the MMORPGs game *World of Warcraft*. In particular, they consider first, how gamer 'role-playing' introduces several intersecting, sometimes conflicting, frames into *World of Warcraft*. Second, they discuss gaming styles, the use of game guides and walkthroughs. Third, Glas *et al.* consider the complexities of categorizing and understanding game play when this intersects with, or even becomes, work. Finally, they consider when game frames bleed into frames outside of the game, and hence, carry over consequences into other frames, such as when in-game items start to carry out-of-game monetary value. What all of these cases clearly highlight is the complexity of conceptualizing game play, and illustrate how the relationship between gaming and other frames is always a constantly shifting one, with frames intersecting and building upon each other.

The first chapter of Part III of the book is by game designer, author, researcher, teacher, curator and artist, Celia Pearce. Here, Pearce discusses players of the MMORPG *Uru: Ages Beyond Myst*, and considers how this case provides important insights into, not only gaming practices, but the meaning, creation and maintenance of community in contemporary society. *Uru* was one of the final and only online games in the popular *Myst* series. It was released as beta version in 2003, and was quickly populated by up to 10,000 players; however (at least at that point), *Uru* did not make it to full release, closing after less than 6 months. What to that point had been primarily a single-person game series, and Pearce suggests, played by many who would describe themselves as 'loners' became an online communal

technical skills and providing them with the opportunity to socialize with others. To this end, Delamere argues that *Second Life* may lead to a blurring of social capital between 'virtual' worlds and 'real' worlds.

Next, Keith Massie examines racial and gendered representations of avatars in the online video game *EverQuest.* Although some past research has highlighted the gendered nature of female characters, little research has been conducted into the racial stereotypes that are contained in video games. From the beginning to the end of the gaming process within *EverQuest*, characters are constructed with such stereotypes in mind. From the character construction phase, to the gaming tutorial, to the actual game-play, gamers are presented with limited, marginalized and stereotypical representations of women and black characters. Massie highlights the highly sexualized and derogatory images of female avatars throughout the game. In addition, he draws a correlation from the lack of black characters to choose from, the fact that black avatars are labelled as a non-human species and the geographical exclusion of black 'races' within the game to ghettoes. Such stereotypical roles result in what he refers to as cyber segregation between white and black avatars. Such images mirror and reinforce societal representations of black as 'other' and women as weak, sexualized objects. As gaming has developed as arguably the most popular leisure pursuit today, enjoyed by men, women and children across the world, Massie argues that it is time that game designers offered less gendered and radicalized representations for its players to engage with.

Finally, in Part III, Kate E. Taylor provides an important chapter in her consideration of the role of the video game *Wordslinger* in aiding and empowering women who have experienced domestic abuse. The game was developed as part of a self-help website 'You Are Not Crazy' aimed at women experiencing domestic abuse and aims to help women to identify and overcome domestic violence. In particular, Taylor explores the ways in which the game interacts with the embodied experiences of the player to aid them in their empowerment. Though the game is rather simplistic in design and usability, the impact of the game for the player is profound. The video game consists of an 'embodied narrative' which acts as a tool aiding the personal development of the player in the 'real' world. Taylor suggests that much of the game's success may be due to the fact that it is a home-based activity, just like domestic violence. As domestic violence is more likely to occur in the home, the Internet is one space that may open up the possibility of support and communication for women experiencing domestic abuse. The game is unique in its approach combining the game-play format of more traditional video games with the aim of empowering women. As the body is key to our experiences of the world around us, the chapter examines the ways in which the human embodied experience can be shaped and influenced by the virtual world. As the game engages the female player in narratives of abuse, it encourages the player to challenge the passivity often experienced in domestic violence to become an active participant. *Wordslinger* therefore succeeds in producing a counter-narrative to women's own views of themselves as victims, thus changing an oppressed identity to an active and developing identity. Therefore, the physical and emotional interaction with the game allows women to connect with their own physical and embodied narrative of violence while encouraging them to construct new ones.

References

Aarseth, E. (1997) *Cybertext: Perspectives on Ergodic Literature*, London, John Hopkins University Press.

Avedon, E. M. and B. Sutton-Smith (1971) *The Study of Games*, New York, Wiley and Sons.

Bauman, Z. (1998) *Work, Consumerism and the New Poor*, Buckingham, Open University Press.

Bauman, Z. (2001) *Community: Seeking Safety in an Insecure World*, Cambridge, Polity Press.

Boellstorff, T. (2008) *Coming of Age in Second Life*, New Jersey, Princeton University Press.

Bourdieu, P. (1984) *Distinction: A Critique of the Judgement of Taste*, London, Routledge.

Bryce, J. and J. Rutter (2006) Digital games and the violence debate. In J. Bryce and J. Rutter (eds) *Understanding Digital Games*. London, Sage: 297–322.

Burn, A. and D. Carr (2006) Motivation and online gaming. In A. Burn and D. Carr (eds) *Computer Games: Text, Narrative and Play*. Cambridge, Polity Press: 103–18.

Caillois, R. (1962) *Man, Play and Games* (Trans. M. Barash), London, Thames and Hudson.

Castronova, E. (2005) *Synthetic Worlds: The Business and Culture of Online Games*, Chicago, University of Chicago Press.

Clark, A. and D. Chalmers (1998) The extended mind. *Analysis*, 58(1): 7–19.

Cogburn, J. and M. Silcox (2009) *Philosophy through Video Games*, London, Routledge.

Coupland, N. and H. Bishop (2007) Ideologized values for British accents. *Journal of Sociolinguistics*, 11(1): 74–93.

Crawford, G. (2009) Forget the magic circle (or Towards a sociology of video games) – Keynote. *Proceedings of the Conference: Under The Mask 2*, Luton.

Crawford, G. and J. Rutter (2007) Playing the game: Performance and digital game audiences. In J. Gray, C. Sandvoss and C. L. Harrington (eds) *Fandom: Identities and Communities in a Mediated World*. London, New York University Press: 271–81.

Durkheim, E. (1982 [1895]) *The Rules of Sociological Method*, New York, Free Press.

Fine, G. A. (1983) *Share Fantasy: Role-Playing Games as Social Worlds*, Chicago, University of Chicago Press.

Fox, M. (2006) *The Video Games Guide*, London, Boxtree.

Gere, C. (2008) *Digital Culture*, 2nd edn, London, Reaktion.

Goffman, E. (1968) *Asylums*, Harmondsworth, Penguin.

Goffman, E. (1969) *The Presentation of Self in Everyday Life*, Harmondsworth, Penguin.

Goffman, E. (1974) *Frame Analysis: An Essay on the Organization of Experience*, New York, Harper and Row.

Gosling, V. K. (2008) "I've Always Managed, That's What We Do": Social capital and women's experiences of social exclusion. *Sociological Research Online*, 13(1). Retrieved from http://www.socresonline.org.uk/13/1/1.html.

Huizinga, J. (1949 [1938]) *Homo Ludens: A Study of the Play-Element in Culture*, London, Routledge.

Juul, J. (2001) Games telling stories? *Game Studies*, 1(1). Retrieved from http://www.gamestudies.org/0101/juul-gts/.

Juul, J. (2005) *Half-Real: Video Games between Real Rules and Fictional Worlds*, Cambridge, MIT Press.

Juul, J. (2008) The magic circle and the puzzle piece. *Proceedings of the Philosophy of Computer Games: DIGAREC Series 1*, Postdam.

Keegan, M. (1997) A classification of muds. *Journal of MUD Research*, 2(2). Retrieved from http://brandeis.edu/pubs/jove/HTML/v2/keegan.html.

Longhurst, B., G. Smith, G. Bagnall, G. Crawford and S. Osborne (2008) *Introducing Cultural Studies*, 2nd edn, London, Prentice-Hall.

McMahan, A. (2007) Second Life: The game of virtual life. In B. Atkins and T. Krywinska (eds) *Videogame, Player, Text.* Manchester, Manchester University Press: 131–46.

Neiborg, D. B. and J. Hermes (2008) What is game studies anyway? *European Journal of Cultural Studies*, 11(2): 131–46.

Newman, J. (2004) *Videogames*, London, Routledge.

Nielsen, S. E. (2010) Three Screen Report, 7 (4th Quarter). Retrieved from http://in.nielsen.com/site/documents/3Screens_4Q09_US_rpt.pdf.

Nielsen, S. E., J. H. Smith and S. P. Tosca (2008) *Understanding Video Games: The Essential Introduction*, London, Routledge.

Pargman, D. and P. Jakobsson (2008) Do you believe in magic? Computer games in everyday life. *European Journal of Cultural Studies*, 11(2): 225–43.

Perron, B. and Wolf, M.J.P. (2009) 'Introduction', in B. Perron and M.J.P. Wolf (eds) *The Video Game Theory Reader 2*, London: Routledge, 1–22.

Salen, K. and E. Zimmerman (2004) *Rules of Play: Game Design Fundamentals*, London, MIT Press.

Taylor, T. L. (2006) *Play between Worlds: Exploring Online Game Culture*, Cambridge, MIT Press.

Turkle, S. (1995) *Life on the Screen: Identity in the Age of the Internet*, New York, Simon and Schuster.

Wright, T., E. Boria and P. Breidenbach (2002) Creative player actions in FPS on-line video games: Playing counter-strike. *Game Studies*, 2(2). Retrieved from http://www.gamestudies.org/0202/wright/

Part II

Production and play

marketplace, and particularly with online games, the form and nature of the relationship between the professional producer and game players is crucial to the ongoing success of the game. Thus, we need to see production as an ongoing process involving a range of actors, many of who are trying to create various forms of capital (social, cultural and economic) and value (exchange value, use value and sign value) from the process. These actors and values often come into conflict and while these production networks are sometimes conceptualized as participatory or co-creative production networks, this chapter explores whether these concepts are adequate enough to capture what is going on here.

The chapter cautions against taking an overly optimistic or overly negative approach to such developments. It asks us to consider what is behind the increasing tendency to encourage user productions and what do empirical examination of such productions reveal? Is what we are observing actually co-creation or is it that the professional media industries are finding new ways to encourage player production and to appropriate and extract value from this labour? Finally, are the relationships between professional game producers and game players governed and regulated in the most effective and fair manner to recognize and reward each actor? Online games provide a fertile, but diverse, ground to empirically and conceptually explore the concept of player and community production and innovation.

Drawing upon both primary and secondary research on amateur and professional player productions and user knowledge accumulation, this chapter hopes to contribute further empirical and conceptual insights into the changing role of the player in relation to game production and innovation.[2] In what follows I will first explore the rise of the 'user' in innovation and media theory.[3] The chapter continues by examining some contested forms of player production and their potential contribution to value generation including data generation, game governance, user-generated content and cheating. The chapter then moves on to explore how these micro-player practices reflect and link into broader political economic trends in the professional games industry, namely outsourcing of certain aspects of production and training. The chapter concludes by examining the implications of the implicit and explicit ways by which professional producers facilitate and accumulate amateur productions and by calling for new rules to govern these new forms of production.

The turn to the user, unfinished designs and user production in games

Elsewhere I have written about the process of game design in a small games company and how game design teams tend to design games that they like themselves and how this is further influenced by the needs and wishes of publishers, marketing departments and other external actors (Kerr 2002). For many game companies who produce off-the-shelf boxed products, the engagement with users is primarily mediated by publishing companies who tend to want to replicate previous successes and market to existing markets. This approach is still prevalent in the cultural industries despite the rise of strategies to explicitly and implicitly involve users in the design process in other sectors, such as product design, and there is often very little contact between designers and their users in boxed off-the-shelf media products even if the games are play tested with actual game players (Oudshoorn *et al.* 2004). Over 10 years ago, Silverstone and Haddon (1996) noted that designers' knowledge of users was often tacit, contradictory and untested, and in this uncertain environment, powerful sub-groups compete to determine design. By focussing only on upstream production processes, we are in fact falling back on the old linear innovation models that placed user involvement at the end of the process and discussed them mostly in terms of consumers, impacts and effects. Indeed, Stewart and Williams (2005) argue that many theories still conceptualize design and production as an upstream process that delivers a finished product to the market place which contains particular conceptualizations of 'the user' and 'user activities' and which the user encounters and must adapt to.

By contrast, the growth of Web 2.0 and online games in the last decade have opened up new opportunities for professional producers to engage with game players: explicitly, through online forums and community support and implicitly, through tracking and monitoring player behaviour. While professional development companies design the environment and attempt to configure how the player engages with the game, they are also faced with new community and network challenges and ongoing production issues. If we view production as a social process involving multiple cycles of design and use, then we have to recognize the unfinished nature of artefacts that are launched on the market, the fact that technical artefacts change over time and that part of this change over time is induced or produced by users and/or their knowledge, or knowledge about them, and their labour. This line of thinking takes seriously concepts from the sociology of science and technology which argues that we should view technology as malleable and as something whose meaning, use and interpretation change over time and only stabilize as networks of human and non-human actors coalesce (Bijker 1995; Callon 1987; Mackenzie and Wajcman 1999). It also accepts that the design of technological artefacts involves political decisions and that the design of artefacts may exclude certain users and actions (Wajcman 1991; Oudshoorn and Pinch 2005). Digital networks, in general, and online games, in particular, enable us to research and explore the extent to which production and innovation occurs throughout the lifecycle of an artefact, and the Science and Technology Studies (STS) perspective alerts us to the degree to which users are constrained, negotiate and indeed translate technological artefacts in the marketplace. This perspective highlights how the role of game players may move beyond beta testing, focus groups, purchasing and game play and how conflictual issues may signal the need for alternative rules and practices.

A crucial outcome of this line of thinking is that game players are more than consumers and the relationship between producers and users needs to be respected, symmetrical and governed in such a way so as to reflect this important role. In other words, professional producers (i.e. developers/publishers) need to be transparent in their relationship with game players and conceive of new ways publishes/develops the games and in some cases third parties. By installing and accepting the license agreement to a game, players submit to both the external and the internal game rules and they must agree to ongoing monitoring of their play behaviour and their machines. Most hardware and software interventions in games are introduced under the guise of increasing security and improving the game play experience but the impact is that all players are monitored and are open to automatic disciplining and in some cases mistaken disciplining (Taylor 2006, 2009). Unfortunately, most players have little knowledge about the gathering of such data until they infringe upon certain rules and regulations and discover that they have been flagged for 'cheating' or deviant play and punished in some way. Indeed, it is clear that the relationship between producer and player in MMOs in particular is not symmetrical and while there is much monitoring, most games lack channels and procedures whereby the player can defend them. This is not to say that game players have no agency, but they are aware that some of their activities may contravene either the explicit rules of the game or more informal community rules, and in the absence of means to defend themselves, they develop a range of strategies from self-surveillance to purchasing third party software to protect their avatars and accounts (De Paoli and Kerr 2010).

The move to what Poster (1990), Taylor (2006) and Albrechtslund (2008) would call 'participative surveillance' certainly introduces a new element to the producer–user relationship that may or may not be to the benefit of the user. It also signals the degree to which surveillance is not all top down, but that game companies rely on players to co-regulate and report on deviant behaviour. In line with contemporary neo-liberal regulatory systems in Western economies where regulation has moved from state forms of control to co- and self-regulation by industry, it is informative to observe the ratio of game masters to players in games and the degree to which game producers rely on automatic tools and player co-operation and reporting to police their games (de Zwart 2009; Humphreys 2008, 2009). One example of direct player involvement in game governance and design can be found in the democratically elected committee of nine players who comprise the 'The Council of Stellar' in the MMORGP, *Eve Online*. However, depending on your perspective, these players are engaged to 'govern' other players in Eve and have no formal power or are 'empowered' players who are elected to negotiate future design directions with the professional development team. However, the role comes with some rather onerous requirements including that they should reveal their true identity on the forums and be prepared to monitor and report back on player concerns. It remains to be seen how this works in practice.

An area that has received some attention from game studies scholars in the recent past and again provides useful empirical material for thinking about information gathering and monitoring, producer–player relationships and how player practices can be productive is the area of cheating. Cheating in games is a negotiated practice, and some activities, which may be defined as cheating, are accepted by player communities, while others may not be. Again different publishers take different approaches to cheating but many now rely on automatic tools to implement their End User License Agreements (EULAs) and to supplement player reporting. Consalvo (2007) has provided the most comprehensive investigation of the practice to date and provides useful detail on the extent to which the industry capitalizes economically on the practice (through providing cheat codes, walk-throughs and so forth) while needing to appear to regulate it also. However, we can also view cheating as a practice that provides forms of knowledge that may result in further innovation by the producer (who acts to exploit and regulate the behaviour), by third parties who offer cheating services to players and by players who attempt to deal with the practice (De Paoli and Kerr 2010).

In ongoing research with De Paoli in the MMORGP *Tibia*,[4] we have found extensive online negotiations between players and third party companies who produce commercial cheating software and a number of responses from *Tibia's* publisher that have included mass bans, updating licenses and the development of an automatic cheat detection tool (De Paoli and Kerr forthcoming). Each new iteration of the automatic cheat detection tool results in a new process of research and development for the software cheating companies in consultation with their lead users; players who cheat in this game. Indeed, the cheating players must provide initial data from their playing experience as to how they believe the cheat detection tool works and need to test in-game any new cheating technologies. So the cheating players are performing much of the initial research and indeed the testing of the tool. They then provide feedback via online forums. It is again evidence of the circular nature of innovation, the value that players, and the knowledge they have of a game, can provide to professional producers and the economic benefits that different companies can derive based on this free labour. It is unclear at this point to us as researchers, to the cheaters and to the cheating software companies how exactly the cheating detection tool operates but evidence from other games, especially *World of Warcraft*, would suggest that such tools are very invasive and operate as spy ware sending back regular screenshots of one's computer to the monitoring centre and scanning a player's hard drive for offending software and copyright infringements. Again much of this activity is not transparent to the player and while they willingly agree to the 'consent to monitor' clause in a game's EULA, the manner in which this is executed would appear to be well beyond what is necessary in an entertainment product.

One of the key rights asserted in EULAs relates to ownership of intellectual property. We can see the struggle for ownership quite explicitly in relation to user-generated content and more specifically game modding. In game studies scholars have been attempting to critically assess the economic value and social capital created by volunteer game players. They also highlight the contentious area of ownership over the content created and document how publishers have penalized copyright infringements in user-generated content while at the same time benefiting themselves economically from the productions (Kücklich 2005; Nieborg and van der Graaf 2008; Postigo 2003, 2007, 2008; Søtamaa 2007; Taylor 2006). The focus in this work has been on the immaterial labour of game players (de Peuter and Dyer-Witheford 2005) or playbour (Kücklich 2005). As Postigo (2003: 597) notes 'hobbyists' leisure work is converted from gift to commodity, what results is the circumvention of the initial investment risk for the are locating outside their home market and close to their target markets (Kerr and Cawley 2009). Over half the jobs in Ireland in the games industry are in customer support with a focus on the European and Asian markets. Blizzard, for example, has a large 'European customer support' branch in the Republic of Ireland which employs a range of European nationals and provides support mainly for *World of Warcraft*. In an interview about their recent expansion in Ireland, the chief operating officer stated that costs in Ireland were very competitive and that employment laws provided a good balance between staff and employer rights (Collins 2009). Most companies who responded to our survey stated that access to labour was a key reason for locating in Ireland and the diverse range of nationalities employed and relocated to Ireland points to the international dimension of their work and perhaps to the less than favourable balance in employment laws elsewhere in Europe. It is also notable that there would appear to be, across the industry, a very low participant rate in unions. Less explicit, but nevertheless important, has been the significant support, financial and otherwise, of the local industrial development authority.[6] The current system of supports is not game specific, but Ireland has a corporation tax rate of 12.5 percent, which is highly attractive to companies who are making a profit.

When we look at the type of work being undertaken by small game companies in Ireland, both with Irish and foreign ownership, the focus is on work for hire, on porting games across platforms, and on acquiring and aggregating intellectual properties. For successful companies, the trend has been for them to be acquired by multinational companies. For example, in the past 3 years, Demonware, a network middleware company, was acquired by Activision while Havok, a physics middleware company, was acquired by Intel. Both companies were successful in commercializing what were initially university-based research projects. At least in the case of Demonware, it is clear that their technology is now only available to Activision projects, much to the dismay of their former clients (Fahey 2007). Company acquisition is of course a key strategy to acquire successful intellectual properties and to restrict access to them without having to invest in the initial risky idea. Overall, the past decade for the Irish games industry has seen good growth in employment in support functions in multinationals and in middleware but also tremendous flux in the content creation part of the value chain.

In other countries, offshoring and outsourcing of elements of the game production process has been encouraged by game-specific supports and would appear to have included more professional production work as a result. Canada, for example, at both a national and regional level has been successful in attracting elements of the global games industry using tax incentives and employment supports (Dyer-Witheford and Sharman 2005). In 2008, France introduced a tax credit system that enables game producers (regardless of nationality of employees and including outsourced work) to apply for a significant 20 percent tax credit on production costs (excluding testing and support) of particular types of games.[7] In the United Kingdom, there is ongoing lobbying of the government to provide tax credits and other financial supports to game developers (in particular) to help them to compete with Canada, South Korea and other rapidly emerging centres of game production (Oxford Economics 2008). It would appear that countries are keen to encourage, to relocate or to keep production activity in their countries and are receiving significant public funding or reductions on tax to locate in particular areas. Thus, production activity and labour are moving to lower cost sites – a fact borne out when we see that more German nationals are employed in the Irish games industry than Irish citizens (Kerr and Cawley 2009).

Programming/modding competitions are another example of this tendency to outsource, but these relate more to training and the provision of a constant supply of suitably trained staff. In an industry where passion and portfolio are as important as a university qualification, the rise of industry-sponsored amateur game production competitions requires some critical investigation. On the surface game production, competitions appear to be a great way for students and 'indie' developers to gain experience of working on particular technologies, under conditions similar to those in a professional company and in some competitions to obtain advice and mentoring from industry professionals. What is clear, however, is that the sponsors in some cases, and affiliate companies in others, use the competitions as a way to introduce and market their tools to students – a type of outsourced training – which is used to supplement the more general education obtained by most students on games and related courses. As with user-generated content, the games produced often become the property of the competition organizers and sponsors and participants must agree to their image and information being used in media promotions during and after the event. Indeed, participation in the competitions is strongly regulated with extensive terms and conditions. An added bonus for the students and companies is that the 'hothouse' competitions lead to a final 'beauty pageant' where the winning teams and participants meet companies who are recruiting and receive free sponsored hardware and software. When these activities are placed alongside the relatively low level of expenditure by game companies on in-house training of staff (Grantham and Kaplinsky 2005), the high costs of using recruitment agencies, the rapid pace of technological change and the requirement for staff to keep up to date with new software and hardware, mostly through learning by doing on the job, and the 'brain drain' of talented and experienced staff from the industry, one suspects that these competitions require more critical investigation (IGDA 2004; Kerr forthcoming).[8] As Søtamaa (2007: 12) notes 'mod competitions bring together a variety of industry practices . . . to enculturate the free modder labour' and amateur gaming competitions would appear to work in a similar way.

What is clear from the examples provided is that professional game production and related functions are moving to lower cost locations and that they are becoming increasingly active in the development of gaming competitions and related events, which is yet another strategy through which they encourage amateur productions, acquire intellectual properties and recruit amateur developers. These competitions would appear to operate as a public–private partnership given that countries forego taxes, industrial development bodies provide space and money to run competitions and lecturers/teachers give up time and curriculum space to facilitate such competitions. In addition, amateurs and students give up their

Technological Systems. New Directions in the Sociology and History of Technology. Cambridge, MIT Press: 83–103.

Charles, D., J. Kücklich and A. Kerr (2007) Player-centered game design: Player modelling and adaptive digital fames. In S. de Castell and J. Jenson. (eds) *Worlds in Play. International Perspectives on Digital Games Research.* New York Peter Lang: 249–65.

Chesbrough, H. W. (2003) *Open Innovation: The New Imperative for Creating and Profiting from Technology*, Watertown, MA, Harvard Business Press.

Collins, J. (2009) Blizzard Confirms 500 Extra Jobs in Cork. *Irish Times* (May 29). Retrieved from http://www.irishtimes.com/newspaper/finance/2009/0529/1224247670056.html.

Consalvo, M. (2007) *Cheating: Gaining Advantage in Videogames*, Cambridge, MIT Press.

De Paoli, S. and A. Kerr (forthcoming) "We will always be one step ahead of them" A case study on the economy of cheating in MMORPGs. *Journal of Virtual Worlds Research.*

de Zwart, M. (2009) Piracy vs. control: Models of virtual world governance and their impact on player and user experience. *Journal of Virtual Worlds Research: Technology, Economy, and Standards,* 2(3). Retrieved from http://jvwresearch.org/index.php?_cms=default,0,0.

Dyer-Witheford, N. and Z. Sharman (2005) The political economy of Canada's video and computer game industry. *Canadian Journal of Communication,* 30: 187–210.

Fahey, R. (2007) Activision Confirms Demonware Acquisition. Multiplayer Technology Specialist Snapped up for an Undisclosed Sum. *gamesindustry.biz,* (6 March). Retrieved from http://www.gamesindustry.biz/articles/activision-confirms-demonware-acquisition.

Grantham, A. and R. Kaplinsky (2005) Getting the measure of the electronic games industry: Developers and the management of innovation. *International Journal of Innovation Management,* 9(2): 183–213.

Hesmondhalgh, D. (2008) Neoliberalism, imperialism and the media. In D. Hesmondhalgh and J. Toynbee. (eds) *The Media and Social Theory.* New York, Routledge: 95–111.

Humphreys, S. (2008) Ruling the virtual world. Governance in massively multiplayer online games. *European Journal of Cultural Studies,* 11(2): 149–71.

Humphreys, S. (2009) Discursive constructions of MMOGs and some implications for policy and regulation. *Media International Australia,* (130): 53–65.

IGDA (2004) Quality of Life in the Game Industry: Challenges and Best Practices.

Jarrett, K. (2008) Interactivity Is Evil! A Critical Investigation of Web 2.0. 13(3). Retrieved from http://www.uic.edu/htbin/cgiwrap/bin/ojs/index.php/fm/article/view/2140/1947.

Jenkins, H. (2006) *Convergence Culture. When Old and New Media Collide,* New York, New York University Press.

Kerr, A. (2002) Representing Users in the Design of Digital Games. *Proceedings of Computer Games and Digital Cultures Conference.* Tampere, Finland.

Kerr, A. (2006) *The Business and Culture of Digital Games: Gamework/Gamplay,* London, Sage.

Kerr, A. (forthcoming) The culture of gamework. In M. Deuze. (ed.) *Managing Media Work.* Thousand Oaks, Sage Publications.

Kerr, A. and A. Cawley (2009) The Games Industry in Ireland 2009. Maynooth and Limerick.

Kerr, A., S. De Paoli and C. Storni (2009) Rethinking the Role of Users in Ict: Examples and Reflections from the Internet *The Good, the Bad and the Challenging. The User and the Future of ICTs,* Copenhagen, Denmark.

Kline, S., N. Dyer-Witheford and G. De Peuter (2003) *Digital Play. The Interaction of Technology, Culture and Marketing,* Montreal, McGill-Queens' University Press.

Kücklich, J. (2005) Precarious playbour: Modders and the digital games industry. *Fibreculture*, (5). Retrieved from http://journal.fibreculture.org/issue5/kucklich.html.

Lash, S. and J. Urry (1994) *Economies of Signs and Space*, London, Sage.

Mackenzie, D. and J. Wajcman. (eds) (1999) *The Social Shaping of Technology*. Buckingham, Open University Press.

Mactavish, A. (2008) Licensed to play: Digital games, player modifications, and authorized production. In S. Schreibman and R. Siemens. (eds) *A Companion to Digital Literary Studies* Oxford, Blackwell: 349–68.

Nieborg, D. and S. van der Graaf (2008) The mod industries? The industrial logic of non-market game production. *European Journal of Cultural Studies*, 11(2): 177–95.

Oudshoorn, N. and T. Pinch. (eds) (2005) *How Users Matter. The Co-Construction of Users and Technology*. Cambridge, MIT Press.

Oudshoorn, N., E. Rommes and M. Stienstra (2004) Configuring the user as everybody. Gender and design cultures in information and communication technologies. *Science, Technology and Human Values*, 29(1): 30–63.

Oxford Economics (2008) *The Economic Contribution of the UK Games Development Industry*. Oxford.

Poster, M. (1990) *The Mode of Information. Poststructuralism and Social Context*, Cambridge, UK, Polity Press.

Postigo, H. (2003) From pong to planet quake: Post-industrial transitions from leisure to work. *Information, Communication and Society*, 6(4): 593–607.

Postigo, H. (2007) Of mods and modders: Chasing down the value of fan based digital game modifications. *Games and Culture*, 2(4): 300–13.

Postigo, H. (2008) Video game appropriation through modification: Attitudes concerning intellectual property amongst modders and fans. *Convergence*, 14(1): 59–74.

Ralph, N. (2008) Sony Killing Questionable Littlebigplanet Levels, without Warning. *Wired*, (November 10). Retrieved from http://www.wired.com/gamelife/2008/11/littlebigplan-1/.

Søtamaa, O. (2007) On modder labour, commodification of play and mod competitions. *First Monday*, 12(9). Retrieved from http://www.firstmonday.org/issues/issue12_9/sotamaa/.

Stewart, J. and R. Williams (2005) The wrong trousers? beyond the design fallacy: Social learning and the user. In H. Rohracher. (ed.) *User Involvement in Innovation Processes: Strategies and Limitations from a Socio-Technical Perspective*. Wien, Profil: 39–71.

Taylor, T. L. (2006) *Play Between Worlds: Exploring Online Game Culture.*, Cambridge, MIT Press.

Taylor, T. L. (2006) Does wow change everything? How a Pvp server, multinational player base, and surveillance mod scene caused me pause. *Games and Culture*, 1(4): 318–37.

Taylor, T. L. (2009) The assemblage of play. *Games and Culture*, 4(4): 331–9.

Terranova, T. (2000) Free labour: Producing culture for the digital economy. *Social Text*, 18(2): 33–58.

van Dijck, J. (2009) Users like you? Theorizing agency in user-generated content. *Media, Culture and Society*, 31(41): 41–58.

van Oost, E., S. Verhaegh and S. Oudshoorn (2009) User-initiated innovation in wireless leiden. From innovation community to community innovation: *Science, Technology and Human Values*, 34(182): 182–205.

von Hippel, E. (2005) *Democratising Innovation*, Cambridge, MIT Press.

Wajcman, J. (1991) *Feminism Confronts Technology*, London, Polity Press.

Wayne, M. (2003) Post-fordism, monopoly capitalism, and hollywood's media industrial complex. *International Journal of Cultural Studies*, 6(1): 82–103.

3 Conflict, thought communities and textual appropriation in MMORPGs

Esther MacCallum-Stewart

> Importantly, the relationship between player and system/game world is not one of clear subject and object. Rather, the interface is a continuous interactive feedback loop, where the player must be seen as both implied and implicated in the construction and composition of the experience.
>
> (Newman 2002: 410)

This chapter builds on the rich corpus of work available in this field to consider how players negotiate the social experiences of play and some of the tensions that this provokes. It does so by examining how multiple readings of the game text are articulated by player behaviour. Through the creation of new social directives, reinterpretations of existing rule structures, group activities and fan produced texts, players attempt to order the unruly nature of the online world through their own activities.

Online worlds are so broad in narrative and playful in scope that users are able to adopt multiple approaches in play styles. This means that they often have a relationship with the game world that is at odds to other simultaneous users. In MMORPGs, a web of rules, interpersonal codes, conflicting expectations and attitudes of complicated social interactions play as exploration. Furthermore, the need by the player to appropriate the text of the game for their own pleasure provides a final, vital element in the composition of online societies. As Hills (2002) argues, acts of fan appropriation create a 'dialectic of value'; however, this cannot be cleanly separated between public and private needs. Because online games are dynamic environments in which users are constantly interacting with each other, this disharmony is evident as individual aims collide with those of the game and of other players.

This chapter examines how players' appropriation of the game text's meaning can affect play. First, I outline how different social rule structures impinge on understandings of the game text and start to redefine it as a complex social structure. I then investigate different ways in which the game world can be interpreted and discuss how these understandings can conflict, or lead the player into 'fannish' practices in order to reinforce their personal view of the game world. The last part of this chapter examines ways in which designers and producers of MMORPGs have begun to use these complex social expectations to positive effect, manipulating players as well as their worlds to take advantage of social interaction and fannish appropriation as a valid part of game design.

Interpreting play

Huizinga's magic circle (1938, trans. 1955) would posit the MMORPG as a totally enclosed environment. Interactions would remain within the world and be recognized as 'distinct from ordinary life' (Huizinga 1955: 9). This distinction problematizes the MMORPG as a valid 'real' space; however; in a recent speech, Jane McGonigal estimated that players spend the equivalent of part time jobs playing online games and approach them with a similar work ethic (McGonigal 2010).

Instead, this chapter sees online games as more permeable – places where players undertake social activities similar or identical to those in a non-virtual space, but do so with an awareness that they are in a realm that includes ludic codes of practice. This is closer to Juul's understanding of the game space as both fictional and real (and therefore 'Half-Real') (Juul 2005), but avoids the troublesome use of the word 'real', instead simply designating the online game as a space that has additional ludic rule systems. This distinction allows participants to behave in ways that are unexpected or might be seen as unacceptable within social groups, where play is not the foremost activity. It also means that players may bring to the game their own social norms from more familiar groups such as family, home and work. Frans Mäyrä asserts that games do not exist in separation from their players (Mäyrä 2008; Mäyrä *et al.* 2009), rejecting the totality of ludology, and also perhaps suggests that the game is the player and the player is the game. Social activity by players is seen as a core part of the game experience – players join because their friends do, migrate across games together and play because it allows them to socialize (Ducheneaut *et al.* 2007a, 2007b; Taylor 2006). The implication that players in online worlds are usually gaming *and* engaging in social activity at the same time has interesting consequences for the development of these spaces. The player's negotiation of the different codes and rules for each type of behaviour can also be one of the main causes of social friction within the online space, most obviously described by Richard Bartle's classification of player types in 'Hearts, Clubs, Diamonds, Spades: Players who suit MUDS' (Bartle 1996) (expanded in *Designing Virtual Worlds* (2004), and taken up elsewhere by Yee (2006).

Social activity in online worlds demands that players in online worlds necessarily move between multiple levels of socialization both within and exterior to the game world (Klastrup and Tosca 2010; Mortenson 2007; Taylor 2006; Watkins 2009). Commonly shared activities required by users of MMORPGs, such as organization of groups, allocating tasks and roles, or simply chatting and gossiping whilst grinding, means that players are constantly fluctuating between game and social activity. If a player is imagining a set of real rules and placing them within a fictional world, she is also taking part in this world. Constance Steinkuehler describes this as a social 'mangle':

> . . . the game that's actually played by participants is not the game that designers originally had in mind, but rather one that is the outcome of an interactively stabilized (Pickering 1995) "mangle of practice" of designers, players, in-game currency farmers, and broader social norms.
>
> (Steinkuehler 2006: 199).

The dragon does not get up and slay itself – the player does that all by herself – or perhaps she asks her friends to come along and help if it is a particularly large, fierce dragon. If the dragon really is tough, she might have to also ask people she does not know, but have appropriate dragon-killing skills. With a large group of people she now has several social concerns that stretch across both real and 'half-real' spectrums – who gets the dragon's treasure when it dies? What time should everyone meet? Who is going to coordinate the fight when it is in progress? What happens if xx player, who does not like yy player, shows up and asks to join in and so on. Very quickly, social criteria have started to affect what might have seemed a very simple act, complicating and changing the process.

Although the social relationships players form most often take place within the game, they rely on pre-existing 'real-world' strategies and behaviour in order to succeed. Players mediate their game experience of the game world as a unique environment, tempered by an awareness that social situations within it have their roots in the same types of behaviour deployed in the rest of their lives. This can be exemplified by the 'hints' for good play that Blizzard Entertainment show players as they log into *World of Warcraft* (*WoW*) (2004 – present). Some of these are simply social guidelines such as 'It's considered polite to talk to someone before inviting them into a group, or opening a trade window' and 'If you enjoyed playing with someone, put them on your friends list!' Neither have anything to do with functional play in the game, but instead facilitate good relationships with others and give an indication of what type of behaviours might be expected within the game world. These early signifiers alert players to the type of ideology one might discover within the game, as well as suggesting modes of expected behaviour (MacCallum-Stewart 2010). In this case, the emphasis on 'polite', 'talk' and 'friends' all posit social activity without, for example, linguistic or role-playing pointers that might encourage additional modes of interaction.

Finally, players must use emergent strategies in order to comprehend the game world. Players experiment in online worlds as part of emergent play (Smith 2001). These activities can be physical – such as testing the limits of an avatar or the world around it, as well as social – what happens when a player behaves badly or well towards others, or technical – is it possible to alter code in order to make MMORPG gold more quickly. These behaviours have fascinated games scholars for some time (Consalvo 2007; Mortensen 2006; Taylor 2006), and emergent play is now often a central aspect of game design; see, for example, the DS title *Scribblenauts* (2009).

Most importantly, players regard rule-breaking and deviant (or perceived deviant) activities as an expected part of the MMORPG experience, even if they do not always appreciate it occurring around them. Cheating is also a flexible term, originating from the confusion between the types of rule structures, both formal and informal, within the game (Consalvo 2007). So, for example, in *WoW*, Ganking (repeatedly killing another player's avatar when they are helpless), is not cheating, but *is* considered both a social annoyance and an amusing form of upsetting others. At the same time, it is so common that it is not really even deviant behaviour (Koster 2007). The wowhead discussion thread (http://www.wowhead.com/forumsandtopic=105453.4) 'To gank or not to gank' is rife with players (denoted here by the first letter of their names) excusing their actions:

> . . . About the only time I intentionally gank is sometimes when I am really bored I like to dress my belfadin up like a SW guard and pose as a quest gver in northshire. If you are dumb enough to think some guy with a horde logo and? level is an alliance quest gver you deserve to get killed. (O)
>
> . . . i love it when people call in there guilds etc to get me, its like i really caused such a disturbance that i now have 5–20 people running around giving me all the attention.' (S)
>
> . . . I almost never gank, only time I ever corpse camp is for the icecrown PvP Daily, because frankly, I don't feel like hunting down 10 separate allies. All other times, unless first ganked, I will /challenge, wait for them to eat/drink, the beat them in fair battle. (F)

Attempts to justify aberrant actions such as these reveal a fundamental clash between players and play styles (Castronova 2007; Griffiths and Light 2008; Grundy 2008; Steinkuelher 2006). Players find themselves working at cross-purposes to each other within the same place. They therefore try to evolve coping mechanisms. In order to do this, players become textual poachers, appropriating and sometimes subverting narrative, ludic and social patterns within the game. The rest of this chapter examines why this conflict cannot be resolved simply by the 'rules' of the game, and then goes on to show how it takes place within social communities.

Playing nice – different types of behaviour in MMORPGS

Before examining the ways in which players act to change or appropriate game texts in order to contain and personalize their meaning, it is important to review the ways that players may approach MMORPGs. A player's approach to a game varies in one critical way from that of most other social structures. Play dominates or drives all other activities.

Juul defines what are here called ludic laws as forming 'Game Space', and argues that these rules define the fictional world within a game (you cannot kick a football 'outside' of the screen in *FIFA*) (Juul 2005). Whilst this is true to an extent, a player in an MMORPG is also aware of the other structures listed here. In MMORPGs, the narrative is so detailed and so integral to the Game Space itself that it cannot be ignored. MMORPGs encourage players to emotionally engage with the world around them on an ideological level that encourages them to consider the text on

> This charter is similar to a contract between us as a community and you as a member of said community. You can count on us to organize 25-man raids and assure all the logistics implied by that, while making the whole process as much fun as we can.
>
> (TF 2009)

Players frequently encode these social codes of practice formally in documents on forums or elsewhere in an attempt to give them legitimacy, but it is essential to realize that these tenets do not extend across the entirety of a server, world or realm and are not usually available outside of the guild formation. These tenets also prompt re-interpretation on a more localized level. The two charters above demonstrate this – they are player-based constructions placed upon the text in order to manage it within a social context that only partially exists within the game.

What is important about each set of laws or practices laid out above is that they are often conflated by players, who confuse them, or simply appropriate the most useful system at the time to justify their actions. Thus, players interviewed by Consalvo (2007) were adamant that creating walkthroughs, gold-farming for their own characters and assisting other players financially within the game were not 'cheating', but instead argued that cheating consisted of specifically violating the End User License Agreement. Modders and hackers often express a desire to reconfigure the game, often to improve it (Brown 2010; Moshirnia 2007); walkthrough and wiki contributors are seen as vital elements of participatory culture (Burn 2006; Jenkins 2010); and griefers often defend their acts as challenging by either player or world sphere *status quo* (Myers 2008) or performing valid protest against unfair or broken parts of the game. Activities such as stealing from and repeatedly killing ('ganking') other players are also seen by some as fun, even though these latter are considered the more anti-social activities.[1] However, all of these activities are seen as play, as valid activities within the context of game-as-place-of-exploration that are then mapped onto the world.

The example below is from a weblog post discussing this rocky territory. Noticeably, although the player includes a useful diagram, it is filled with jargon that in fact renders 'the proverbial gray area' nonsensical. Their own statement that this system is usually internalized also highlights the difficulties that players feel when delineating the differences between exploiting a game's existing content, cheating and play.

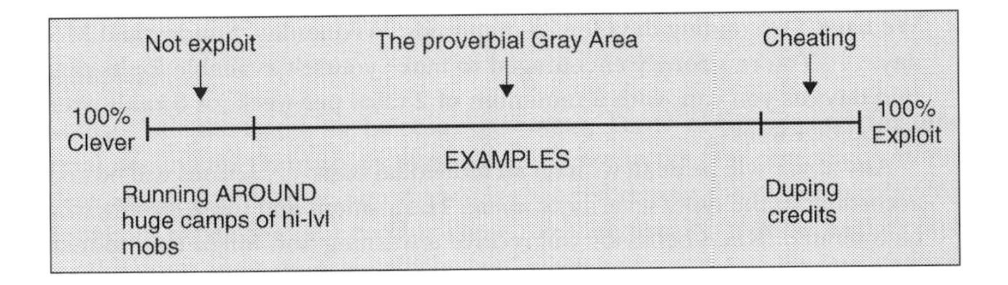

> Anyways, I have my own internal system for deciding if I'm exploiting or if I'm just being incredibly clever. I have made a visual aid for the discussion.
> (The AFK Gamer 2005)

Overall therefore, the diversity of behaviours that exist in online worlds clashes dramatically with the 'rule' structures that are layered upon this. This chapter investigates how some of these acts change the perception of MMORPGs for the players.

Forming groups: thought communities

MMORPGs have selective roles built into the games' ludic[2] structure. Players choose attributes, professions, tasks and skills for their avatars, which mean that they automatically specialize within the game. In many MMORPGs, this structure is furthered by making certain in-game content only available to specific groups, such as specific classes or races, meaning that content in this differentiation of types is important. However, it is also likely that players will come to these groups with different information about the game itself. An average group might not only contain avatars with a selection of abilities, but it might also include social roles within this such as the person who knows the dungeon/game/area best, a task leader, a quartermaster who allocates looted objects and characters who might know when best to use their own skills or items. These informal groupings form thought collectives as described by Ludwig Fleck. The general structure of a thought collective consists of both a small esoteric circle of close members and a larger exoteric circle of acquaintances or other people less directly connected to the central group, each consisting of members belonging to the thought collective and forming around any work of the mind, such as a dogma of faith, a scientific idea or an artistic musing. A thought collective consists of many such intersecting circles. Any individual may belong to several exoteric circles but probably only to a few, if any esoteric circles (Fleck 1979).

If we see the game itself as the structure around which the thought collective forms, then the esoteric structures can be mapped onto guilds and small groups. More sustained groups such as guilds can be seen as communities of practice (Wenger 1999), whereby players form more permanent groups with a consistent allocation of roles and the additional understanding that knowledge is also viewed collectively.

The specialization of play has re-percussions in both gaming and social domains. Users are encouraged (or forced) to play diverse characters, since group play often requires a variety of abilities to succeed (e.g., the small group formation in many MMORPGs consists of a healer, a 'tank' who takes damage, and damage dealing classes who either melee or fight from range),[3] and they tend to carry this thinking into the ways that they manage information and relationships relevant to the game. The more complex a game is, the more likely it is that players will specialize in certain areas of knowledge or ability. This begins in a very ludic manner, relevant only to the game itself, but gradually extends into social, organizational and both types of community remain internally compatible, albeit operating under very different communicative criteria. From the previous guild charters, despite their difference of length, there are specific references to social behaviour. For the TF charter, the justification between practical and social is often crossed. It is, for example, not allowed to talk when a raid leader is speaking because it is important to listen to instructions but also because it is 'bad manners' (TF 2009). In the following incident from the FC guild, where one player killed a cat NPC (Mr Bigglesworth) in front of another player whose real-life pet had just died, the cross-over between game world and real-world social behaviour is again made explicit in the resulting comments on the forum (again players are named by their initials):

> I agree it was pretty insensitive to [kill Mr Bigglesworth]. I'm particularly irritated because the Raid Leader did specifically instruct the Raid not to do so. Quite aside from distressing N, it was possible that it could have had plot/Raid content issues. If the Raid Leader asks you to do or not to do something, please behave, or you will not be permitted to Raid with FC.
>
> There are plenty of nasty things in the game; some of these will have echoes of things in our real lives, and we have to balance being aware of that with interacting with the game environment as presented.
>
> So please be sensitive.
>
> (D, FC 2009)

Finally, breaking guild or charter rules is penalized, but this is often instigated through play itself. The author has seen occasions where players have 'blocked' another from sight with smoke bombs, awarded a low-quality 'dunce' hat for the raid's most ineffective/annoying/aberrant player or teased players for consistently being late until they mended their lackadaisical ways. More serious infarctions have involved banning players from attending raids for a period of time, requesting 'official' apologies on forums, and 'three strikes and you are out' warnings. Overall however, there is direct correlation between the moniker of a 'good' player who is able to conform to the group dynamic (in whatever form that may take) and a 'bad' player who is seen as a rule breaker.

So far, this chapter has argued that players exist in a liminal space in which they negotiate a series of different ludic, legal, social and narrative structures. The confusion that arises over which discourse dominates means they have to moderate social interaction accordingly, using a composite framework of social signifiers taken interchangeably from all of these structures. Players are appropriating a system of best fit to validate their activities and in an attempt to control their situation when they feel it is moving beyond their control. This is epitomized in their frustration with the higher authorities and often articulated through public channels within each world such as forum or chat channels. Google produces more than 48,000 hits for the keywords 'Blizz should ban', where disgruntled players air their complaints about everything from unfair advantages in some game areas to specific players. Importantly, their comments are aimed at the institution ('Blizz', or Blizzard Entertainment), which they perceive to be largely anonymous and unreachable, and they often confuse the various figures in authorial control:

> With all the time that you say you spend on the patch and the 4+ Weeks it spent in the PTR why the HELL CAN"T YOU GET IT RIGHT! and heres one for ya if the main part of this patch was the new raid/Heroic changes why the hell are all the instace mobs gliching?! what outstanding change did you make to make them glich that bad when all you change what 1 badge drop? Also to all and any other pissed players feel free to add to this list.
>
> (*WoW* forums 2009)

This rant assumes that game developers are also games masters/moderators (paid employees who moderate disputes reported in game), and that other players will be 'pissed [off]', constructing a clear 'them and us' situation (which is in fact refuted in later posts), and transposing personal frustration to more ephemeral targets. The post typifies the feeling by players that they lack agency within MMORPG worlds. This last section examines how players take autonomy for themselves, using a case study whereby players have altered an existing narrative to their satisfaction and demonstrating a new trend towards player empowerment.

Player appropriation, or *the craft of war*: how onyxia fell from power

> Readers [players] are travellers; they move across lands belonging to someone else, like nomads poaching their way across fields they did not write, despoiling the wealth of Egypt to enjoy it for themselves.
>
> (De Certeau 1984: 174)

De Certeau's early definition of textual poaching posits that readers [players] are no better than thieves (poachers) when they steal literature for their own use and change its meaning for their own ends. However, if we see these poachers as the players I have already described, their actions make a great deal more sense. Players are already negotiating complex meanings within the text, and they need the world around them to conform to these when required. If, as Newman (2002) suggests at the very start of this chapter, the player is part of a 'continuous feedback loop', both 'subject and object', then it stands to reason that they will change the narrative for their own ends. This case study demonstrates how players have helped to change an unsatisfactory plotline in *World of Warcraft* to suit a variety of different ends.

Elsewhere, I have argued the narrative of a world is interpreted differently according to need by players. This includes, for example, their appreciation of objects, their understanding of narratives and their engagement with other players through actions such as role-play (MacCallum-Stewart 2010; MacCallum-Stewart and Parsler 2008). Whilst players experience the same quests, conquer the same dungeons and carry out group tasks together, there is often a very little understanding of the stories behind each situation. Many players, more focused on the playable content than its underlying narrative, are happy to consume the text passively without really paying attention to the storylines. Options in many MMORPGs (which are often themselves more focussed on *ludus* than *paidia* (Caillois 1961)) allow them to scroll quickly through quest dialogue, or to 'tab out' of cut scenes and continue playing immediately. These curtail the interaction between narrative and player. In *World of Warcraft*, some players pay so little attention to the back story that they are unaware that the two sides, Horde and Alliance, are sharing an uneasy truce.

However, at the same time, there seems to be an implicit understanding that a player's narrative in relation to the individual quests completed or goals they aspire to is a private dialogue between themselves and the world around them. Players may construct stories that describe their relationship to this world, but because most MMORPGs retain players by containing many quests players can perform alone, it is an individual's choice as to how much attention they pay these narratives. This can be seen during the relative bafflement of some players to cut scene interjections during raids in *World of Warcraft* – some will ignore them, some talk over them and some remember them for the future. It is rare, however, that discussion of what is actually happening takes place *between* players, or that they role-play with them as interactive artefacts as events unfolds:

Players have adopted this passive attitude because their agency in the overall MMORPG world plot is usually minimal. Like more traditional fan communities, they must usually respond to the narrative through creative means outside

Figure 3.1 Players watch the descent of Nightbane during a cut scene in *World of Warcraft*. Although they are able to interact with each other during this event, the players depicted watch the animation, point out things that other players should notice or prepare themselves for combat.

the text. Since history continues around them without much direct intervention, players are the 'powerless elite' (Tulloch 1995). The necromancer Abercrombie in Darkshire will always remain uncaught in *WoW*, the heroes of Lord of the Rings will still destroy the Ring in *Lord of the Rings Online* and so on.

However, this has begun to change. The vast appropriation of the game text by players through modding, machinima, wiki construction, forums, beta testing, role-playing stories, events and so on, has caused a real shift in the power dynamic of the player as an active agent in their world. Although changes do not happen quickly, games designers are relying on their players to help them sustain their products. One of the ways they are doing is by allowing players a greater autonomy in their worlds. This includes the increasing use of public test realm beta testing by players and attention to the artefacts that esoteric groups such as modding communities are producing to compliment the game. *World of Warcraft*, for example, has integrated mods such as the addition of a threat metre into the game code (patch 3.02), and *WoWwiki* was recently estimated to be the second largest wiki online (after Wikipedia) (Dybwad 2008). As a result, game narratives, often a neglected element of these design changes, have also started to gradually change due to pressure by players. The following case study shows how a significant alteration in *World of Warcraft* has changed a storyline that players have had issues with since its inception.

Onyxia, aka Lady Katrana Prestor, is the subject of a lengthy quest within *WoW*, and as such players on both Horde and Alliance side have extended contact with the narrative. In 'vanilla' WoW (before the expansions), Onyxia's Lair was a difficult encounter for 40 players. Players wishing to undertake this raid had to 'attune' to the instance by getting a key, comprising a quest chain of over a dozen different tasks. In Stormwind (the Alliance main city), players would often congregate during a part of the quest that involves marching through the city in order to confront Onyxia in her human disguise. This stately march made it a cinematic and popular event. When 'unmasked' Lady Katrana Prestor transforms into a smaller version of her dragon form and attacks the players, in the first incarnation of the game, this was a difficult encounter requiring a group. Onyxia then flees to her lair and the quest chain transforms into the raid encounter. Ultimately, players who succeeded in slaying Onyxia were highly rewarded. One person in the raid would get her decapitated head as a prize. When this head was presented to the leader of each faction, 'Onyxia's Warcry' gave surrounding players a substantial statistical boost, and the head could be seen hanging from the battlements of each city.

However, this also meant that Onyxia could co-exist in many places at once: inside the raid instance in her dragon form, at the top of Stormwind, sequestered next to the boy king Arduinn Wrynn, masquerading as his advisor, as an item within a player's inventory and displayed in both Ogrimmar and Stormwind. Players would metagame by keeping a head from a previous raid, congregating in the city, using the head to boost their performance and heading off to kill the dragon whose bloody head already adorned their city gates. Since the quest also lead to Onyxia's unmasking and transformation, players were often frustrated by the fact that Onyxia could be repeatedly 'outed', yet would perpetually remain in the throne room. Obviously, those who had completed the quests to this point were also well aware of her treacherous persona:

> That the game is delivered in real-time does not mean however that the game is linear and nor does it have complete temporal continuity: players can kill the dragon Onyxia over and over again – just because a group of players kill her once doesn't alter the game world itself other than temporarily.
>
> (Kryzwinska 2008: 19)

> Onyxia dies, the bosses will all be killed. And the next day everything will be back to normal. Nothing will ever change. Seeing the endings happen for other players shows us our own potential endings and reassures us that the end of the story does not mean loss: everything will still be here. True, if you kill Onyxia, she'll be dead *for you* for five days. But she'll be back.
>
> (Walker-Rettbur 2006: online)

When Krzywinska (2008) and Walker-Rettburg (2006) observed that Onyxia always returns, an act that can destroy the mythologies of the self as an agent of destiny within *World of Warcraft*, they did not anticipate the impact of fannish disapproval. I have used Onyxia as a focal point within this chapter since she is clearly problematic; her troublesome storyline and ludic function have also been discussed by Aarseth (1997) and Mortensen (2006); a coin belonging to Anduinn Wyrnn (who stands next to Prestor) in the Dalaran fountain of *WoW* reads, 'I wish I would grow up, it feels like I've been ten for years'; and it seems that even the players with little investment in the lore of *World of Warcraft* have a great fondness for the Onyxia questline (Holisky 2008; Wilhelm 2007), enough that her discordant manifestations were cause for irritation.

In 2008, the *Wrath of the Lich King* (*WotLK*) expansion pack advanced the Onyxia story. The figure of Onyxia (as Lady Prestor) was removed from the throne room and although princeling Arduinn Wyrnn remained, his father Varian took the place of his former mentors. It was assumed that Wrynn the elder had returned and denounced the dragon-in-disguise. This implicit understanding was coupled with the removal of the attunement requirement for Onyxia's Lair. Players could now simply walk into most of the early raid instances without having to undertake the lengthy prior quests.

This type of narrative development is unusual, but is indicative of the huge amount of fan texts that surround the figure of Onyxia implying that a change should be made. The machinima *The Craft of War: BLIND* (Percula 2008) is typical of this; a re-inscription of the Onyxia story whereby a blood elf assassin tries to kill Onyxia. The Alliance warrior (possibly Mathias Shaw, an NPC involved in several Alliance quests) who tries to defend her is blinded at the crucial moment by a spell, failing to see Onyxia grow a massive pair of wings, and thus remaining unaware of her true identity.

BLIND is regularly recommended as a 'must see' machinima, and when host Vimeo accidentally deleted it in May 2009, there was an outcry. In 2009, the EU

Figure 3.2 Lady Katrana Prestor and guards prepare to defend themselves in *BLIND: The Craft of War.*

section of Blizzard also ran a lore competition in which one of the questions was 'Who slays Onyxia?', and a version of the story has in the accompanying *World of Warcraft* comic book series (DC Universe 2009: #9). As peripheral to the game itself, these texts all qualify under Jenkins' depiction of fan text, and all re-write what is considered an aberrant piece of material. This re-writing to a more satisfactory version is also typical fan poaching (Jenkins 1988).

The success of Onyxia-related fan texts is, however, moderated by another incident which takes place regularly in the throne room. The Onyxia confrontation is one of the most visible narrative events in the game – not least because the ultimate confrontation happens in an area where players are often queuing for another aspect of the game – the PvP battlegrounds. Interestingly, after the necessary attunement for Onyxia was removed in *WoTLK*, very few players were still aware of the quest chain and its resultant effect, and would often express surprise when hordes of dragons suddenly appeared. Whereas in earlier stages of the game, Onxyia's elite guard were a force to be reckoned with, three years later in 2008, they were merely a surprising, and rather easy annoyance. The fact that so few players knew about the narrative reasons for this sudden incursion into their gaming points to a lack of attention by players to storylines, it was a collective impetus by a more narratively involved group of players that changed the situation. This indicates very clearly that information about narrative does not always transmit itself between groups, and in particular, that it can go virtually ignored until something dramatic that requires a reaction occurs. If fans are appropriating texts, they are still also forming themselves into knowledge groups where the transmission of information stops occurring, or it simply is not of interest to other esoteric circles.

the Social Dynamics of Massively Multiplayer Online Games. *Proceedings of the SIGCHI Conference on Human Factors in Computing Systems.* Montreal, Qubec, Canada: 407–16.

Ducheneaut, N., N. Yee, E. Nickell and R. J. Moore (2007) The Life and Death of Online Gaming Communities: A Look at Guilds in World of Warcraft. *Proceedings of the SIGCHI Conference on Human Factors in Computing Systems.* San Jose, CA, USA: 839–48.

Dybwad, B. (2008) How Gamers Are Adopting the Wiki Way, Massively. Retrieved 8 March, 2008, from http://www.massively.com/2008/03/08/sxsw08-how-gamers-are-adopting-the-wiki-way/.

Eladhari, M. and C. Lindley (2003) Player Character Design Facilitating Emotional Depth in MMORPGs. *Proceedings of DiGRA and Level Up 2003*, Utrecht, The Netherlands.

FC Forums. (2009) Retrieved 14 April, 2010, from http://finalchapterguild.com/index.php.

Fleck, L. (1979) *Genesis and Development of Scientific Fact* (Trans. F. Bradley and T.J. Trenn), Chicago, University of Chicago Press.

Gee, J. P. (2004) *Situated Language and Learning: A Critique of Traditional Schooling*, New York, Routledge.

Griffiths, M. and B. Light (2008) Social networking and digital gaming media convergence: Classification and its consequences for appropriation. *Information Systems Frontiers*, 10(4): 447–59.

Grundy, D. (2008) The presence of stigma among users of the MMORPG RMT: A hypothetical case approach. *Games and Culture*, 3(2): 225–47.

Hills, M. (2002) *Fan Cultures*, London, Routledge.

Holisky, A. (2008) Onyxia Attunement to Be Lifted. Retrieved 24 May, 10, from http://www.wow.com/2008/09/16/onyxia-attunement-to-be-lifted/.

Huizinga, J. (1949 [1938]) *Homo Ludens: A Study of the Play-Element in Culture*, London, Routledge.

Jenkins, H. (1988) Star trek reread, rewritten: Fan writing as textual poaching. *Critical Studies in Mass Communications*, 5(2): 85–107.

Jenkins, H. (1992) *Textual Poachers: Television Fans and Participatory Culture*, New York, Routledge.

Jenkins, H. (2010) Learning in a Participatory Culture: A Conversation About New Media and Education. Retrieved 13 May, 2010, from http://www.henryjenkins.org/.

Juul, J. (2005) *Half-Real: Video Games Between Real Rules and Fictional Worlds*, Cambridge, MIT Press.

Klastrup, L. (2003) A Poetics of Virtual Worlds. *Proceedings of the Fifth International Digital Arts and Culture Conference*, Melbourne, Australia.

Klastrup, L. and S. Tosca (2011) When Fans Become Players: Lotro in a Transmedial World Perspective. In T. Krzywinska, E. MacCallum-Stewart and J. Parsler. (eds) *Ringbearers, Lord of the Rings: Online as Intertextual Narrative.* Manchester, Manchester University Press: In Press.

Koster, R. (2007) Ganking Meaning and Playing as You Are. Retrieved 24 May, 2010, from http://www.raphkoster.com/2007/02/06/ganking-meaning-and-playing-as-you-are/.

Krzywinska, T. (2006) Blood scythes, festivals, quests and backstories: World creation and rhetorics of myth in *World of Warcraft. Games and Culture*, 1(4): 383–96.

MacCallum-Stewart, E. (2011) The place of role-playing in MMORPGs. In T. Krzywinska, E. MacCallum-Stewart and J. Parsler. (eds) *Ringbearers, Lord of the Rings: Online as Intertextual Narrative.* Manchester, Manchester University Press: In Press.

MacCallum-Stewart, E. and J. Parsler (2008) Role-play Vs gameplay. The difficulties of playing a role in World of Warcraft. In H. Corneliussen and J. Walker-Rettberg.

(eds) *Digital Narrative, Identity and Play: A World of Warcraft Reader.* Cambridge, MIT Press: 225–46.

Mäyrä, F. (2008) *Introduction to Game Studies: Games in Culture*, New York, Sage Publications.

Mäyrä, F., J. Paavilainen and J. Stenros (2009) The Many Faces of Sociability and Social Play in Games. *Proceedings of MindTrek*, Tampere, Finland.

McGonigal, J. (2010) Gaming Can Make a Better World. *TED Talk.* Retrieved 30 April, 2010, from http://www.ted.com/talks/lang/eng/jane_mcgonigal_gaming_can_make_a_better_world.html.

Mortensen, T. (2006) Blizzard and the Gold Sellers. Retrieved 22 November, 2006, from http://torillsin.blogspot.com/2006/11/blizzard-and-gold-sellers.html.

Mortensen, T. (2007) Me the other. In P. Harrigan and N. Wardrip-Fruin. (eds) *Second Person, Roleplaying and Story in Games and Playable Media.* Cambridge, MIT Press: 297–307.

Moshimia, A. (2007) Emergent Features and Reciprocal Innovation in Modding Communities. *Proceedings of the World Conference on Educational Multimedia, Hypermedia and Telecommunications.* Vancouver, Canada.

Myers, D. (2008) Play and Punishment: The Sad and Curious Case of Twixt. *Proceedings of The [Player] Conference*, Copenhagen, Denmark.

Newman, J. (2002) In search of the video game player: The lives of Mario. *New Media and Society*, 4(3): 407–25.

Percula. (2008) The Craft of War: Blind. Retrieved 2 February, 2010, from http://www.vimeo.com/5241163.

Runescape. (2010) How Do I Report Abuse? Retrieved 25 May, 2010, from http://www.runescape.com/kbase/guid/how_do_i_report_abuse.

Smith, H. (2001) The Future of Game Design: Moving Beyond Deus Ex and Other Dated Paradigms. *Proceedings of IGDA*, Copenhagen, Denmark.

Steinkuehler, C. (2006) The mangle of play. *Games and Culture*, 1(3): 199–213.

Taylor, T. L. (2006) *Play Between Worlds: Exploring Online Game Culture*, Cambridge, MIT Press.

TF. (2009) Forums. Retrieved 30 April, 2010, from http://www.temptingfate.eu/forums/index.php.

Tulloch, J. (1995) We're Only a Speck in the Ocean: The Fans as Powerless Elite. In J. Tulloch and H. Jenkins. (eds) *Science Fiction Audiences: Watching Dr Who and Star Trek.* London, Routledge: 143–72.

Walker-Rettberg, J. (2006) The End of Endings. Retrieved 13 November, 2006, from http://jilltxt.net/?p=1815.

Watkins, C. (2009) *The Young and the Digital: What the Migration to Social Network Sites, Games and Anytime, Anywhere Means for Our Future*, Boston, Beacon Press.

Wenger, E. (1999) *Communities of Practice: Learning, Meaning and Identity*, Cambridge, Cambridge University Press.

Wikipedia. (2010) Thief (Character Class). Retrieved 25 May, 2010, from http://en.wikipedia.org/wiki/Thief_%28character_class%29.

Wilhelm2451. (2007) Marshal Windsor and Lady Prestor. Retrieved 25 May, 2010, from http://tagn.wordpress.com/2007/12/28/marshal-windsor-and-lady-prestor.

World of Warcraft Europe. (2009) Comic Quiz Contest. Retrieved 14 May, 2010, from http://www.wow-europe.com/en/contests/comicquiz-09/index.html.

WoW Forums. (2009) Question to Any/All Blizz GMs'. Retrieved 25 May, 2010, from http://forums.worldofwarcraft.com/thread.html?topicId=1910995271 8andsid=1.

Yee, N. (2006) Motivations for playing in online games. *CyberPsychology and Behavior*, 9(6): 772–5.

4 Thrift players in a twisted game world?

A study of private online game servers

Holin Lin and Chuen-Tsai Sun

This chapter considers massively multiplayer online role playing game (MMORPG) player motivations for using 'private game' servers and considers their gaming experiences. Here, the term 'private server' refers to unauthorized MMORPGs video game servers that operate in parallel to official game servers. Private servers are set up and operated by individuals who do not pay licensing fees and use leaked, stolen or hacked source code to run games. A private server may support anywhere from less than 10 to several thousand users. Many private server operators offer access for free, or charge much lower fees than official servers, while some rely on a donation system for fee collection.[1] Statistical data on this 'informal gaming sector' are unreliable, yet game companies and authorized local distributors believe that they are losing substantial amounts of revenue due to private server activity. They are working closely with law enforcement agencies to crack down on illegal servers and the Internet cafes that provide access to them.

Game companies and even some players who have no experience on private servers generally perceive private server players as either trying to save money or as super-achievers who want to get around the player level requirements of specific games. Game companies and law enforcement agencies have a very clear perception of private game servers as illegal substitutes for official servers, similar to the selling of pirated copies of music or software. However, there are some important differences: first, the owners of private servers do not simply copy server software, but frequently bend or alter official game environments. Second, to a significant proportion of private server operators, profit seeking is not their primary goal.

This chapter considers evidence gathered from Web-based sources (mostly in blogs and game forums) in an effort to: (a) understand the motivations and experiences of players who use private game servers; (b) determine what meaning private game servers provide to the players; and (c) identify the distinctive meanings of private servers to the general game culture. Our primary assumption is that private game servers and players should not be considered parts of a marginal phenomenon but as factors that support our understanding of the gaming experience.

Background

There are at least two reasons why it is difficult to accurately estimate the number of private game servers in operation at any time. First, due to their illegal status, many operators purposefully avoid drawing attention by limiting their announcements to forums that require membership registration or by attracting users via interpersonal communication. With a few large-scale exceptions, most private servers are used by small and relatively short-lived player communities. Second, owners frequently change server names to avoid detection by law enforcement authorities. Nevertheless, by monitoring private server-related Web sites and interviewing players, we were able to find hundreds of private servers for popular MMORPGs such as *World of Warcraft*, *Lineage*, *Lineage II*, *Ragnarok Online* (RO), and *MapleStory*. Significantly, whether they are pay-to-play or free-to-play, almost all popular games have private servers. This single phenomenon may suggest that saving money is not the primary reason for private server players.

Because of the geographic locality and familiarity, we limited our search to Chinese-version servers in Taiwan, Hong Kong, China, and Southeast Asian countries; not surprisingly, the largest number of servers we found were operating in China. The technical requirements for setting up a private server (also known as an *emulator*) are not complex. One interviewee with such experience told us that a '2D game such as *Lineage* requires only about an hour to set up. A more complex game such as *Lineage II* needs some research time, but can be done in two days.'

Most private server operators eliminate all original game quests to save themselves capacity and to simplify operations for players. At the same time, they almost always increase levelling speeds and virtual money-making return rates to levels that are tens or even hundreds of times higher than those offered by official game servers, thereby supporting players' monster-slaying and resource collection efforts. Some servers even give new members immediate access to top-ranked characters and the highest-level game features.

Small private servers generally do not collect payment of any sort. Donations are often welcomed. Sometimes private server operators encourage donation by giving high-ranked gears as gifts. Many players consider such voluntary donation a legitimate charge as the cost of maintenance and operation of private servers needs to be covered. Only a small number of the largest private servers offer 'virtual malls' for buying equipment, using the above-mentioned donation mechanism to generate revenue.

Perspectives

Our analysis is based on two pivotal perspectives: (a) private servers should be collectively interpreted as a parallel culture to mainstream game culture, and (b) 'fun' is the central motivation for players to use private servers. When addressing these issues, it is important to remember that MMORPGs are ongoing worlds in which player roles are both active and productive; therefore, their gaming activities are central to creating game value. As Jenkins (2006) notes, media users consume and participate in the production of media commodities, a process he calls 'participatory culture'. He has observed that some Japanese animation fans in the United States are adding their own original English subtitles (known as 'fansubbing') to imported products – an activity that he views as an example of unofficial production that promotes Japanese anime in the United States. Whether legal or not, such examples of media user participation are shaped by social and cultural consensus and they are generally controlled more by consumers than producers.

In a similar manner, Taylor (2006) and Kerr (this book) point out that the game activities of MMORPG players produce savings for game producers and add considerable value to games. In the absence of economic incentives, players participate in public testing and debugging prior to the official launching of games, and later participate in the production of game objects and exchanges of virtual goods and services. To achieve more efficient and rewarding gaming experiences, player communities frequently initiate and support borderline illegal products such as mods, plug-ins, and user interfaces. Products that make game worlds more colourful or efficient are sometimes incorporated into the upgrades of official game versions, thus saving companies the costs involved in market surveys and product research and development. Taylor, who uses the term 'productive players' when discussing gamers who create value in this manner, has commented on another aspect of this relationship: some game companies try to exert too much control over game ownership, thereby running the risk of putting unwanted constraints on the creativity of individual users. The complexity of the progressive implications of player contributions to gaming culture resists simple interpretations based on intellectual property rights or End User License Agreements.

As stated above, there is a fun or pleasure aspect to using private servers that we believe is at the core of understanding all gaming experiences. As MMORPGs inherit and remix various features such as action, adventure, role-playing, strategy, and social game genres, they increasingly provide a multidimensional framework for pursuing fun. McMahon (2003), Buckingham (2006), and Yee (2006) are among researchers who have identified three categories of MMORPG gaming elements that are closely related to game enjoyment:

1. *Ludic* factors include rules, goals, missions, and constraints established by game systems. Players interacting with these elements and overcoming challenges enter an attentive state called 'engagement' (Buckingham 2006), and those maintaining a level of successful involvement over time are motivated by a sense of accumulative progress (Rieber 1996). According to Csikzentmihalyi (1990, 2002), this sense of fun exists in isolation from external rewards because players maintain a balance between game system challenges and their own skills when entering a 'flow' state. Player anxiety increases when skills cannot meet current challenges and no alternatives exist; player boredom increases when challenges fall below skill level. Both scenarios result in a decreased sense of fun. Game companies therefore try to provide missions with finely

calculated degrees of challenge to allow players to develop their skills incrementally during a prolonged flow state consisting of optimal gaming experiences. At a certain point, game missions become less interesting and engaging, and players lose a sense of challenge, such as when facing the repetitive monster slaying and resource collection tasks that game companies promote as necessary for achieving promised endgame rewards. The rewards may appear worthwhile, but the majority of MMORPG players consider the process of achieving incrementally higher gaming levels as either tedious or, at best, a necessary evil.

2. *Narrative* fun elements can be divided into two categories based on 'passive' or 'active' player participation. Passive players who enjoy background scenery, character appearances, and storylines feel effortless involvement, immersion, or pleasure that is akin to reading a novel or watching a movie (Buckingham 2006). In the active category, Murray (2004) notes that virtual worlds are story-rich, with an abundance of narrative elements. When events take shape in MMORPGs, players can actively put together narrative elements in conjunction with other players for stories such as castle sieges or improvised cyberdramas. Narrations and discourses solidify experiences and memories that form a sense of game history. Such deep immersion is an important reason for strongly identifying with a game. Ideally, MMORPG players should be able to freely navigate, explore, and immerse themselves in these worlds either passively or actively, but the emphasis of game companies on player level has resulted in the division of game worlds into zones. There are serious consequences for players entering maps or dungeons they are not qualified for, and this design feature seriously affects the preferences of players for narrative or immersive fun.
3. The *interactive* dimension of gaming consists of social features such as chatting, teaching and learning, cooperation, and organization, among others. Players can collectively construct social networks, gaming cultures, and community identities. Game companies are fully aware of the importance of social bonding to game success in terms of popularity and profit, but such interactions are often restricted by game system design and character level. Once again, playing level is at the centre of this dimension: the highly interdependent structure of player roles/occupations and the collective requirements of many quests demand that players cooperate with each other in order to achieve higher levels. Furthermore, MMORPG designs generally discourage the mixing of players at different levels, meaning that players at lower levels risk rejection when they try to join an achievement-minded community. But on the other hand, when players are accepted into social-oriented communities where everybody knows your name (Steinkuehler and Williams 2006), they may feel pressure from the community when they want to just play alone with everyone else in the background ambience (Ducheneaut *et al.* 2006). In certain cases, players who want to maintain distance from others and play less seriously may find it hard to do so because their characters have established reputations.

Other interviewees made specific comments about what they felt was the unnecessarily long process of moving up the player-level ladder on official servers – 'levelling hell' was a phrase we came across frequently during data collection. Some players just want to have some casual fun without getting involved in competitive levelling, others are willing to compete but are tired of repetitive monster slaying and other levelling-related tasks, and still others complain that they do not have enough time to follow the normal levelling process due to school or work responsibilities. When players hear reports from friends about immediately participating in high-level game activities via private servers, their efforts on official servers can lose meaning. One interviewee told us:

> At one time I was practicing to become a sorcerer.[4] As you know, you have to do it alone. Every day, practice, practice, practice – two hours a day, alone, automatic. It is really outrageous, why should it be so painful, so repetitious? Why should I go online and torture myself? Really, that dark thought of deleting my account would emerge sometimes . . . When I went to play on that private server, I was actually very disappointed and annoyed because my good online friend, she left the official server for a while and came back to tell me how nice the private server was, how easy and fun it was, blah, blah, blah. To me, the private servers are there to mock people like me who practice very slowly, very obediently . . . I felt a little betrayed.
>
> (Yoshi, female, age 27)

Another time and effort-consuming task in gaming is organizing teams for major quests, a process that entails collaboration with unknown players. In order to promote acquaintance networks and formal player organizations, many gaming companies purposefully add a strong interdependence component into their game designs, usually in the form of character occupations. Many quests need specific combinations of occupations, thus limiting opportunities for solo play, or forcing players to spend a great deal of idle time waiting for chances to form teams. Team play also increases the potential for individual failure and criticism from team mates. For these players, private servers allow them to play whenever they want without being dependent on the availability of other gamers.

To a certain extent, the private server phenomenon can be interpreted as resistance to a rigid 'one-size-fits-all' game system on the part of players who come from different backgrounds and social contexts and who possess different technical skills and cultural preferences. Players generally move back and forth between official and private servers, and if our interviewees are representative of larger gamer populations, the majority clearly understand their distinctive qualities. This leads us to suggest that most players use private servers to satisfy needs that are not being addressed by official systems – in other words, players use private servers to express dissent towards game operators and managers. Private servers thus represent a third option to remaining completely loyal to a game or leaving it due to a sense of frustration.

Third, *explorative gamers* describe themselves as searching for gaming fun that they cannot find on official game servers; they view levels, competition, and equipment as conditions for achieving other types of fun and not as goals in and of themselves. Our interviewees frequently described fun on private servers in terms of visiting places they are denied access to on official servers because of their current levels. They enthusiastically described the places they visited, the bosses they had pictures taken with, the magic creatures they rode, the cuteness of non-player characters they saw, and the breathtaking landscapes they 'flew over'. To have the same experiences on official servers, players must have top ranks, wear the best equipment, and have experienced team mates. Even then there is no guarantee of success, and in many cases the best that players can hope for is to take quick glance at their victorious surroundings before 'running for their lives'; posing for pictures is out of the question. In contrast, private server players can easily obtain the best armour or weapons, and testing those pieces of equipment at a leisurely pace is viewed as a form of recreation.[5] One interviewee described this activity as similar to 'playing with paper dolls'.

In the same manner as 'real-world' tourist travel, picture taking plays an important role in MMORPG explorative experiences. In addition to serving as records of 'I was there,' pictures taken in rarely visited places on private servers add to a sense of ongoing immersion in a game. As described by one interviewee:

> It compensates for something I cannot do on the official server. For example, I can do the quests and read the stories on the official server, but I cannot take a good look at certain scenes, I can only do that on other [private] servers. On the private server I also tried on clothes that I usually won't wear. I started with the lowest-ranked clothing, one piece at a time. I created them, tried them on one-by-one, and took pictures.
>
> (Shita, female, age 25)

In addition to exploring game worlds, private servers help players explore game systems. Developing characters on official servers is costly, making it difficult to experiment with various character occupations, skills, and roles. Private servers thus play very useful roles as inexpensive laboratories for testing new characters. Two interviewees specifically addressed this positive aspect of private servers. First, Shita stated:

> On the official server I can only play as a Prophet. If I want to play using other characters, I need to invest a lot of time in developing them. On private servers I can learn the characteristics of each occupation. With that knowledge, when I come back to the official server I will know what others need for their occupations . . . I have deliberately created various characters on private servers and researched all of their skills.

Similar sentiments were expressed by Yoshi (female, age 27):

> [On the official server] I did not have sufficient time to develop so many characters, yet I envied others for their good-looking Burst Shots or high-powered

Backstabs. I can get these characters very quickly on private servers, take a couple of shots, and enjoy the instant killing of monsters or other characters. So private servers sooth my frustration . . . It was only after I levelled up rapidly on a private server that I learned what occupations I was really interested in, and I may develop them after I go back to the official server. In *Lineage II* there are so many occupations, the first one you pick may not necessarily be your favourite. I know many Bladedancers who have reached level 79, then found that their second occupations were their favourites. If you haven't tried them, you won't know what you really prefer, you don't have that much time to try every occupation to its limit. Private servers give you experimental space. It's like a practice field, telling you that 'you may become that strong!' It is a space for imagination, for dreaming. Sometimes it turns out to be a motivation for going back to the official server.

To me, private servers are like short dreams, because at the end of the day you have to go back to the official server. Otherwise – like on my private server – everyone is acting alone. The monsters stink, you can reincarnate yourself, you don't need any team, so if you want to have interactions with others, you need to go back to the official server. I feel that private servers are only there to tell me what skills I can have when I grow up, or what places are out there for me to explore, so I can do that later on with a team on the official server.

The meanings of private servers

Private servers represent player efforts to control their own gaming experiences and to satisfy their individual needs. MMORPG players have long-expressed interest in having stronger and more active control over their gaming experiences. Evidence of this is the number of bots, cheats, and cash trades for virtual goods and higher ranks that are used in response to the repetitive 'grind' of monster slaying and other levelling tasks. Private servers support more extensive modifications of environmental settings and fundamental game structure, especially for players who establish close relationships with private server managers. This explains why many players compare multiple private servers with various settings before choosing one to use on a regular basis, after which they regularly interact with the game manager to bend rules, adjust parameters, create more convenient ways of transportation, and so on to fulfill their individual gaming needs. In comparison, game managers on official servers are impersonal or indifferent, with most strictly adhering to established game rules. It is not unusual to find private server game masters regularly chatting with players on public channels, exchanging MSN messages with them, and holding spontaneous activities from time to time.[6] The next two excerpts attest to the more personal nature of player–manager relationships on private servers. The first is from Evilshadow (male, age 30) who expresses the greater level of connection felt with game controllers on private servers:

You feel isolated from the GM [game master] on the official server, you don't have very much interaction with him. He is more like a robot, starting an

> activity as part of his job, not as something that he wants to participate in himself. On my private server, whenever the GM wants to create something, he calls for everyone to create it together.

And similarly Kitty (female, age 22) highlights the personalization and personal contact possible with private server, significantly adding to her pleasure in gaming:

> What attracts me most about private servers is . . . all prayers are answered . . . On my birthday, the server owner built a guild house on the ocean, and gave it to me as a present.

Flexibility also has a down side: instability. Private server operators feel constant pressure from game companies and law enforcement authorities; therefore, players must be prepared to lose their servers and characters without warning. Instability is also increased by the above-mentioned tendency among many players to jump between different private servers in the belief that the next one may fulfill all of their needs. One interviewee pointed out that one of the best features of private servers also represents a reason for leaving them:

> Gaming experiences on official servers are cultivated at a slow tempo, but on private servers they are very fast. At the beginning it surprises you, and when you have reached your goals, you don't have so much desire to stay. And private servers are temporary businesses, you never know if tomorrow morning you will still find them running. When they say we'll close it, it is closed. But official servers, with all that money we pay them, can they do that to us?
>
> (Shita, female, age 25)

Hence, it is evident, that for players, private servers provide multiple functions; and we can identify at least five of these: (a) private servers act as 'photo albums' for remembering past game lives and histories; (b) private servers provide spaces for flexible gaming schedules, thus helping players avoid the pressure they feel to spend large amounts of time working on skills as well as the feeling that they 'must get their money's worth' from official servers; (c) private servers provide buffers for leaving specific games by offering players a mechanism for moving on; (d) private servers act as inexpensive laboratories for serious gamers who want to experiment with new game features that they want to use on official servers; and (e) private servers are playgrounds for leisure players who want to enjoy the pleasures of exploration without having to endure the grinding process of performing repetitive levelling tasks.

Economic relationship between official and private servers

It is evident that official and private servers have a complex economic relationship. This relationship cannot be reduced to a formula in which one side completely suffers from the success of the other. There are four aspects of this relationship that must be considered when determining whether one side or the other receives benefits. First, the game designs and degrees of player participation on some private servers occasionally benefit game companies that want to meet player demands for new and innovative game features.[7] Second, a significant number of gamers use official and private servers concurrently, some simply refuse to join official servers under any circumstances, some player communities established on private servers move intact to official servers, and sometimes private servers help keep jaded players 'in the game' until new versions are released. Third, while money is a factor, it does not appear to be a primary motivator for players to join private servers. Player cost-benefit calculations are influenced by the payment mechanism – as stated above, a monthly fee creates pressure to play more to get value, but that can also create a sense of being controlled by a game. The 'donation' payment system used by many private servers gives them an advantage in this regard. This dual system also encourages simultaneous play on official and private servers, at a rigorous pace on the former and at a leisurely pace on the latter. Finally, in terms of gaming experience, private servers act as supplements to, rather than replacements, for official servers.

In summary, private server players are usually strong fans of gaming in general and of specific games in particular. Almost all of them have experienced play on official servers in the past, and very few limit their play to private servers only. When choosing a free game or a private server, they are not making arbitrary decisions, but are actively selecting a gaming source with certain affective or social qualities that they understand based on their past experiences with official servers. Also, the stability factor is most likely a primary reason why more players do not jump to private servers. MMORPGs are considered ongoing worlds, and participant identities and memories of past gaming events are representative of the large amounts of time they have invested.

Conclusion

As commercial products, MMORPGs are designed to keep mainstream players in game worlds as long as possible so as to prolong the lives of game titles. Low flexibility in terms of rules and experimentation is one result of these marketing goals. In a limited sense, private servers are versatile mutants of official games for players interested in finding more customized gaming experiences. In a broader sense, they can be viewed as new incarnations of Coleman and Dyer-Witheford's (2007) *digital commons* (in which everyone voluntarily contribute to the production and exchange, with no one claiming ownership), a disappearing aspect of gaming due to digital game commercialisation. However, the potential loss of game company revenue means that private server operators will almost always be considered illegal outsiders. In this study, we offered evidence indicating that: (a) despite their illegality, private servers offer some benefits to game companies; and (b) most players move between the two worlds for reasons tied to the current characteristics of commercial gaming and not simply to save money. We have highlighted that gamers have a variety of motivations for using private servers, but these are mostly as an addition to, and to enhance their game play, on official servers. Although private servers are invariably cheaper for players, this does not appear a primary motivation, and while private servers may be used as a counter to issues encountered in the official game version and server, this often ensures that gamers remain in the official game. It is our hope that this information can be used to support further discussion of related issues.

Acknowledgement

The authors would like to thank National Science Council of Taiwan, ROC for their financial support of this project. This research was sponsored through Project NSC 97-2410-H-002-061-MY3. We also want to thank Yu-Hao Lee, who was our research assistant at the time of this project, for his significant contribution.

Notes

1 This system is very similar to that used for popular 'free-to-play, but pay-for-equipment' MMOs, except that all payments are used to support server maintenance.
2 http://lolopo416073.pixnet.net/blog/post/24389410 (Retrieved 1 August 2009).
3 Available http://koukaki.pixnet.net/blog/post/11174203 (Retrieved 1 August 2009).
4 The formal name for a sorcerer in *Lineage II* is 'Storm Screamer'.
5 Certain private servers offer objects (e.g., clothing that makes players invisible) that are not available from official servers.
6 One interviewee told us that a 'GM suddenly announced that we are going to raid a certain boss 10 minutes from now, and he set a transmitting crystal in the castle so that everyone could join this event. Then the GM set free the boss for the players to attack, and he just stepped aside and watched. He did not join us in the attack, but he did revive dead characters for us.'
7 In November 2006, NetDragon Websoft, Inc., published their *JY2* online game, in which they offered game features commonly found on private servers – for instance, accelerated experience point rewards and exceptionally high drop rates for high-level equipment. They called their product a 'high' server. Within a few days, a large number of players reached top scores that would have taken them several weeks on a regular server. Another example can be found in Blizzard's new version of World of Warcraft and Wrath of the Lich King, released in November 2008. Players who have a character of at least level 55 on the account they play are able to create a new level-55 death knight and start levelling from there. This design considerably saves the players from repetitive work for developing a second character.

References

Buckingham, D. (2006) Analyzing digital games. In D. Carr, D. Buckingham, A. Burn and G. Schott (eds) *Computer Games, Text, Narrative and Play*. Cambridge, Polity Press: 88–102.

Coleman, S. and N. Dyer-Witheford (2007) Playing on the digital commons: Collectivities, capital and contestation in videogame culture. *Media, Culture and Society*, 29(6): 934–53.

Csikszentmihalyi, M. (1990) *Flow: The Psychology of Optimal Experience*, New York, Harper and Row.

Ducheneaut, N., N. Yee, E. Nickell and R. J. Moore (2006) "Alone Together?": Exploring the Social Dynamics of Massively Multiplayer Online Games. *Proceedings of the SIGCHI Conference on Human Factors in Computing Systems.* Montreal, Qubec, Canada: 407–16.

Jenkins, H. (2006) *Convergence Culture. When Old and New Media Collide.* New York, New York University Press.

McMahan, A. (2003) Immersion, engagement and presence: A method for analyzing 3-D video games. In M. P. Wolf and B. Perron. (eds) *The Video Game Theory Reader.* New York, Routledge: 67–86.

Murray, J. (2004) From game-story to cyberdrama. In N. Wardrip-Fruin and P. Harrigan. (eds) *First Person: New Media as Story, Performance and Game.* Cambridge, MIT Press: 2.

Rieber, L. P. (1996) Seriously considering play: Designing interactive learning environments based on the blending of microworlds, simulations and games. *Educational Technology Research and Development,* 44(2): 103–7.

Steinkuehler, C. and D. Williams (2006) Where everybody knows your name: Online games as "Third Places". *Journal of Computer Mediated Communication,* 11(4): 885–909.

Taylor, T. L. (2006) *Play between Worlds: Exploring the Online Game Culture,* Cambridge, MIT Press.

Yee, N. (2006) Motivations for playing in online games. *CyberPsychology and Behavior,* 9(6): 772–5.

5 The only (end)game in town

Designing for retention in *World of Warcraft*

Douglas Brown

Many of the elements that make online gaming unique, including but not limited to socializing, group play and inhabiting online worlds can be found within massively multi-player online role-playing games (MMORPGs). This branch of online gaming is rooted in multi-user dungeons developed in the 1980s. The majority of these games share similar designs inspired by multi-user-dungeons and their early commercial incarnations in *Ultima Online* and *EverQuest*. These are online social worlds where character growth and reputation are gained through repetition, and where progress bars are linked to a multitude of actions. Players are given options to manipulate their progress, while being offered the benefits of social interface features, such as chat rooms and instant messaging. Gamers can team up against the game, known as 'player versus environment' (PvE) play, or take one another on within the game, 'player versus player' (PvP) play. These features facilitate a long-term immersive experience. King and Krzywinska (2006, p. 28) note that 'regular visits to a MMORPG world enable players to keep up contacts with friends made in the game, but also create pressure to keep on playing or risk being left behind as others advance in capacity.' Although the genre is a newcomer compared to other video game forms, it has made a heavy impact quickly and is now one of the most ubiquitous ways of gaming online.

MMORPGs have evolved from their original non-commercial model into one of the most profitable video game industry genres.[1] As PC gaming's MMORPG user base grows, various business models have developed to take advantage of these new players. While they range from the innovative in *EVE Online's* embracing real money trading within its virtual economy to facilitate customer retention through to new takes on old formats in games funded by micro-transactions such as *Rappelz* or *Runescape*, charging monthly subscription fees as part of or the whole of the service remains popular. This subset of MMORPGs is dominated by one game that dwarfs not only the rest of the genre but any video game ever in terms of user base, profitability and global reach. Blizzard Entertainment's *World of Warcraft* (*WoW*) is a phenomenon, boasting in excess of eleven and a half million subscribers;[2] a number that has consistently grown since its 2004 release. It has seen off all challenges to its dominance of the sector since. This is far from a one-horse race; Blizzard has faced threats to its hegemony from studios with equal pedigree in Square-Enix's *Final Fantasy XI*, competition with more genre experience in Sony's *EverQuest II* and heavyweight intellectual property in Turbine's *Lord of the Rings Online*. These titles and many more have put up the extremely high initial cost of entry[3] to challenge *WoW*, yet all now maintain market shares at a fraction of their rival.[4].

This chapter examines the success of *WoW* from a design angle and critiques a specific part of the experience, which will be defined as the 'endgame'. In doing so, the chapter offers a deconstruction of the success of *WoW*, focusing specifically upon the centrality, complexity and incompleteness of its endgame, and the approach Blizzard entertainment has taken to the trajectory of the game's development. The chapter considers the role of Blizzard in unashamedly appropriating player developments and features of *EverQuest* particularly in this respect. Further it points to the 'first mover' advantage, around the player experience and path dependency associated with this strategy, which has enabled *WoW* to remain, at least until now, ahead of the competition. This chapter argues that due to Blizzard Entertainment's approach, competitors' products appear to be under-designed in comparison to the *WoW* experience. In conclusion, this chapter suggests that radically new approaches to game design will not work in this field, where older is better and repetition is no bad thing.

WoW's success is unquestionable, but the actual extent of its subscriber base is difficult to verify. Press releases trumpeting subscriber numbers were plentiful after release, but are now drying up since milestones are less frequent and competition is not so threatening. Statistics for this commercially sensitive information are hard to confirm, and public studies are updated infrequently, often relying on anonymous sources. Subscriber numbers are not even the most efficient way to register how many people actually play the game, since players' holding multiple accounts is a fairly common occurrence. Nevertheless, it is clear that if anything *WoW*'s market share has increased beyond the 63 percent the MMOGchart project suggested in 2008.[5]

To maintain this dominance, *WoW* did not remain static. The game has evolved over its history from the breakout title that overtook *EverQuest*. At the time of writing, two retail expansions have been released, with a third imminent, and the game has been frequently updated or 'patched' with new areas, 'content' to explore and character abilities as well as re-balanced, debugged and updated. Graphical quality options, for example, have gradually increased, although the game still looks dated compared to its competition, as Aarseth (2008: 113) observes:

> Significantly, it does not up the ante in terms of more, or nicer. The technical graphics and physics details are significantly less ambitious than with competitors such as *EverQuest 2*, which came out a few months before.

This development strategy of patching and expansions is common for the genre, so it is unusual that *WoW* has had so few retail expansions over its history. The original *EverQuest*, by comparison, released 10 expansions over the same length of time. *WoW*'s two expansions saw the core game updated to new version numbers, so alongside the original game's version 1.0, early 2007's *The Burning Crusade*, version 2.0 and late 2008's *Wrath of the Lich King*, version 3.0 make up three distinct 'ages' of *WoW*. While the focus of this chapter is on the game in its current 3.x state,[6] a holistic approach is necessary in order to understand how the endgame has changed, grown and evolved.

Blizzard's title is not just dominant within the MMO sector but also within the games industry as a whole, eclipsing the sales of any other single game and the vast majority of game franchises or series. US sales of 8.6 million units[7] at $30 equates to $258 million gross from retail alone. Each of *WoW*'s 2.5 million US subscribers[8] pays $15 a month, yielding an additional $450 million annually. Figures are close to the same in Europe, where Blizzard has boasted of 2 million subscribers.[9] WoW, a single game, makes up at least $1 billion of the global games industry's approximate $40 billion value.[10] *WoW*'s business model makes it difficult to pirate successfully for very long, and its low subscription price compared to other MMORPG games makes a WoW account feel like less of a luxury. Even within game theory and Game Studies, *WoW* reigns supreme: 'I am deeply indebted to Espen Aarseth for coining the term "World of Warcraft studies", because this is exactly what Game Studies 2.0 is about' Kücklich (2006) quips in a Digital Games Research Association column. In this book, *WoW* is the focus of two chapters, and at least referred to in most of the contributions. Academic studies have been based around *WoW* to the point where holding it up as shorthand for MMORPGs or online gaming is almost reductive. Navel gazing with regard to WoW is particularly dangerous since much of the game's popularity stems from its utilitarian design philosophy, which generates a level gamescape. Competition has been able to gain a foothold by emphasizing the dimensions WoW tries so hard to level out. *EVE Online* creates a single cohesive world rather than one shattered into separate server incarnations and mirrors this cohesion in its dog-eat-dog design philosophy, where as long as player interactions are within the rules and programming of the game, any conduct is permitted. This philosophical seeding allows a different, more brutal, world to emerge. In contrast, WoW goes to great lengths to ensure that wherever possible players' interaction is consensual. The danger is that *WoW*'s allowances could in this and so many other areas be seen by critics as *the* way online gaming works. It would be both disappointing and dangerous if WoW were to be adopted by critics as well as the industry as the last word on, and blueprint for, online RPG genre gaming. Taylor (2008 p. 187) is right to worry: 'we have a fair number of studies that have focused on a very small number of MMOGs through which we are beginning to get an implied generalized theory of online games'.

Part of this problem could be seen as a result of academic objectivity resulting from some motivations common to traditional textual analysis. Seeing *WoW* as 'whole cloth', a fair text fit for fair analysis in this regard may actually exacerbate the issue, as this assumption runs afoul of designers' commercially motivated attempts to improve or maintain only the profitable elements of the user experience. Textual analysis mindful of the commercial element of the game (Rettberg 2008) is one way to begin to mitigate this worry. *WoW* is certainly an impressive, successful game, but its success is such that it has shifted from a revolutionary or evolutionary project to one concerned primarily with maintaining the status quo with itself at the top. How far financial motivations affect Blizzard's design strategy is arguable, but the consensus among players and commentators is that *WoW*'s designers are well aware of the competition and tune some of their decisions accordingly. It is the maintenance of this state of affairs and how the designers go about keeping players in their multitudes attached to *WoW* rather than migrating to other games, which will be focused upon using 'endgame' analysis.

This chapter addresses the question as to why *WoW* has remained such a dominant force, and addresses some of the worries above through textual analysis. After exploring just what *WoW*'s success consists of, and how it came to such prominence, we observe how 'endgames' work and are defined. Comparing the design of *WoW*'s levelling game, endgame and patching priorities with other MMORPGs will highlight the key innovations in *WoW*'s endgame model and finally, through a return to the commercial angle on design strategy, this chapter attempts to highlight how *WoW*'s longevity and competitive edge rest squarely in this design element. This study builds and draws upon the work of Paul (2010) who highlighted some similar signs of the shifts in design strategy evident in *WoW*'s endgame.

My methodology is primarily that of textual analysis that involves making use of my own experiences as a long-term MMORPG player and *WoW* denizen with multiple maximum-level characters who saw all of these changes first-hand.[11] Krzywinska's (2005) encouragement of the merits of textual analysis also motivates this methodology.

To understand design, the way it seeks to shape the player's experience, and to evaluate the values of a game, it is important to conduct a detailed textual analysis. Being up close and personal with a given game forces you to think through its specificity, helping thereby to insure against the temptations of over-generalization and testing the validity of top-down analyses of 'games' as a general category.

What's in an endgame?

When looking at modern MMORPGs, one could be forgiven for thinking these games, although designed to be played for a long time, do have a finite end. A maximum level, an ultimate weapon, a culminating story arc, or similar, these things could happen and effectively bring down the curtain. There is only so much content in a game, and players will inevitably progress through it faster than the developers can put out more of similar quality. Even given a positive dedication to extending the game and the things it is possible to do within it, surely by the time players reached the top of the ladder proffered by the original game, they would have enjoyed most of what was on offer? It is becoming uncommon to see threads on the *WoW* forums from players asking what to do now their character has reached the maximum level, primarily since endgame content has evolved to be more welcoming, but also since the question has been answered so many times before: 'Life begins at [level] 80', since this is where the endgame (currently) starts.

To be clear, an 'endgame' is the options and activities available to players once the levelling game is complete and their characters cannot grow any further through experience points. In games such as the original *EverQuest*, it took so long to reach the final couple of levels that endgame activities could be said to begin earlier. Other games such as *EVE Online* do not use levels at all but do still have an endgame, although it is more difficult to define. In *EVE Online*, the endgame stems from mastery of particular roles and could be located in the lawless '0.0' regions where player-run groups have freest rein. Much of the seduction of *WoW*'s design rests in the constant, gradual growth of character capabilities and thus player agency, with new moves being unlocked at regular intervals and opportunity to specialize increasing as players go up the ladder of levels. The actual features of endgames, the things available for players to do, vary widely between the different games on the market.

The gap between the values and play styles of levelling game and endgame is so great that there is a strong case for the analysis of endgames and levelling games as entirely separate entities, as this chapter goes on to consider regarding *WoW*. Reaching the endgame begins a new experience for the gamer that might be short lived or, as seems to be the case with *WoW*, the 'hook' (Klastrup 2003) that leads to a much longer relationship beyond the average 15.5 days of online time needed to generate a maximum level character (Ducheneaut *et al.* 2006).[12] Because all that it takes to reach *WoW*'s endgame is time, any given *WoW* player's can eventually reach the endgame, and endgame design is an important element of the total MMORPG blueprint which may occasionally cause friction with the design of the levelling game. The ideal levelling game is a progressive reveal, gradually unveiling elements of the endgame which are useful then and there, but really shine once levelling is over. However, the dangers are that either endgame features are so situational or complex that they serve little to no purpose in the levelling game, or so simplistic and presented with such broad strokes that the resultant endgame lacks depth. This design challenge encompasses an array of features, most of which are beyond the scope of this chapter but addressed in Bartle (2003), Mulligan and Patrovsky (2003) and Koster (2005) among others. Character classes, roles in groups, group cohesion, specializations and other systems which can feed into characters' abilities such as professions or crafting systems all have to be balanced out in this slow reveal and then turbocharged or somehow complicated at the end.

Once the levelling game is completed, the endgame begins. Challenges emerge to replace levelling that are characteristically long-term endeavours. In *WoW*, game modes which have been inaccessible are unlocked: hard mode small group play or 'heroic mode', large group play or 'raiding' and competitively ranked PvP combat in 'arena matches'. Others become more accessible. PvP sub-games or 'battlegrounds' have been largely awkward encounters when available at all during the levelling game, but reaching the endgame allows a more even footing to emerge. Levelling up, character progression and new areas of the world opening all slow down or cease entirely, with the only unexplored areas left being the large group play 'raid dungeons' and solo or small group content designed specifically for the endgame. Various factions in the game offer desirable rewards of items for maximum level characters but require a certain amount of 'reputation' to be earned before the player can receive them, generally via repeating particular actions or quests on a daily basis. The game even offers the option of replacing the experience bar used in the levelling game with a reputation bar for a specific faction, smoothing the transition. Paul (2010) details these rewards and endgame itemization more thoroughly, but it is their positioning as transition from levelling game to endgame activities that is relevant here.

A key difference between *WoW*'s endgame and levelling game is that the items a character wears gain emphasis, as their capabilities to endow better statistics and, by extension, agency, continue to grow when levels cease to increase. Blizzard uses an equivalent level system to distribute statistics on an item, so if an item has a level of 60, the budget of statistics available for it is roughly those of a level 60 character. To demonstrate the discrepancy between player levels and item levels in the endgame, during *WoW*'s original incarnation, the maximum level was 60, while item levels available from raiding ranged from 65 to 92. When the level cap was raised to 70, this range increased to 115–164 and currently at two-thirds of the way through the third expansion with a maximum level of 80, raiding gear ranges from 200 to 272.[13] Thus, emphasis has gradually been shifted from a character's level – her progression through the levelling game – over to her gear, effectively her progression through the endgame.

In *Guild Wars* by comparison, limited access to the endgame is offered to players right from the start in the form of pre-made maximum level characters designed to allow instant PvP action. Fighting against other players is only possible in the endgame, where characters who have completed the levelling game are able to explore dungeons for better gear and take part in PvP that influences the availability of PvE content. While it is possible to play the endgame entirely with pre-made characters, the accrual of points to grow these characters takes much longer than it would with a levelled character.

In *EverQuest* as a third example, the same ladder of raid encounters as the PvE side of *WoW* exists, but in a more regimented fashion, since PvP competition is a rarity. Reaching the maximum level replaces experience with 'alternate advancement experience' that can be used to buy statistics, skills and character abilities, some more fundamental than others and none absolutely necessary to tackle raid content. Raids form gates that need to be cleared in order to allow access to new areas (known as 'flagging') for solo and small group hunting.[14]

In none of the above examples, nor MMORPGs in general, is the distinction between the levelling game and endgame as clear as this analysis is suggesting it should be. To some extent, this is because few games make the distinction as rigid and absolute as *WoW* does. It would be foolish to completely divorce the two since the presence of the endgame strengthens the levelling game significantly. One of the most tempting options presented to the player upon reaching the endgame is reprisal of the levelling game on a new character. *WoW* makes this even more desirable by splitting the world up into two factions whose paths to endgame vary, particularly during the first half of the levelling game. The recently announced *Cataclysm* expansion revamps the *WoW* 1.0 world significantly in an attempt to further encourage this reprisal, but still focuses the majority of its content on an expansion of the endgame.

A high-level character's resources can smooth out the levelling game for a second character, showing in microcosm one of several synergies between levelling game and endgame. Those who have reached the endgame may contribute to others involved in the levelling game voluntarily and speed up their levelling. They also involuntarily make the levelling game easier for everyone by stimulating the economy (Castronova 2001). Endgame players generate larger amounts of in-game currency than players in the levelling game. This wealth feeds into the values of lower level commodities since endgame players will be purchasing these for other characters on the same account or in order to change their character's specialization, and trickles down to create a net increase in value of the economic contributions of those in the levelling game. This makes it less problematic for levellers to accrue the large sums necessary for static costs that speed access to the endgame such as mounts for faster travel. Castronova (2006: 197) touches upon this effect when discussing the wider ramifications of 'mudflation':

> as the cash flows upward and the equipment flows downward, the net result is that the most advanced players are often sitting on very large amounts of cash, while the least advanced players often have very good equipment.

Although Castronova is referring to specific sorts of game designs, these observations hold true for *WoW*. Other advantages a strong endgame brings to a levelling game are less quantifiable. A large community of players in place at the top who do not leave the game creates a more positive environment for players starting out, with access to knowledge and expertise in the ways of the game world readily available. Partly as a counterweight to the ruthlessness of some players' activities targeting new players in *Eve Online*, the 'Eve University' training guild was founded. A large endgame player base assures an easier and more enjoyable experience of learning the game for new players, and lubricates the in-game economy to enhance the new players' agency within the levelling game.

While learning the game may be eased by the presence of a large endgame community, the experience of play can potentially be far less pleasurable. Endgame items can increase characters' potency to the point where a character still in the levelling game would have little to no chance of a fair fight, yet maximum level characters are in some games free to kill off anyone they like. Some of these actions, colloquially known as 'ganking' arguably make for a healthy, competitive environment, while others stray into the realm of grief play (Myers 2009). *Warhammer Online* takes a somewhat extreme solution to this problem by causing any player a significant number of levels above their target to turn into a giant chicken when attempting to attack, reinforcing the game's commitment to 'skill'. *WoW*'s approach to these actions is laissez-faire by comparison, letting players decide on the rule-set and PvE/PvP bias of a particular server before they create a character, and allowing transferral of characters between these servers for a nominal fee.

Endgames inevitably split developer focus both during the initial game design and when additional content is being released. As will become apparent, this resource management hurdle is one of the primary ways Blizzard is able to maintain *WoW*'s market dominance. Blizzard often justify their focus on the endgame with a desire to keep looking forward rather than looking back at 'old' content. Prior to the announcement of the forthcoming *Cataclysm* expansion, which does just that, lines such as this from Blizzard representatives were commonplace: 'We are nostalgic about many of our old instances, but sometimes it's best to continue evolving and producing new content for players to enjoy'.[15]

While this primarily manifests in content patches adding things for endgame players to do rather than diversifying or improving the levelling game experience, sometimes changes made to re-balance the endgame can have an impact on the levelling game. Particularly, in *WoW*, it is not uncommon for characters' various moves and abilities to be edited with every patch. Changes made to balance how classes use these moves in endgame encounters with specifically released content or against one another may echo down to how the skills are introduced in the levelling game, or unbalance classes at lower levels. A recent patch reduced the hit points that druids natively received when they shape shifted to bear form, since this class was doing too well in endgame PvE encounters. Aiming to re-balance the endgame, the developers unintentionally made the levelling game more difficult. Similarly, a recent change was made to Warlocks' voidwalker pets' 'sacrifice' ability. Whereas previously the pet could be sacrificed in order to shield the character, now the shield activates in exchange for a portion of the pet's health. The change was made to make Warlocks more competitive in endgame PvP, and resulted in the ability making less sense when introduced early on in the levelling game. Endgames are thus a design challenge, containing a vocal player-base clamouring for content, whose demands must be judiciously met by the developer, also mindful of their duty to the levelling game which brought these players to this point. In short, it has a secondary function as a retention tool, to which we now turn.

The endgame as a retention tool

The player base for MMORPGs is one of the most diverse in the industry (Taylor 2006). Part of the challenge *WoW* faces when attempting to safeguard its position is appealing to this player base. *WoW*, as the dominant game, has shifted from attempting to accrue more subscribers into aiming at retaining those it already has through patching and expansions. This is a fundamentally different aim to its competitors and places even more emphasis on *WoW*'s endgame. Most of the criticisms gamers level at *WoW* will be at the endgame since it generally remains static, while issues with the levelling game tend to go away swiftly since they can be avoided or levelled through. This is also true for *WoW*'s competitors, which mostly ape its construction. In fact, this may be the best way of defining the endgame: the position in the game where the majority of the community is located, both within the game itself and within discussion of the game externally. The endgame is where the 'meta-game' resides. In the case of *WoW*, the meta-game is huge. Some *WoW* blogs focus on fishing or novelty pet collection, while others cover the depths and intricacies of specific classes. Arena matches are treated like an E-sport complete with breathless commentaries, star players and prizes, while specialist sites analyze game mechanics in depth with communities of programmers, game industry members and statisticians to formulate the endgame strategies. This meta-game infrastructure helps retain players by expanding resources available to allow deepening of their preferred endgame playing style (Taylor 2006). Blizzard reached out to the meta-game from the outset by allowing their interface code to be modified, enabling players to create applications enhancing particular play styles, from tools allowing role players to display back-stories in-game through to meters allowing raiders to calculate the damage their team is doing. Blizzard later renewed this focus by providing data covering all characters' stats and gear on a website, the *WoW* armoury.[16] Blizzard has not only fostered the meta-game by providing it with sufficient tools to grow and develop but also somehow satisfied the community with endgame content provided at a glacial pace compared to how often competitors, in the form of whole new levelling games, spring up.

The other element of the endgame's role as a retention tool is its position in the expansion schedule. It is important that at its core the endgame remains static once players enter it in order to best cultivate the community, yet it also needs to grow in order to keep players interested. The initial series of *WoW* patches up to 1.12[17] (September 2006) represents growth and accrual of new customers. New dungeons for small groups are added, first Maraudon and then Dire Maul, new quest locations for the levelling game are inserted and the whole face of the game is changed with major additions of new systems to attract more players; a battlegrounds system streamlines PvP, and while the 'honour' system makes PvP attainments more visible by associating them with titles, characters can put in front of their names (Paul 2010). Endgame content was added too, but this mainly took the form of large raid dungeons being released at long intervals. The model eventually shifts from accrual to retention, focusing development on the endgame in the patches from 1.12 through to 2.4. The current situation sees a shift from a game that attempted to impress and attract new customers as well as challenge them with exclusive, difficult content to a game whose prime function is keeping everyone entertained with sufficient content to allow the *WoW* endgame to compete against other levelling games. Blizzard has moved from pushing a levelling game onto potential customers to tending existing customers within its endgame. Every patch since the initial set includes more endgame content for raiders and either major changes to or new content for the PvP aspect of the game. Sporadic releasing of different types of content was replaced by a controlled schedule when *Burning Crusade* was released. Each shop-sold expansion comes with a new levelling game (increasing the level cap by 10 levels and providing content equivalent to about half of the original levelling game) and the new endgame will have 3 or 4 major content updates before the next shop-sold expansion extends the levelling game again. This is the route Blizzard has taken since the first retail expansion was released in 2007. The 2.x era of *WoW* shows the shift from the original subscription accrual strategy towards the current retention angle, with endgame content for raiders focused upon initially in patch 2.1 (May 2007) then gradually dialled down in 2.3 (November 2007) which focused on providing a broader spectrum of content. The first expansion culminated in the 2.4 patch (July 2008), an attempt at providing endgame content for everyone with the release of the extremely difficult 'Sunwell' raid alongside a slew of single player and small group content. Patch 3.0 continues the focus on accessibility and breadth of content at the cost of exclusivity in the three patches so far, each of which has been swiftly followed by minor patches reducing the difficulty of the content. While Blizzard may well have wanted to switch from accrual to retention earlier, as is implied by how drastically they changed the endgame prior to the release of the third expansion in the 3.0 patch, detailed below, their hands were effectively tied by the expansion schedule. The endgame can only be altered in the slew of new content that comes with a shop-bought expansion, since otherwise major revisions will harm the community.

The endgame player base

But who makes up this player base, and who needs to be appealed to? This question has been the topic of much research among games academics, with Bartle's (1996) archetypes of 'Explorer', 'Achiever', 'Killer' and 'Socializer' player types forming the basis of most criticism of MMORPGs. The Bartle types are reliably found in multi-player online gaming yet, I would argue, not necessarily the best targets for designers to aim their content squarely at when designing endgame content. Looking at *WoW*'s patching schedule and history, a cursory assessment would make it appear that those who would become dissatisfied with *WoW*'s endgame most quickly would be Socializer and Explorer types. Socializers could be frustrated by *WoW*'s awkward chat system, slow implementation of voice chat and lack of social features compared to newer games. Explorers could be bored by the slow addition of new terrain and relative inaccessibility of new places to explore outside of raid groups. Features explorers enjoyed such as the ability to get to places that should be inaccessible via techniques or items in the game were removed, although this could be seen as balanced out by the addition of flight in the endgame of *WoW* 2.0, which allows sections of the world to be explored anew. As a design, *WoW* is a compromise, palatable to all four Bartle types, but not a panacea to any one of them, so enjoying any of these elements would not be sufficient to explain why *WoW* has maintained its popularity so long.

Most patches seem aimed squarely at Achievers through the regular addition of raids and Killers through the constant expansion of PvP combat options. Socializers and Explorers would be pleased, for a while, by the bi-annual expansions and the occasional major update such as the achievements system and the eventual arrival of graphical upgrades or voice chat, but not to the point where they would not desert the game. Hegemony's side effect of the creation of a very social meta-game could be expected to slow this process further, as could elements of the initial design such as the horde/alliance split and the instancing of most PvP combat areas, essentially corralling Killers and Achievers by giving them more reason to play together than roam the world alone, thus minimizing their impact on Explorers and Socializers' gaming experience. Despite this, the focus on Achievers and Killers should have generated a gradual exodus from *WoW* to other games.

Blizzard's shift from designing for accrual to designing for retention can be seen as an attempt to cater for Socializers and Explorers in an attempt to make the work they had done on content for Achievers and Killers more accessible. The 3.0 patch, the last major content patch in *WoW* 2.0, marks the point where the developers realized they were trying to serve diametrically opposed ends in the form of accrual *and* retention of customers, and embraced the retention model by dramatically reducing content difficulty. The specific line detailing these changes released soon after patch 2.4 to prime the game for the forthcoming second expansion reads as follows:

> All raid dungeons from Karazhan to Sunwell Plataeu as well as the outdoor encounters of Doom Lord Kazzak and Doomwalker have been adjusted and now all creatures and bosses have less health points and do less damage.

This major shift in the focus of the PvE endgame basically turning what had been a complex series of difficult encounters into a much easier free-for-all occurred alongside the early implementation of the achievements system, which had been touted as a feature of the second expansion. This occurred much later than the controversy Paul (2010) addresses when discussing 'welfare epics' from a purely PvP perspective, but worked to cement the 'rhetoric of rewards' that he isolates as core to the future of the game, a strategic shift of the whole design to the 'fine-tuned incentives and rewards structure' that Duchenaut *et al.* (2006: 314–315) isolated as the game's great strength, proofing it against the chief weakness they also highlight: an inaccessible endgame. This philosophical shift has galvanized *WoW* as the situation their study predicted came to pass: 'the situation could devolve dramatically, should (the) supply of newcomers be exhausted. Of course, it could also be that this market is growing so fast that *WoW* will never have to face this issue' (ibid).

Paul describes a PvP element of exclusivity, dialling back some rewards in favour of the 'needs' of Achievers and Killers, and Blizzard's aiming towards 'potentially increasing the allure and perceived legitimacy of these epics' (Paul 2010: 172), but even in light of this, the non-stop commercial dominance of *WoW* both in its accrual and retention guises shows that in its specific case, a twist on the Bartle types may be necessary to explain the designers' thinking. Even more so, reference to the Bartle types alone does not explain why players continue to prefer *WoW* over the multitude of other entirely new MMORPG experiences with content for all four player types which challenge it still and have attempted so often to seduce its subscribers. Surely, all four player types, but particularly explorers and socializers, should prefer a whole new world with new challenges to face, continents to explore and ways to take one another on over a gradual drip-feed of new content in an old world which *WoW* inevitably offers much of the time.

Some of the answers to this have been given in studies that deal with player psychology (Yee 2005). Taylor's (2006: 28) observation that 'what you do in *EQ* is immerse yourself in a space . . . people live and, through that living, play.' is still valid and could explain the attachment many gamers have to the world Blizzard has provided for them to inhabit. Other approaches to player types have been tried either in isolation or building upon Bartle's research (Callios 1961; Edwards 2001; Kiersy 1998). These mainly reinterpret the model through statistical studies or philosophical perspectives, but generally steer clear of the commercial sharp end of game design. We can find an exception outside of academia somewhat by making use of articles written by Mark Rosewater, lead designer of collectable card game *Magic: The Gathering*, about the kinds of players he felt he needed to appeal to when coming up with new cards. He gave his three different 'psychographic profiles' the nicknames of Timmy, Johnny, and Spike (Rosewater 2002). Timmy represents the 'power gamer' who likes big, impressive cards that emphasize one particular game mechanic, looking for the extremes of the system and instant gratification these pyrotechnics bring. Johnny was the somewhat more social, combination-oriented player who found the web of the game design intriguing and enjoyed unexpected combinations or strange takes on familiar mechanics, essentially seeking novelty and depth, deferred gratification. Spike was the tournament player or single-minded gamer, looking to win at any cost and seeing the game as a tool by which games against others could be won regardless of style or the 'spirit of the game'. Rosewater later added another type to his mix, the more story or lore focused 'Vorthos/Melvin' role-player type, and revised his article to bring it more in line with the Bartle types (Rosewater 2006, 2007). However, he defends the commercial imperative that underlies his player types in an update, discussing the commercial dimension of his research candidly:

> If the designer is the marketer, then it is his responsibility to understand his product and his consumer. If I want players to buy the first pack and then want to buy the second, I have to understand what I can create in that first pack that will inspire the desire to buy the second. This is where Timmy, Johnny and Spike came from. I wanted to understand who Magic's consumers were. Once I figured it out, I gave them names so that R&D could talk about them.
>
> (Rosewater 2007, online)

Rosewater is first and foremost a game designer and was writing as such, while Bartle wrote from a more academic perspective. More importantly, elements of Rosewater's role and position when he wrote these articles make his work directly relevant to the material we are focused on when looking at *WoW*. Rosewater inherited *Magic* after the game attained dominance of its market sector. Aside from a brief spell in the doldrums during the *Pokémon* fad of the late 1990s, *Magic* was the dominant game-producing regular expansions that balanced, expanded and added new mechanics while crucially maintaining gamers' interest and keeping them buying cards. It gradually moved from releasing very different expansions to 'cycles', groups of three expansions over the course of year that shared a similar narrative focus before switching to a new set. The original core set of cards was also updated on a biannual basis. This had been built into tournament rules, where the most prestigious tournaments allowed play only with the core set and the last two cycles. This allowed designers to dial up the power and extremity of cards as the end approached, emphasizing the mechanics focused upon in the cycle. The celebrated 2002 'invasion' cycle showcased this new focus and brought a lot of players back to the game. It is no coincidence that Rosewater's writing about player types emerges after this renaissance. There are similarities in terms of the endgame or expansion design of *Magic* at that point in its development and current generation *WoW*'s retention model. When Rosewater wrote about making cards for players, he was talking about retaining interest in a very similar model to the challenge *WoW* faces now. The three archetypes differ from the Bartle types precisely because of this difference of focus. Rosewater did not want to analyze how or why these gamers played that was incidental, rather he wanted to know which buttons to press in order for them to keep buying more cards. While Rosewater was selling packs of cards, *WoW* is selling months of game subscription, but otherwise the two games' situations are strikingly similar. While the initial design of a levelling game should be aimed at the Bartle types, the expanding of a successful endgame will focus more on satiating the Rosewater profiles. Ironically, *Magic* is currently no longer at the top of its market, having been eclipsed by the spin off *World of Warcraft Trading Card Game.*

Elements of the Bartle types are emphasized in Rosewater's three archetypes, pinpointing aspects that need to be massaged in order to keep gamers hooked. This targeting combined with the *WoW* endgame's unusual maintaining of focus on the primary game play mechanic of its levelling game could go some way towards explaining not only the retention but also why the game can in some cases be so addictive. Viewed from this angle, the lack of appeal, through content at least, to the more socially focused players becomes more explicable. An analysis of the current endgame options available to *WoW* players at the 3.1 point may help cement these theories on how to effectively analyze endgames.

The *WoW* endgame in depth

Currently, *WoW* gamers reaching level 80 have access to three broad strands of endgame content. This has evolved from the initial endgame where, as Paul (2010: 159) observes: 'Raiding was one of few play options available to maximum level characters . . . the game that shipped offered almost exclusively group based content to players when they reached 60'.

The first option is solo or small group content. Quests unlock in the central city of Dalaran as well as in the zones towards the end of the levelling game. These tend to be daily repeatable quests requiring one to three players to complete, and recent patches have added many more. In addition, all the previous dungeons players could have encountered in the levelling game are re-tuned to 'heroic' versions designed for maximum-level characters. Alternatively, players can carry on experiencing levelling game content and receive gold and achievements instead of experience, since the amount of levelling game content in both 2.0 and 3.0 is considerably more than necessary by design, or try to complete other achievements, which generally involve collecting items or repeating actions to fill a progress bar.

The second route is raiding, where 10-player content (double the usual number of players allowed in a dungeon) or 25-player content is available. The main reason to raid is to receive items that make characters more effective at raiding. The dungeons are also where the major villains of the storyline reside and story arcs encountered during the levelling game are concluded. Raid dungeons reset every week and dozens of trawls will be necessary for a character to receive all the items they could possibly use. Raid content gradually gets harder and a new dungeon is released approximately every quarter, requiring items from the previous raid for characters to be most effective tackling it. Raiding in *WoW* 3.0 is much more forgiving than in earlier versions, with the developer's mantra 'bring the player, not the class' yielding the possibility for raid teams to run the gamut from committed gamers who play five nights a week through to more socially oriented groups who raid sporadically.

The third route is arena matches and PvP combat in the form of battlegrounds and the public PvP arena of Lake Wintergrasp. Arena matches take place over a season that generally concludes around the same time as the next major raid dungeon is released. At the end of a season, prizes are distributed, and over a season players earn points weekly which they use to buy equipment that makes them more effective at PvP combat. The latter is also true for the other PvP formats, although the gear rewarded is less potent. New seasons launch with upgraded items available, so that PvP gamers need to start gathering gear again to be at their most effective. Gradual improvement regardless of the path chosen is characteristic of the *WoW* endgame, with all the different paths allowing characters to advance in power, albeit at different rates as they accomplish different goals.

Players tend to dabble in all three routes and generally specialize in one or two. Being even moderately effective or just spending time taking on the content rewards tokens that can be turned in for relevant rewards. Taking up multiple routes is implicitly encouraged through overlaps built into the system. It may be easier, for example, to supplement PvP with items from raiding, especially weapons, which require very high ratings to purchase from the PvP strand. Moreover, PvP items designed to enhance combat, particularly those which remove effects that cause loss of control of a character, can make raid encounters easier. Both strands require consumables for maximum efficiency, and these are bought with gold easiest accrued through solo or small group content. Reputation items function as either minor rewards for other strands which can be accessed more quickly (without the weekly raid re-setting or doling out of arena points) or enhancements for other items which remain useful for a long duration. Narrative exposition is split between solo content and raid content, both often bleeding into each other. The 3.1 solo content hub, a coliseum, evolved into a fully constructed raid arena in 3.2. Narrative is drip-fed to players, functioning as a reward for completing a series of quests or defeating milestone raid encounters. Solo content and raid content need to be played through if players want to see the whole conclusion of the story begun during the installation sequence of *Wrath of The Lich King*.

While the content certainly provides options for all the Bartle types, looking at how they are provided for tells us little about why *WoW* is more appealing than the alternatives. Simultaneously, the Rosewater types, designed for a collectable card game, fit awkwardly into the MMORPG model. There are no 'combos' to isolate exactly, nothing so simple as a card's attack/defence statistics to function as an indicator that it would favour a particular player type. Paul's (2010: 173) conclusion regarding 'welfare epics' and the life of the game may answer half of the question:

> The moment players cease finding entertaining things to do within a game they are likely to leave, but by integrating rewards to solo and small group players, (game designer) Tigole's 'welfare epics' are likely to assimilate more players into the never-ending treadmill to keep subscription dollars rolling.

However, when we observe the designer as the marketer and hybridize Bartle's players with Rosewater's consumers, a different picture begins to emerge. The endgame caters for those seeking instant action with solo content and PvP battlegrounds and those seeking more diverse, complex, delayed gratification with long-term raid dungeons and skill-testing arena matches as well as hefty daily 'grinds'. It supports players' meandering through the content just to see its extremes in action in the easier 10 player raid instances or the all-out PvP battle in Wintergrasp. At the same time, it is also easy to slice through the content and find a direct way to maximize a character's effectiveness over time.

WoW provides for the different player types its drip-feed of improvement and agency is controlled by the weekly resets rather than how much time or effort players put into the system. Alongside the constant tweaking of PvP balance and raid content difficulty, Blizzard retain control of their endgame and allow players to grow at pace with the speed of content delivery, or at least make accurate predictions over when players will run out of direct things to do in a particular strand, by calculating the average number of lockout iterations necessary for a player or group of players to attain a majority of the rewards on offer. By retaining control over the speed of the treadmill by rationing everything weekly, Blizzard to some extent sidestep the problem of players essentially wresting control of the game from designers in the endgame by eating through content at a pace which developers cannot maintain. Simultaneously, they cater not for the play styles of their gamers as players but rather the buying decisions of their gamers as consumers, offering a product which their rivals cannot match because they might as well be competing in different markets. The *WoW* endgame can only be challenged by another endgame, not by a levelling game, because the two are fundamentally different entities. As such, the only competitor *WoW* should fear would be another entrenched endgame, and the market sees to it that other games which get that far are inevitably niche titles with an endgame focus distinct from that of *WoW*'s utilitarian philosophy.

It is the relatively static nature of the endgame that allows this edifice to keep functioning. We have established that endgames act as retention tools to keep players playing. This is accomplished in a number of ways. Some games provide infinite scope to some of their systems. *EverQuest's* 'alternate advancement

history is built upon. All these help contribute to the sense of worldliness a trial subscriber can feel in *WoW* (Krzywinska 2006), affecting the initial buying decision. The founders of *WoW*, many of whom are still its senior designers today, began as smart, vocal endgame *EverQuest* players approached by a Blizzard executive and given the chance to build a better mousetrap. Unashamedly, re-using many of *EverQuest*'s features and dealing with issues *EverQuest* players had encountered allowed *WoW* to distinguish itself. The textual poaching (Jenkins 1992) was pretty much literal in this case, and the poachers became the gamekeepers of the new franchise. Blizzard's existing global infrastructure from its earlier strategy games was brought to bear to ease any growing pains. Lucky timing, strong infrastructure and a long development time meant that *WoW* was able to take over, but in the current market this is extremely unlikely to happen again. While *WoW* did not invent the genre, it has successfully redefined it since the game moved away from competition and towards retention in the 3.0.2 patch a month before the second expansion. Through this, *WoW* has essentially proofed itself against a coup similar to the one it managed to pull off over *EverQuest*. *WoW* is pervasive now; much like a social networking system it defines an aspect of gaming culture. We have established that radically new approaches will not work in this field, where older is better and repetition is no bad thing. Time will erode *WoW*'s hold on the market and it will eventually relinquish dominance to a successor, and this element, Blizzard's 'exit strategy', could come to define and demarcate this particular avenue of online gaming. But until then, *WoW* remains the only (end)game in town.

Notes

1 Western consumer spending on MMORPGS has been estimated in excess of US$1.5 billion. (Harding-Rolls 2009).
2 Activision-Blizzard's February 2010 fourth quarter earnings call: http://investor.activision.com/results.cfm.
3 WoW cost in excess of US$50 million to develop and release, and the barrier to entry is always high, even for more modest efforts, as detailed in Castronova and Falk (2009).
4 WoW's Western market share is reported as more than 58% by Crossley (2009) citing Harding-Rolls (2009).
5 http://www.mmogchart.com/Chart7.html.
6 Patch 3.2.2 at the time of writing. Inevitably, *WoW* will be a very different game upon the publication of this analysis, as many of the differences between its state when Paul (2010) was published and the present have motivated this piece.
7 NPD research data provided to Gamasutra, 29 July 2009: http://www.gamasutra.com/php-bin/news_index.php?story=24518.
8 Blizzard Entertainment Press Release 22 January 2008: https: //eu.blizzard.com/en-gb/company/press/pressreleases.html?080122.
9 *ibid.*
10 Crossley (2009) citing Harding-Rolls (2009) claims WoW generated $2.2 billion in subscriber fees over 4 years in the west. The figures used above are, if anything, conservative. They utilize older data for the subscriber numbers of individual regions, while total subscriber numbers have risen. Blizzard has diversified the services they offer to WoW account holders, always charging a fee, and begun selling in-game vanity items directly. They also do not account for Blizzard's far-east operations since there has been controversy and murkiness in this area ever since early 2008. Chinese WoW players do not

subscribe monthly, but instead buy game time on a pay-as-you-go basis. Before figures became unreliable, it was claimed that there were five million WoW players in China.

11 This can all be verified on the EU wow armory page for my main character at the following URL, but only while my account is active: http://eu.wowarmory.com/character-sheet.xml?r=Emerald+Dreamandcn=Henchbeard.

12 Although this study was written using data from original 1.0 WoW, the experience system has been patched since to return the amount of playtime necessary to reach maximum level in 3.0 to the same amount necessary to reach in 1.0. The levelling game still takes around the same amount of time even though it encompasses more content.

13 Blizzard recently added functionality to WoW allowing the viewing of item levels, so this information is now confirmed from in-game sources. The final level 272 gear referred to is Trial of the Grand Crusader Tribute loot.

14 The complexity of these endgame labyrinths of content and flags is showcased in the 'planes of power' expansion mapped out at http://www.rising-dawn.com/EQProgression/PoP/Images/pop3.png. The initial design of WoW 2.0's endgame looked similar, if a little more forgiving before Blizzard turned away from this approach to endgame design. They produced their own map which, while now redundant since all these restrictions have been removed, remains archived at http://www.wow-europe.com/en/info/basics/bcattunement/.

15 A forum response by Blizzard community manager Zarhym from 7 September 2008, archived at http://blue.cardplace.com/newcache/us/7903651737.htm.

16 http://www.wowarmory.com.

17 A full, detailed archive of major WoW patches is online and provided the material for this analysis: http://www.wow-europe.com/en/info/underdev/implemented/.

18 Information from archived news posts on Gamasutra and Wikipedia since game companies attempt to draw as little attention to these sorts of actions as possible after informing affected players.

References

Aarseth, E. (2008) A hollow world: World of warcraft as spatial practice. In H. G. Corneliussen and J. Walker-Rettberg (eds), *Digital Play and Identity: A World of Warcraft Reader.* Cambridge, MIT Press: 111–22.

Bartle, R. (1996) Hearts, Clubs, Diamonds, Spades: Players Who Suit Muds. Retrieved 25 May 2010, from http://www.mud.co.uk/richard/hcds.htm.

Bartle, R. (2003) *Designing Virtual Worlds*, New York, Pearson Educational.

Castronova, E. (2001) Virtual Worlds: A First-Hand Account of Market and Society on the Cyberian Frontier. *CESifo Working Paper Series No. 618.*

Castronova, E. (2006) *Synthetic Worlds: The Business and Culture of Online Games*, Chicago, University of Chicago Press.

Crawford, C. (2003) *Chris Crawford on Game Design*, Indianopolies, New Riders.

Crossley, R. (2009) World of Warcraf Dominates MMO Market. Retrieved 24 March 2009, from http://www.edge-online.com/news/world-warcraft-dominates-mmo-market.

Ducheneaut, N., N. Yee, E. Nickell and R. J. Moore (2006) Building an MMO with mass appeal. *Games and Culture*, 1(4): 281–317.

Harding-Rolls, P. (2009) Subscription Mmogs Life Beyond World of Warcarft. Retrieved 24 March 2009, from http://www.screendigest.com/reports/09subscriptionmmogs/SD-09-03-SubscriptionMMOGs/view.html.

Klastrup, L. (2003) A Poetics of Virtual Worlds. *Proceedings of the Fifth International Digital Arts and Culture Conference*, Melbourne, Australia.

Koster, R. (2005) *A Theory of Fun for Game Design*, Scottsdale, Paraglyph Press.

Krzywinska, T. (2006) The Pleasures and Dangers of the Game: Up Close and Personal *Proceedings of DiGRA*, Vancouver, Canada.

Krzywinska, T. (2008) World creation and lore: World of warcraft as rich text. In H. G. Corneliussen and J. Walker-Rettberg (eds)., *Digital Play and Identity: A World of Warcraft Reader*. Cambridge, MIT Press: 123–42.

Krzywinska, T. and G. King (2006) *Toob Raiders and Space Invaders – Videogame Forms and Contexts*, London, IB Taurus.

Kulklich, J. (2005) Game Studies 2.0. *Proceedings of DiGRA*, Vancouver, Canada.

Mulligan, J. and B. Patrovsky (2003) *Developing Online Games: An Insiders Guide*, Indianapolis, New Riders.

Myers, D. (2008) Play and Punishment: The Sad and Curious Case of Twixt. *Proceedings of The [Player] Conference*, Copenhagen, Denmark.

Paul, C. A. (2010) Welfare epics? The rhetoric of rewards in World of Warcraft. *Games and Culture*, 5(2): 158–76.

Retteberg (2008) Corporate ideology in World of Warcraft. In H. G. Corneliussen and J. Walker-Rettberg (eds), *Digital Play and Identity: A World of Warcraft Reader*. Cambridge, MIT Press: 19–38.

Rosewater, M. (2002) Timmy, Johnny and Spike. Retrieved 25 May 2010, from http://www.wizards.com/Magic/Magazine/Article.aspx?x=mtgcom/daily/mr11.

Rosewater, M. (2006) Timmy, Johnny and Spike Revisited. Retrieved 25 May 2010, from http://www.wizards.com/Magic/Magazine/Article.aspx?x=mtgcom/daily/mr258.

Rosewater, M. (2007) Melvin and Vorthos. Retrieved 25 May 2010, from http://www.wizards.com/Magic/Magazine/Article.aspx?x=mtgcom/daily/mr278.

Taylor, T. L. (2006) *Play Between Worlds: Exploring Online Game Culture*. Cambridge, MIT Press.

Taylor, T. L. (2008) How a Pvp Server, Multinational Player Base, and Surveillance Mod Scene Caused Me Pause. In H. G. Corneliussen and J. Walker-Rettberg (eds), *Digital Play and Identity: A World of Warcraft Reader*. Cambridge, MIT Press: 187–202.

6 The boardgame online

Simulating the experience of physical games

Neil Randall

While sales of video games have thoroughly dwarfed sales of boardgames – 2009 sales in the United States of $11.43 billion for digital games versus 2008 sales of $791 million for boardgames[1] – boardgames show no signs of going away. Classic games such as *Scrabble*, *Risk*, *Pictionary*, *Scattergories*, *The Game of Life*, *Clue(do)*, and *Monopoly*, not to mention Chess and Checkers, continue to thrive on toy store shelves, and tie-in boardgames for such media franchises as *Twilight*, *Harry Potter*, and *CSI* appear whenever the popularity of the original movie or television show creates viability. Over the last two decades, German-style boardgames (also known colloquially as Euro games) have also claimed significant market share, with titles such as Klaus Teuber's *The Settlers of Catan*, Klaus Jürgen Wrede's *Carcassonne*, Andreas Seyfarth's *Puerto Rico*, Alan R. Moon's *Ticket to Ride*, and also numerous products based on *The Lord of the Rings* having garnered large followings. Boardgames have received little in the way of academic attention during the digital game era, with the vast majority of journal articles, monographs, and scholarly anthologies covering video games exclusively. Of course, game studies as a field has shown sustained growth only in the last decade, and many (if not most) game-playing scholars come from a video game background, so the dominance of video games among scholarly consideration is understandable. Boardgames offer play experiences far different from video games, with well-known topics in game scholarship such as immersiveness, narrative creation, player identity, playing spaces, and simulation techniques in need of re-consideration in light of these experiences.

This chapter focuses on two of those topics, playing spaces and simulation techniques, within the context of the playing of boardgames. Specifically, I examine the means by which programmes designed to allow online boardgame play, seek not only to replicate boardgame components but also to simulate the experience of face-to-face play. To this end, the chapter is not about simulation techniques in the more typical game-related sense – i.e., the means by which the game itself imitates specific elements of the real world (or the fictional world it represents). Instead, I examine the simulation of the physical components and social environments in which the physical games are designed to be played by exploring computer programmes produced specifically to allow this play to occur. The chapter concludes with a semiotic analysis of some of the underpinnings of these simulative elements, drawing specifically from the social semiotic theory of 'modality', and an application of Charles Pierce's concept of 'indexicality' – concepts I will explain in more detail later on. My purpose is to situate boardgame play within the growing study of digital game play whilst demonstrating the need for a sustained exploration of their differences. Playing games online is of course a digital activity, but playing boardgames online, even when the boardgames are presented purely digitally, combines the virtual and the physical in a challenging hybrid manner.

The boardgame and games studies

Before the introduction of video games, game researchers included boardgames in their analyses to varying degrees. Caillois (1961) touches on chess, checkers, and card games in his wide-ranging discussion, but the boardgame as such does not undergo extensive analysis outside of his famous classification of games. Suits (1978) similarly treats games more or less as a whole and does not attempt to distinguish between analyses of boardgames and numerous types of other games. What Suits does, however, is pave the way for an exploration of how rules function – their purpose, their volatility, and their agreed-upon authority – and I take up these ideas briefly here. Dunnigan *et al.* (1977) provide a practical look at how board war games are designed and produced, and in the process address issues surrounding simulation and iconography in games, but it is a largely non-theorized set of guidelines that offers little to game studies *per se* beyond its historical interest and the guidelines as genre-specific design suggestions. Murray (1969) demonstrates the huge range of themes, styles, and game types that boardgames have encompassed over many years, but it is essentially a database, similar in that regard to Freeman's (1980) compendium on war games, the specific genre of boardgames designed as battle or campaign re-enactments. Elsewhere, Palmer (1977) and Perla (1990) provide examinations of the history of war games and details of their play. Closer to the needs of game studies is Parlett's (2003) detailed examination of boardgame types, including a useful, if brief, look at game boards and their representational functions (Parlett 2003).

Recent scholarship on boardgaming does exist, but it is quite rare. Of particular interest to the play of boardgames is the detailed psychological study by Gobet *et al.* (2004). Another recent example is Kirschenbaum's (2009) study on the creation of narrative generated by play simulated military battles, a narrative construction previously argued against by Costikyan (2007). Specifically, Kirshenbaum takes issue with Costikyan's denial of story in games such as chess, *Monopoly*, or [the board war game] *Afrika Korps*, claiming that the written genre known as the after-action report, common to war game players as a means of outlining for interested audiences how their game played out, creates a kind of 'war story' that serves to refute Costikyan's premise. Kirschenbaum contributes to this chapter in its recognition of the importance of extra-game communication to the boardgame experience and in its understanding that the study of boardgames requires a different focus from the study of video games – a point I will discuss later. As Kirschenbaum (2009: 369) argues boardgames offer 'a qualitatively different kind of ludic artefact than an electronic game'. Also recently, the first issue of the *International Journal of Role-Playing* has appeared, which has role-playing games (RPGs) of all kinds, from digital to paper-and-pencil, as its overarching topic; boardgames and paper-and-pencil role-playing games informed each other early in the life of the RPG industry (*Dungeons and Dragons* began as a miniatures war game, called *Chainmail*, played with pieces on a table representing terrain), and role-playing of various types appears in the systems of numerous boardgames, so a journal covering RPGs must rightly feature boardgames. Indeed, the first issue contains Jason Pittman and Christopher Paul's 'Seeking Fulfillment: Comparing Role-Play in Tabletop Gaming and the World of Warcraft', and while the tabletop gaming of the title refers to paper-and-pencil RPGs, scholarly issues surrounding the RPG tabletop can certainly apply to boardgame settings as well. A sustained effort to cover boardgames is *Board Games: The International Journal for the Study of Board Games* (www.boardgamestudies.info/studies/), with articles covering analyses of game boards (Yuhara 2004; Bock-Raming 1999) and other game equipment, but primarily focused on the history of boardgames (Rothöler 1999; Whitehill 1999).

A highly promising event in boardgame scholarship occurred in late 2007: the publication of Philip Sabin's *Lost Battles: Reconstructing the Great Clashes of the Ancient World*. It is promising, in fact, for two reasons. First, it offers a model, grounded in military history scholarship, for simulating ancient battles that could be delivered in a video game but that is presented specifically as a board war game system. Second, this system expressly (if all too briefly) critiques existing board war games that simulate ancient battles at the tactical level,[2] particularly the *Great Battles of History* series from GMT Games (designers Mark Herman and Richard Berg). Obviously, anyone engaged in scholarship of a particular type of artefact welcomes academic scholarship that helps to inject the field with its own ethos – Sabin (2007) calls conflict simulation games an 'academic neglected literature' –, especially when, as in Sabin's case, the scholarship is strong. Sabin's issue with Herman and Berg's simulations, despite his assertion that they are the 'best existing simulations of ancient battles', is that they are 'too detailed and complex for my own taste, incorporating the kind of fine-grained precision regarding orders of battle and battlefield geography that may be appropriate for later, better-documented periods but which in this case means that many evidentiary gaps and uncertainties are filled by sheer guesswork' (Sabin 2007). Within the context of games scholarship, Sabin's book offers a highly useful historical analysis of board war games that simulate ancient battles, but it is rare in the fact that it examines boardgames at all and even rarer in the fact that it considers board war games.

Online-enhanced boardgames

While Sabin (2007) addresses boardgames from the standpoint of simulating their topics, here I address the simulation of their play, including their physicality, in emerging online space. Through the use of networked game presentation programmes, players can compete against each other in simulated boardgames while online. These tools offer limited programming capabilities for creating ease-of-play enhancements that are not available to those playing across a table from one another, and in that respect differ from the physical games they represent, but in most respects they simply provide a digital and networked version of the board and pieces that come in the box. The point of these tools is to provide an environment in which players can 'face' each other across a simulated board in a simulated way. In other words, they simulate not only the game, but the room in which the games are played – the gamespace.

I term these games 'online-enabled boardgames' (which I initialize as OEBs) and the programmes on which they run 'online-enabled boardgame platforms'. They are known in the hobby as online play-aids, but for several reasons they need not be considered only as aids to existing boardgames. First, some designers have begun looking to these platforms as a means of publishing games without the physical product – e.g. Dan Versson Games, (www.dvg.com). Furthermore, in many instances in which a digital boardgame has been prepared from an existing physical boardgame, designers of OEB versions frequently enhance the OEB with displays, documents, and automation commands that attempt to simulate the boardgame experience but that do not exist as explicit items in the physical game itself. Further still, designers and developers of 'physical' boardgames (PBGs) are increasingly using OEB platforms to test the designs, sometimes exclusively, throughout the game's development. Ed Beach tested his *Here I Stand: Wars of the Reformation, 1517–1555* using only the OEB platform Cyberboard (cyberboard.brainiac.com) – which is one of the earliest OEB platforms, which often involved gamers constructing 'DIY' versions of boardgames. There are various other OEB platforms including, VASSAL (www.vassalengine.org), Wargameroom (www.wargameroom.com), and SpielByWeb (www.spielbyweb.com). Each of these platforms functions as a simulation – or at least an attempted simulation – of the face-to-face boardgaming experience. And a designer of an OEB uses the tools of the specific OEB platform to create a *module*, i.e., a data file containing the elements (graphics, text, macros, and so forth) that players load into the platform in order to play.

Designer and developer Adam Starkweather has declared the VASSAL platform the sole means of testing his IGS (International Game Series) products for Multiman Publishing (www.multimanpublishing.com), although he does make graphics files available for those who wish to print them and construct physical playtest kits. Other developers and I have followed suit. Thus, even games that are designed for eventual print are currently being played throughout the development period – typically a year or more – exclusively or nearly exclusively as OEBs. And to judge from comments posted to the ConsimWorld (www.consimworld.com) and BoardGameGeek (www.boardgamegeek.com) discussion sites, players of numerous types of strategy boardgame are increasingly finding OEB platforms either a necessary method for playing their games, because they have no opponents available close to where they live, or a preferred method because they allow for short sessions at convenient times.

A boardgame can have any number of physical components: the board itself, the rules documents, play aids such as charts and tables, player mats for tracking various game-related functions, playing cards, storage containers, and of course dice and playing pieces (the latter variously called tokens, counters, markers, and so forth). An OEB can replicate or simulate as many or as few of the items on this list as required, including a randomizer of the kind required in the game under conversion (typically dice). OEBs include renditions of the board and other printed material including cards, consisting of graphics files either self-made, created in the OEB programmes, scanned from the game's original art, or publisher-supplied digital copies of the digital art. Play aids, including charts and tables, player mats, and setup displays, use a separate window or can be included in the graphic file of the board if scrolling to the displays is acceptable.

The iconicity of the board presents OEB designers with the significant challenge of making the game's major component resemble its physical counterpart as close as possible, but providing iconic representations of playing pieces is typically an even greater challenge, because both are unmodified flat graphics files that, without modification, look to be on the same vertical plane. Traditional board wargames with their small cardboard counters have not proven overly difficult to replicate; a simple scan of the originals or the use of publisher-supplied graphics files gives the OEB a look very close to the physical game, with OEB designers often enhancing the counters with simple 3-D effects from a graphics programme (e.g., bevelling or shadowing) in order to provide a representation of the thickness of the piece. For stand-up pieces, plastic or metal miniatures, or other kinds of fully 3-D elements, the representation is much more difficult, with the pieces essentially impossible to scan and with graphics files produced by anyone but an accomplished artist frequently resembling the 2-D flatness of early chess programmes. In such cases, the OEB typically loses its immediate visual appeal, the draw of iconicity, and takes on the status of boardgame aid rather than real or potential boardgame substitute. These presentations do not necessarily discourage players from using the OEB to play the game – indeed, for many players, the OEB is the only available way or even the preferred way to play whether or not it looks like the physical game – but prior knowledge of the physical game becomes more important than it does for games with a stronger iconic representation.

The simulated gamespace

Boardgames demand physical space. They are designed with space in mind, and players judge them in part by their spatial requirements. Furthermore, players use and at times even embrace these spaces. Physical space, the co-location of the people within that space, the relationship between the physical game and the players, and the relationships among the players, all combine to form the physical gamespace. This gamespace is, on the one hand, necessary simply because the game itself requires physical space, but it is also a desired element of the play experience: boardgame players do not just play games; they get together in the same place to play games.[3] Playing boardgames has long been known as a social activity – they were once commonly known as parlour games, after all – and the social element remains a focus of boardgame design and boardgame purchase and reception. Boardgame designers and developers focus significantly (but certainly to varying degrees) on the nature of player interaction in a communal space, and the requirements of the gamespace itself is one of the core elements of this focus.

A physical boardgame occupies a specific and non-negotiable amount of physical space. Boardgames are discussed by players and marketed by publishers according, at least in part, to their physical footprint, the total amount of space taken up by the board and any displays separate from the board (typically game tracks, player mats, and component holders) essential to playing the game. As a well-known example, the footprint for *Scrabble* includes the board, the tile trays for each player, the container for the unused tiles, and whatever players use to record scores. Some board wargames (appropriately called monster games) require table space of at least 7 ft. × 4 ft. (four standard 22″ × 34″ maps), along with associated tracks and displays, and some are even larger. Even for a genre as variable and as capable of generating large games as the board wargame genre, 'footprint' (the space the board occupies) becomes a significant issue for buyer acceptance: bulletin board comments frequently state the need to fit the game on a specific surface, with anything else being too large for the writer of that particular post to play. Publishers, as a result, look for games with a relatively standard footprint for that genre.

But the game's footprint is not the only physical space issue. The room in which the game is being played also matters to the concept of gamespace. Whether that room is the kitchen, family room, bedroom, or unfinished basement of a house, conference room of a hotel, or the table on an airplane, the room's size and ambience factor in the game experience. Obviously, the room must be large enough to contain a surface on which the game is set out, obviously (although gamers have been known to pin game boards and playing pieces on their walls), it must also be large enough to accommodate the players and their interaction needs. In a game of Albert Lamorisse's *Risk*, e.g., players frequently walk away from the board into a private corner of the room in order to discuss deals; in Allan B. Calhamer's *Diplomacy*, discussions in private corners, often with more than two players, are what the game is all about. Both games require a large enough single room, or enough spaces outside but close to the room, to permit players to engage in diplomacy and deal-making.

Therefore, the simulation of communal space is important to the online boardgaming experience. Playing a boardgame usually means being in the same room as at least one other person (unless the game is designed to be played solitaire or is simply played by oneself for convenience), and having people in the room means – usually, at least – talking with them. Conversation during a boardgame session can range from purely social interaction to commentary about the game itself. To function as replacements for the boardgame experience, OEB platforms must provide tools to allow all such discussion.

Three communication technologies currently dominate the attempt to simulate communal space in online boardgaming: e-mail, text chat, and voice chat. Only the two chat technologies offer real-time communication, and as such are the closest to approximating the experience of conversing with people in a shared physical room. Furthermore, voice chat provides the stronger simulative experience, not only for the obvious reason – hearing the other person's voice – but more importantly because it is not necessary when listening to voice chat to view the transaction on screen: as with conversation in a physical space, speech can occur even when one or more parties are using their eyes for non-speech activities (Burrows 1990).

There is technologically no difference between voice chat in OEB platforms and voice communication in networked video games. In both, players use headsets or microphone/speaker combinations to engage in live conversation. While players in online first-person shooters tend to use voice systems as a means of either speaking on behalf of their characters as part of the team of fighters/adventurers, and players of massively multi-player online role-playing games also do this while sometimes using the voice platform for socializing among the players, players on OEB platforms more typically use the voice technology to discuss the rules and procedures or to remind one another of necessary mechanics for specific actions, of the current status of the sequence of play within the game-turn, and of game elements such as the victory conditions. Discussions tend to focus on rules, in other words, a focus necessitated by the collaborative nature of rules comprehension and enforcement. Video games enforce their rules; OEB platforms (with some exceptions such as WargameRoom) do not – and this is a key difference between the game formats.

Part of the boardgame experience, in fact, is precisely the agreement among the participants to abide by an accepted set of rules, whatever that set of rules might be. It might be the printed booklet that came in the box, a file downloaded from the Internet and printed or accessed on a computer, or, in the case of many complex games that undergo frequent updates, both. Or it might be a combination of printed rules and 'house rules', the latter consisting of rules suggested by one or more players and used because those players prefer them for whatever reason (the wide range of house rules covering the Free Parking space in *Monopoly* is just one well-known example of this phenomenon). The players of a boardgame socially construct the rules structure each time a game is begun and indeed continue to construct them socially as the game proceeds. Salen and Zimmerman (2004: 123) argue that rules are 'fixed and do not change as a game is played', but in the same paragraph recognize that in some games the rules are re-negotiated in the process of the game itself. These negotiated changes are 'themselves determined by other, more fundamental rules', and while that restriction is largely true, it need not be the case at all; nothing stops boardgamers from simply using the board, tokens, cards, and dice as the basis for an entirely different game, created on the spot.

Of course, for the majority of boardgame play, the official rules – those published with the game – are the binding guidelines for play. In boardgames, the degree to which they bind is again simply an agreement of the players to be so bound. They are 'accepted for the sake of the activity they make possible' (Suits 1978: 45), but they bind only in so far as players refuse alternatives. The official rules that govern the game, the 'operational rules' (Salen and Zimmerman 2004:

130), are only part of the governing force. Sniderman (1999) comments extensively on the 'unwritten rules' of games, those that guide not only the behaviour of players – the 'implicit rules' in Salen and Zimmerman's terms, but also that determine what players can do in the game if the official rules fail them. He distinguishes between rules and rulings, expressing them as situations in which an extra-game incident affects the play of the game itself, but the rules/rulings distinction can be more usefully applied, I suggest, to boardgame incidents in which players must come to a binding agreement on a rules omission or ambiguity, a ruling that is official for that session and for those players, but not necessarily for any other session or any other players at any time. Such rulings have governed the play of every complex boardgame, and in fact most-simpler boardgames, I have ever experienced.

To capture the crucial process of the negotiation of boardgame rules, OEB platforms must provide the means by which players can determine the rules and rulings under which they will play each particular game during each particular session. Live text chat is one such system, but it is of course limited by the difficulties in discussing complex points by typing. It works, but it does not capture the gamespace experience of talking out a game issue. Voice chat is considerably more effective because the nuances of oral communication come into play, ranging from speed of presentation through rapid repetition, restatement, give and take, and iterations of agreement. No OEB platform has yet incorporated video chat, but of course players may use third-party systems for this purpose if they wish. Video chat has the disadvantage of distraction, of course, although that, too, more fully simulates the face-to-face environment.

Even in their attempts to simulate physical space, of course, OEBs do not dispense with physical space entirely. Players need a physical device to play an OEB, and that physical device requires the space required by that device (desktop, laptop, netbook, smartphone, or similar). In all these cases, this physical space is shared, or at least capable of being shared: the device is typically used for multiple applications at once, so it is not a dedicated playing space. Nor is the space itself designed specific for OEBs; unlike video game consoles, devices used for OEBs are multi-function by default. Of course, most boardgame surfaces also serve other functions, as a desk or table, but in such cases one does not switch from one function to the other at the touch of a few keys or the click of a mouse pointer. Despite the need for device space, we can consider an OEB itself as being without physical space, and since the physical setting and physical components are part of the continuing allure of boardgames, the question then becomes, how do OEB platforms compensate? The answer is that OEBs offer the gamespace beyond the game itself – the space around the table if you will – in a specifically semiotic manner. The gamespace beyond the table is the virtual world the OEB platform creates, but unlike the virtual worlds in video games, it is not a visible virtual world, it is rendered only in a way that points to and extends the table and the room in which the game is played. The semiotic principle at play here is 'indexicality', to which I now turn.

Two semiotic principles in the analysis of oebs

The semiotic dimensions of Cyberboard (an OEB platform) become interesting when considering gameboxes created by that platform's design tools. If a major purpose of an OEB is to represent the physical components of the corresponding boardgame exactly, Cyberboard's DIY gameboxes (made by the players themselves or amateur producers) obviously fail; they are usually low-resolution approximations only. For many players, indeed, this lack of aesthetic authenticity renders Cyberboard a poor choice, at least the DIY versions. For early adopters, and even today for players of games that have only Cyberboard DIY conversions available, the aesthetic limitation is easily and frequently overcome. If you want to play a game and you cannot find (or do not want) a face-to-face opponent, any solution is better than none. This is not a case of players spending weeks or months on a DIY gamebox bemoaning the graphical insufficiencies; indeed, for them the gamebox *becomes* the game, especially for those who play the game exclusively via Cyberboard (or, of course, those who have never owned the physical boardgame). From a social semiotics perspective, the graphics of the DIY gamebox possess a 'modality' different from those of a gamebox that uses company-provided graphics files – where modality refers to claims of truth, fact or reality. Hence, the modality in the former represents approximation, while modality in the latter represents perceived likeness approaching exactness. High modality refers to acceptability, low modality to non-acceptability, with acceptability itself referring to the audience, either intended or obtained.[4] Kress and van Leeuwen (2001) apply this reading of modality to both visual images and multimodal texts, and it is directly applicable to the study of simulations. Simulations are very much about attempted representations of the real, and user acceptance of their mimesis is crucial to their success.

Modality figures in all discussions of simulations (such as video games), because acceptability of and even definitions of simulation differ from audience to audience, often from person to person. Whether something represents a pre-existing object, concept, and/or process in a manner that approximates sensory and functional likeness. Hence, a definition of simulation depends largely on the audience and its needs and expectations. The art of simulation lies in the tailoring of the approximation of the simulateable elements – the objects, concepts, and/or processes being represented – to the projected or known users of that simulation. The audience must accept the simulant elements as sufficiently realistic for the purposes of continuing the use of the simulation *as* simulation. Such acceptance, of course, relies heavily on the audience's knowledge of the reality being simulated, or a willingness to learn of that reality, or both, and they must understand the validity of the simulational elements. The simulational elements must combine, in other words, to produce an artefact possessing high modality for whatever audience it is intended. For some game audiences, a game has high modality if it replicates, in enormous detail, intricacies in weapons, command systems, supply and transport issues, global economic indicators, and so forth; for others, high modality is found in a game that simulates only one or two essential processes in a larger procedural framework.

The issue of modality in OEBs is complex, offering multiple levels of representations of the real. An OEB represents its corresponding physical game, of course, but it also represents the playing of the physical game. In the case of a simulation boardgame, we have several levels of representation – and thus considerations of modality – at work here: (a) the game design's simulation of its events or processes, (b) the physical game's systems of graphical representation, (c) the OEB's depiction of the physical game's graphical representations, (d) the OEB's simulation of the game play mechanism, particularly the manipulation of playing pieces, and (e) the OEB's simulation of the gamespace itself – the room and its participant interactions. We could actually add another set of layers that deal with the representation of the system of representation. Many boardgames (like video games) are part of a series, where a game's success results in more games based on the system of the original, and where the degree to which the sequel reflects the original significantly affects player acceptance. Similarly, OEBs gain adherents in part – although certainly not the same extent – by replicating the look and interface of other OEBs.

The classic semiotic theory I use in this chapter, and one with clear links to the concept of modality, is Peirce's (1998 [1909]) well-known distinction among icon and index, part of his three-part taxonomy (with symbol) of sign types. To be sure, these concepts barely scratch the surface of Peirce's vast contributions to sign theory, but in an age where images and their production have taken center stage in numerous media and in perceptions of truth and the real, they continue to resonate – and because Peirce considered the icon–index–symbol triad an element worth returning to as he elaborated his larger sign typology. Peirce wrote of icon and index in several places, but one passage will suffice, in which he calls icons likenesses 'which serve to represent their objects only in so far as they resemble them in themselves' (Peirce 1998), and in which he writes that indices 'represent their objects independently of any resemblance to them, only by virtue of real connections with them'. Icons are quite easy to understand: they look like, sound like, feel like, smell like, taste like the thing to which they refer. Indices have a somewhat more abstract sense: their sensory apprehension points us to the object to which they are linked. A gunshot heard in a movie is iconic: it is designed by the sound engineers to sound just like a real gunshot, or at least according to how movie audiences expect it sound (in fact, movie sounds are often simulations designed to sound more convincing than the real thing). A gunshot heard in the street outside your window is indexical: it indicates a gun. Of course, it is not as simple as that – the movie gunshot is iconic in one sense (resembles a gunshot), but it can also be indexical within the movie's fictional world, especially if the film portrays the sound but does not show the weapon.

An OEB ionizes the components, converting paper/cardboard playing surfaces into graphics files and playing pieces (cards, markers, tokens, counters, figures) into individual graphics files that resemble or even replicate the look of their physical counterparts. In the case of Cyberboard (and to a much lesser extent VASSAL), a module designer can create iconic elements from scratch, and indeed, in the case of Cyberboard, until recently creating from scratch was by far the most common method of building modules because doing so results in a small file size (important before the days of widespread broadband and still important for some users today). VASSAL modules, by contrast, and increasingly Cyberboard modules as well, are designed by importing either scans of the physical components or, much more frequently today, digital graphics files obtained by permission from the publisher. In either case, the designer's goal is to reproduce the experience of seeing the physical game, thereby quickly orienting players with experience in that game. The game's visual modality is higher when publisher-supplied graphics files are used rather than with user-created graphics because the result is more directly iconic.

OEBs are indexical primarily in their simulation of the gamespace – i.e., just like the gunshot that is not seen, the game 'points to' the existence of this space in a number of ways, rather than showing it. First, OEBs scroll in order to display the full board; an edge of the screen that does not correspond to the edge of the map indicates the unseen portion that lies in that direction – although, since the unseen part does not actually exist until the screen draws it, it could be said to be only symbolically indexical, thereby increasing the semiotic complexity further. Second, other displays that would be visible on the game table, and this includes elements on the physical board that have been removed from the graphics file for the board and placed in separate windows or trays, are similarly indexical, particularly for players who know the physical game and therefore know they must exist somewhere in order for the OEB to function as a usable conversion. Third, while a stack of playing pieces (deck of cards or playing pieces on top of one another) might graphically resemble the three-dimensionality of the corresponding physical combination, seeing the contents of a stack can differ significantly, with the viewing process becoming indexical in and of itself. In VASSAL, *e.g.*, the player hovers the mouse pointer over the stack to cause the stack to expand, with all the pieces lined up on the screen side by side. Each piece in the row indicates both the piece in the stack and its location in that stack.

But beyond the playing pieces, the voice and/or typed chat interaction between the players also functions indexically, as does the act of watching the other player's pieces move on the board (where the OEB platform allows). In these cases, the real-time voice, the real-time chat, and the real-time movement indicate the existence of the other player and thus create indexically derived meaning. Furthermore, the OEB board, through its partial display, not only indicates the full physical board, but also indicates the table on which the physical game would sit and the space beyond the table: the other players, the chairs on which they sit, and the room in which the playing would occur. The indexical extension of the game draws upon a player's experience with playing games in a physical environment and allows an imaginary mapping of that environment onto the borders of the OEB on the screen. The gamespace beyond the game is the fundamental difference between a boardgamer's virtual environment and a video gamer's virtual environment: in the virtual world of a video game there is nothing beyond the boundaries of the board.

The early use of Cyberboard, a use that continues for some players today, demonstrates indexicality in a very specific way. The platform was designed from the beginning to allow the creation of log files in which players would record their moves and send to other players. Many early adopters, however, chose to use Cyberboard as a snapshot tool only, playing the actual game on the physical board with the physical components. They would make their moves physically, then replicate those moves on Cyberboard, sending the saved game file rather than the log file to the opponent(s). Cyberboard functioned as a tool for pointing to the physical game, with each OEB component a separate indexical link to the original. By contrast, as mentioned above, much of today's playtesting of boardgames, particularly board wargames, takes place on OEBs only, in which case there is no physical game for the OEB to point to, except of course the player's mental images, based on a knowledge of similar products, of what the published game will look like. Indeed, in these cases, the physical game, when published, becomes an index of the OEB instead of the other way around; players using the physical game for the first time must, in many ways, re-learn how to manipulate the components. Some playtesters never do play the physical game, in fact, first because by the time of publication they have often moved on to other games, but also because they are so intimately familiar with the workings of the OEB that the physical game feels oddly wrong by comparison.

Next steps

This chapter has introduced a perspective on the understanding of the recent phenomenon of playing boardgames, which are designed to be played in physical communal settings, over computer networks. In particular, I consider the play of primarily board wargames and their adaptation to the online environment, mostly because I am more familiar with this genre of networked boardgame than with any other. Certainly a full study of playing boardgames online would include a more detailed look not only at German-style games (briefly mentioned here) but also the various conversions of *Scrabble*, *Monopoly*, *Risk*, and many other games available online, to say nothing of the endless variations of online chess. Nearly all of these better-known games have found their way online in a manner that includes the automated (computerized) enforcement of rules, and such a system eliminates much of the need for live communication for the sake of rules interpretation. This automation in turn disrupts the simulative value of the play experience, since it is not the same game if the rules are fixed: i.e., where interpretation does not matter because it would make no difference anyway. Or, of course, where rules can be added or changed at will, needing only the agreement of the other player(s). For many boardgame players, of course, the hope is precisely that rules need no interpretation, and that the official rules are unambiguous and therefore binding, but the actual play of a boardgame rarely works this way.

Iconicity, however, does enter importantly in a study of the online conversion of popular games. To be widely adopted by fans of a game, the online version must look almost (if not fully) identical. *Facebook*'s version of *Scrabble* is an example: the virtual board looks like the originals, the virtual letter tiles look like the originals, and even the tile trays recall the original. *Yahoo*'s versions of *Backgammon*, *Go*, and *Risk* are similar. Even with a game as seemingly simple as *Scrabble*, the *Facebook* version – while offering the advantage of playing at your own convenience – fails to simulate the face-to-face *Scrabble* experience in several important ways. A significant example is the dictionary mechanic. In original *Scrabble*, one player can challenge another's word construction by formally stating the challenge and looking up the word in a mutually agreeable dictionary, with a failed challenge resulting in a penalty against the challenging player and a successful one penalizing the initial player. Facebook *Scrabble*'s automated dictionary dispenses with this mechanic and, with it, the sometimes crucial decision of whether or not to challenge. *Scrabble* at its most competitive is not only about making words with well-placed letter tiles, it is also about stopping the opponent from doing so.

A thorough semiotics of boardgames is a much-needed study or, rather, series of studies. Boardgames share many characteristics with video games, but in numerous ways they are vastly different. Many video gamers do not play boardgames, and many boardgamers do not play video games, but even for those who play both, the play experiences of the two bear few similarities. Immersiveness differs, narrative generation differs, simulational value differs, styles of player interaction differ, and social norms of game play differ. The components of boardgames, everything from the rules, to the dice, to the playing pieces, and the game room itself, create meanings to specific audiences for specific purposes through specific types of presentation, and these meanings do not always map onto the meanings produced by the wide range of video games now under a much more intensive scholarly microscope.

Notes

1 The video game figure is from Ortutay (2009) 'October video game sales tumble 19 percent,' Associated Press, November 13. The boardgame figure is from an NPD Group report, cited at *Purple Pawn: Game News Across the Board*, http://www.purplepawn.com/2009/02/board-game-sales-up-poor-analysis-runs-rampant/.

2 Wargame designers and players commonly distinguish among three scales of conflict representation: tactical, operational, and strategic. Tactical refers to individual battlefields, from Cannae to Waterloo. Operational covers specific campaigns within a larger war: Napoleon's 1812 invasion of Russia, for example. Strategic covers entire wars or multiple campaigns.

3 Unless, of course, they desire to play the game by themselves. Some boardgames are designed for solitaire play: examples include Brien J. Miller's *Silent War: The United States' Submarine Campaign against Imperial Japan, 1941–1945* (Compass Games 2005), John H. Butterfield's *D-Day at Omaha Beach* (Decision Games 2009), and many others. In this case, the spatial requirements are nearly the same as for games played by two or more players (the difference is in the space needed for the other players), but in many such cases (*Silent War* being an example) the game itself uses more surface space because such games tend to require additional materials in order to simulate an opponent. Many other players, especially in the board wargame hobby but certainly not exclusively, prefer to play even non-solitaire games by themselves; they simply play both sides, relying on randomization systems within the game (and often added informally) to approximate the multi-player experience.

4 Literary theory such as the work of Genette (1980) teaches us that test will have an intended audience, that is to say, which the text was written or made for, but it also alters us to, that the real audience may differ form that which was intended by the author.

References

Beach, E. (2006) *Here I Stand: Wars of the Reformation, 1517–1555*, Hanford, GMT Games.

Bock-Raming, A. (1999) The gaming board in Indian chess and related board games: A terminological investigation. *Board Game Studies: International Journal for the Study of Board Games,* (2). Retrieved from http: //www.boardgamestudies.info/pdf/issue2/BGS2-complete.pdf.

Burrows, D. (1990) *Sound, Speech and Music*, Amherst, University of Massachusetts Press.

Caillois (1961) *Man, Play and Games*, Glencoe, The Free Press of Glencoe, Inc.

Costikyan, G. (2007) Games, Storytelling and Breaking the String. In P. Harrigan and N. Wardrip-Fruin. (eds) *Second Person: Role-Playing and Story in Games and Playable Media.* Cambridge, MIT Press.

Dunnigan, J. F., R. H. Berg, S. B. Patrick, D. C. Isby and R. A. Simonsen (1977) *Wargame Design: The History, Production and Use of Conflict Simulations*, New York, Simulations Publications, Inc.

Freeman, J. (1980) *The Complete Book of Wargames*, New York, Simon and Schuster.

Genette, G. (1980) *Narrative Discourse (Trans J. E. Lewin)*, Oxford, Blackwell.

Gobet, F., A. de Voogt and J. Retshitzki (2004) *Moves in Mind: The Psychology of Board Games*, Hove, Psychology Press.

Hodge, R. and G. Kress (1988) *Social Semiotics*, Ithaca, Cornell University Press.

Kirschenbaum, M. (2009) War Stories: Board War Games and (Vast) Procedural Narratives. In P. Harrigan and N. Wardrip-Fruin. (eds) *Third Person: Authoring and Exploring Vast Narratives*. Cambridge, MIT Press: 357–372.

Kress, G. and T. van Leeuwen (1996) *Reading Images: The Grammar of Visual Design*, London, Routledge.

Kress, G. and T. van Leeuwen (2001) *Multimodal Discourse: The Modes and Media of Contemporary Communication*, New York, Oxford University Press.

Kress, G. and T. van Leeuwen (2006) *Reading Images: The Grammar of Visual Design*, London, Routledge.

Murray, H. J. R. (1969) *A History of Board-Games Older Than Chess*, Eastbourne, Garners Books.

Palmer, N. (1977) *The Comprehensive Guide to Board Wargaming*, New York, Hippocrene Books.

Parlett, D. (2003) *Oxford History of Board Games*, New York, Oxford University Press.

Peirce, C. S. (1998 [1909]) *The Essential Peirce: Selected Philosophical Writings Volume 2 (1893–1913)*, Indianapolis, Indiana University Press.

Perla, P. (1990) *The Art of Wargaming*, Annapolis, Naval Institute Press.

Pittman, J. and C. Paul (2009) Seeking fulfillment: Comparing role-play in tabletop gaming and the World of Warcraft. *International Journal of Role Playing,* 1: 53–65. Retrieved from http://journalofroleplaying.org/.

Plumb, T. (1997) Low-Tech Gamers Remain Loyal to the Board. Retrieved 22 September, 2010, from http://www.discovergames.com/boston_globe.html.

Rothöler, B. (1999) Mehen, God of the Boardgames. *Board Game Studies: International Journal for the Study of Board Games,* (2). Retrieved from http://www.boardgamestudies.info/pdf/issue2/BGS2-complete.pdf.

Sabin, P. (2007) *Lost Battles: Reconstructing the Great Clashes of the Ancient World*, London, Hambedon Continuum.

Salen, K. and E. Zimmerman (2004) *Rules of Play*, Cambridge, MIT Press.

Sniderman, S. (1999) The Life of Games. Retrieved 25 May, 2010, from http://www.gamepuzzles.com/tlog/tlog2.htm.

Suits, B. (1978) *The Grasshopper: Games, Live an Utopia*, Toronto, University of Toronto Press.

Swink, S. (2009) *Game Feel: A Game Designer's Guide to Virtual Sensation*, New York, Morgan Kaufmann.

Watson-Smyth, K. (2009) The 50 Best Board Games. Retrieved 23 September, 2010, from http://www.independent.co.uk/extras/indybest/outdoor-activity/the-50-best-board-games-1815441.html.

Whitehill, B. (1999) American games: A historical perspective. *Board Game Studies: International Journal for the Study of Board Games*, (2). Retrieved from http://www.boardgamestudies.info/pdf/issue2/BGS2-complete.pdf.

Yuhara, K. (2004) The evolution of Sugoroku boards. *Board Game Studies: International Journal for the Study of Board Games*, (7). Retrieved from http://www.boardgamestudies.info/studies/issue7/article.shtml?yuhara.txt.

7 Games in the mobile Internet

Understanding contextual play in *Flickr* and *Facebook*

Frans Mäyrä

The social and cultural phenomena related to Internet gaming have received their fair share of attention, particularly through numerous studies of massively multiplayer online role-playing games (MMORPGs). In relation to mobile games, they have their own research and developer communities, but the research work related to mobile games has so far been dominated by their technical and design challenges and researchers have not been particularly interested in their social aspects. For example, the recent *Handbook of Mobile Communication Studies* dedicates 1 of its 32 chapters to mobile games and entertainment; see Katz and Acord (2008). This stands in contrast to the numerous studies that have focused on mobile communication through text messages and other means.

This chapter approaches contemporary developments in social and mobile gaming as an Internet phenomenon. Looking back, there has been a noticeable difference between Western usage patterns of the Internet, mostly focused on the personal computer, and Japan, where mobile phone is the predominant access point to Internet services. Recently, Western countries have also been introduced to new generations of Internet-capable smartphones, including the Apple iPhone, Nokia E and N-series, BlackBerry devices and others. Models such as this have reportedly been associated with a noticeable upsurge in mobile Internet usage, and some analysts have claimed that finally 'mobile internet has reached a critical mass'.[1] The relative share of mobile browsers in the Internet usage statistics nevertheless still remains a minority. Why then is there a need to focus on the mobile and playful interfaces of the Internet, and consider them from a social, Internet research and game studies perspective?

The International Telecommunication Union has reported that in late 2008 there were more than four billion mobile phone subscriptions in the world, and that they expected the five billion mark to be passed in 2010.[2] The mobile phone is the most widely available network-enabled terminal device, and one that plays a major role in everyday life, especially now that developing countries are starting to find their way to the Internet. Personal and almost always available a contemporary mobile phone can potentially foster developments of new user cultures, including ones involving the casual creation and sharing of content related to digital photography, the Web (which here mostly means the popular culture of the Internet), music and games. This chapter focuses on the integration of

Internet usage and contents with contextual information. The character of contextual information will be interpreted broadly, as including not only the physical use contexts but also the context provided by social networks. The studies of play and games will provide a particular perspective that helps to understand the main characteristics of this development: what I see as the playfully social character of mobile Internet use.

The phrases 'contextual gaming' or 'contextual play' have so far mostly meant experiments in game design that exploit various sensors and other technical sources of data, used for implementing location-aware games, games that rely on gesture or pattern-recognition, or game worlds that in one way or another reflect the real world, like mimicking its weather conditions or daily rhythms (for a summary, see Tester 2006). Far less experimentation has been dedicated to the development of games that rely on social contexts and information derived from social networking tools. Social motivations nevertheless will remain a major force that influences how many people use or do not use the information and communication technologies. All human contexts are socially determined contexts, as humans are fundamentally social beings. This does not mean that technology use or human behaviour itself would only be determined by social circumstances; rather, I subscribe to the view that human action is never completely determined, nor random – our interactions with and among human actors and technology-rich environments produce complex and deeply dialectical relationships (see Suchman 2007). Since the emphasis on this chapter is on human agency rather than detailed analysis of technologies that currently influence our ways of expressing it, 'contextual play' is primarily used in this chapter to signify distributed and mediated playfulness; the focus is on playful behaviours that are rooted in or that emerge from social relations and exchanges of information that are used to maintain and expand such networks of relationships. It is interesting to note how contextual play gains specific meanings when the location and situation of participants is fluid – as is the case with contemporary online services, which can be accessed in multiple ways. I will focus on a few examples that illustrate this evolving field in the remainder of this chapter. The selection of my brief case studies is based on ongoing work our game research laboratory has been carrying out in areas related to mobile and pervasive gaming, user- or player-created content and online social games.[3] During the 2 years, from 2008 to 2010 when I worked on this study, both mobile and social media went through a period of rapid growth and change, as new services were released and new user cultures emerged, reorganising or displacing existing practices. Thus, this chapter is written also to illustrate a certain kind of transition period towards a more contextually aware and more playful understanding of mobile communications and online media in general.

There are several approaches, including more technically oriented approaches to 'context' in mobile Internet studies that I will not specifically consider here, including those that differentiate between environment context, personal context, task context, spatio-temporal context, terminals context and so forth (Guarneri *et al.* 2004). I consider all these as varieties or aspects of the users' social context. The main reason for such an approach to contextuality lies in the emphatically

measurement system in middle of them. The absolute metrics derived from how users' photos measure up in the interestingness scale, can be used to distinguish 'winners' from 'losers' within the *Flickr* 'gaming community', and thus the quantitative measurement effectively invites more *ludus* style of competitive interaction within the context of playful photo sharing. There appears to be multiple ways, even conflicting ones, to play or game in *Flickr*, and heated discussions on the uses and abuses of the system continue in the discussion forums. The information researchers who have started to pay notice to the workings of *Flickr* have pointed out that the real significance of having a photo appears among the automatically top-ranked. Explore photos is smaller than users themselves typically think. The images actually get most of their attention through the direct contacts among *Flickr* users. This is based on a practice that has been named 'social browsing'; the users look and find new pictures primarily by browsing through one's friends' photo feeds rather than by searching for photos from those who are unknown to them (Lerman and Jones 2006). This does not stop the competitively minded *Flickr* users from trying to tweak their odds in attempts to get their photos into the Explore category.

A number of dedicated gaming applications have also been developed that make use of the *Flickr* Application Programming Interface (API). Typically, these are browser-based small games that rely on a combination of *Flickr* photos with some classic game format, and none of these game designs so far have been as popular as social play within the *Flickr* service itself. Some examples include the following:

My first example is called *Flickr Sudoku.*[12] This is a version of *Sudoku* that replaces numbers with images tagged with certain words from Flickr (see Figure 7.1). The

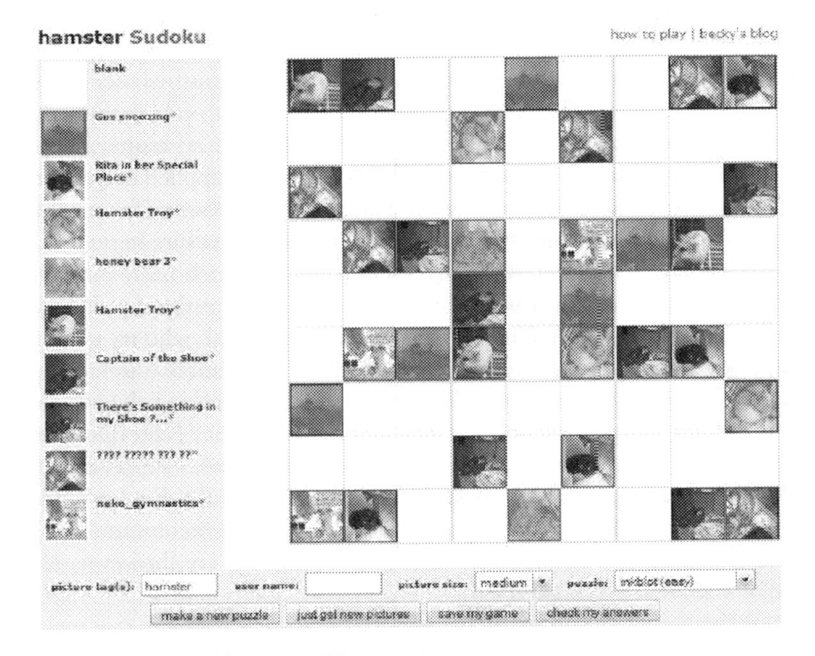

Figure 7.1 Flickr Sudoku with tag word 'hamster'.

aim is to fill the grid with system-generated images so that each column, row and smaller three-by-three grid contains all nine different images. There are differently designed puzzles to choose from, and the player can type in keywords that are used to generate the token images from *Flickr*. Thus, *Flickr* is mostly used to provide a visual flourish to the familiar game mechanics. No multi-player mechanisms or other social game play are apparent in the implementation, but the photos themselves and their surprising combinations may provide a source of humour among the *Flickr* users.

The second example drawn from among these user-created *Flickr* games is named *Fastr*.[13] This is a guessing game which displays 10 images, one by one, and the player needs to type in suggestions of what might be the tag that all these pictures share (see Figure 7.2). A correct guess appears as a blue word and score is given, based on how fast the player was in guessing right. The full version of the game is multi-player and allows chat exchanges among players. Relying on social interaction as well as on playful use of social metadata, this type of game has become relatively popular on the Web – see, for example, the *ESP Game*[14] and *Guess-the-Google*.[15]

As a final example, I introduce here a game called *PhotoMunchrs*.[16] One use for games and the *Flickr* API has been application of them for information research and studies of search technology. *PhotoMunchrs* is a puzzle game that relies on a *Pac-Man* style of navigation through a picture grid, guiding the player character into eating images based on the right tag words, while avoiding meeting the enemy (a red 'Traggle' character) or eating 'wrong photos' (see Figure 7.3). Munching seven correct photos moves the player up one level. The game has been designed

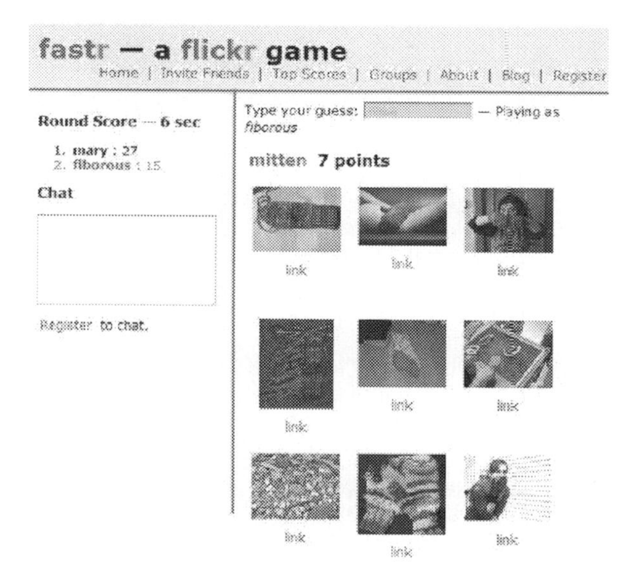

Figure 7.2 The *Fastr* game, with two players.

Figure 7.3 The *PhotoMunchrs* game.

as an experiment in gathering visual relevance data. There is a hi-score list, but no actual multi-player features.

It should be noted that the dedicated gaming applications are in clear minority within the broader field of playful *Flickr* applications. There are literally hundreds of different *Flickr* uploading tools, desktop applications, mobile phone applications, browser extensions and plug-ins for blogging software that are often categorized as 'fun and games', rather than as utility software. The discourse of dedication and passion for extended *Flickr* use also regularly share the rhetorics of addiction with that often associated with games and gamers.[17]

As a showcase of contextual play, *Flickr* is able to provide mixed and diverse lessons. The divided results are no doubt partly due to the current, still rather early state of socially and spatially connected technology, but also to the ways in which the underlying technologies are currently adopted in user cultures. The potential of these kinds of playful and social media services in themselves appears promising. The individual *Flickr* photostreams trace the life and travels of active users in a manner that provides an intimate and detailed view into their daily lives. The service automatically highlights the most recent photos from users' *Flickr* contacts when one logs in to the service account, thereby providing a readily populated social context for interaction.

Yet, the mobile, contextual use of *Flickr* images appears limited, in part due to the laborious process of providing photos with geo-data (something about to change as GPS becomes more integrated in mobile photo devices). It is somewhat common among users to manually insert tags and descriptive texts to photos in order to create some spatial markers or coordinates. The main interface of *Flickr* as a service is nevertheless not organized to bring together users who are co-located or to encourage interactions that are based on mobile use situations. The *paidia* dimension of *Flickr* appears thus primarily contextualized through connections that form the social context to the practices of playful browsing (as analyzed by Lerman and Jones 2006). Thematic or tag-based playful exchanges are also common.

To a certain extent, contextual and spatial information is currently provided for playful uses through *Flickr Places* and *Flickr World Map*, which allow multiple ways of navigating, commenting and other forms of participation in geographically filtered photostreams. The ensuing rich environment of users, social networks, photos and their dynamic interconnections then provides material for that apparent minority whom are interested in engaging in more *ludus* style of goal-oriented or competitive gaming in *Flickr*. The social mini-games, such as *Fastr*, have their potential, but are still rather limited experiments in this realm.

Play in networks

My second example focusses on the contextual access and play in *Facebook*,[18] a popular social networking service and certain related 'social utilities'. The situation is somewhat comparable to that of *Flickr*, even if the services in themselves have been designed with clearly different goals in mind. *Facebook* has its origins as a social networking service for university students – the founders originally designed the service while being students at Harvard.[19] In the web site usage metrics, *Facebook* has challenged *Google* and *Yahoo!* as the most popular destination on the Internet, and the service has grown into one of the most popular web sites in general, with more than 400 million active users in 2010.[20] Even if online sharing of photos is also one of the most popular activities on *Facebook* with more than two billion photos uploaded each month,[21] *Facebook* is a more diverse service, with several distinctly different classes of 'apps' (applications utilizing the *Facebook* API) available.

The spatially contextualized origin of *Facebook* is still visible in the way most of its users are organized into 'networks' that relate to their school, workplace or living area. Some analysts of *Facebook* user data have suggested that allowing users to search each other's profiles for shared city, institution or job type may be an important way to create the 'sense of connection' that, in its turn, facilitates interaction (Lampe *et al.* 2007). Similarly, *Facebook* has been connected with strengthening social capital, such as, by linking students to old school friends, and through this process contributing to their well-being (Ellison *et al.* 2007). Since 2007, *Facebook* has also been made available in various ways as a mobile service.[22] In August 2007, a dedicated iPhone application was provided as a touch-optimized mobile interface into *Facebook* for the Apple smartphone users, and *Facebook* applications for other popular smartphone brands soon followed.[23] All these different forms of access are designed to facilitate making of quick 'status updates' in contexts that are realized away from a more fixed environment such as an office desk and a personal computer.

Games applications are a visible part of the *Facebook* 'ecosystem'. When approached as a playful or gaming environment, the *Facebook* experience is initially focused on acquiring 'friends' to one's contact list. The service has made this easy, by tracking existing social networks and suggesting new contacts. Yet, other social networking services display the number of one's contacts more prominently than *Facebook* (see, e.g., *Linked In.*[24]). I would argue that *Facebook* is more focused on various ways of acting and sharing in the social environment and therefore the service features and add-on applications allow for interacting with one's online contacts. Particularly, since *Facebook* opened their 'Facebook Platform' (a set of APIs) in January 2007, the field has expanded and thousands of different *Facebook* applications have been created and made available through the service.[25] In distributing them, *Facebook* particularly initially relied on an aggressive viral model where each user was encouraged to send an install invitation to their own contacts. This mechanism, while related to the exponential popularity growth of the top applications, also led to a phenomenon called 'application spam', with invitations and notifications rising to such numbers that they were even blamed for drops in *Facebook* user numbers in early 2008. *Facebook* responded by providing users the ability to 'block applications' and the facility to report to administration applications that are forcing its users to invite more friends.[26] The spread of *Facebook* applications such as games nevertheless continues to rely upon/experimentation with different viral mechanisms built in as the core interaction or game play element.[27]

Facebook usage appears closely integrated with the daily media practices of its users; Ellison *et al.* (2007: 1144, 1153) refer to studies according to which the 'typical user spends about 20 minutes a day on the site, and two-thirds of users log in at least once a day', and their own findings among undergraduate students confirms this, adding that their informants reported having between 150 and 200 friends listed on their profile.[28] The significance of social context and the intense socially interactive character of *Facebook* is further underlined by data published by O'Reilly, according to which the three most popular uses for *Facebook* applications are enhanced communication, social comparison and playing a social game.[29] When approached from a game studies perspective, even the communication and comparison applications in *Facebook* appear distinctively game-like or playful. I will illustrate this by briefly highlighting a few typical *Facebook* applications.

For this chapter, I made an informal sampling of the most popular *Facebook* applications, first in August 2008, using the 'Most Active Users' listing in *Facebook* Application Directory, then a year later in August 2009 using the data provided at

Appdata.com web site, which was then again accessed for a sample in May 2010.[30] In August 2008, the most popular application in the *Facebook* application directory was titled 'Slide FunSpace (formerly FunWall)'.[31] As a specimen of the early successful *Facebook* apps, *Slide FunSpace* had in August 2008 more than 21 million monthly active users and was advertised having been used for sharing more than six billion videos and other links. In the blogosphere of Autumn 2008, *FunSpace* and its main competitor, *SuperWall* were also among the most widely criticized applications, sometimes on the basis of their content and sometimes on grounds of being the source of much 'application spam'. Much of the most actively shared content in *FunSpace* was either sexually oriented (see Figure 7.4), humorous or both. Also the sharing of music videos and other media was at the top of the lists of *FunSpace* usage. I would argue that the principal function for 'enhanced communication' applications like *FunSpace* is similar to that of phatic communication – communication that is practiced for maintaining social relations, rather than for its information value. Phatic communication is sometimes considered as a practical synonym for social presence (Rourke *et al.* 1999). *FunSpace* is thus used to construct a shared, pleasantly sociable space among its users – as its name already suggests.

Games applications have got their fair share of visibility in *Facebook*. However, in August 2008, in the *Facebook* application directory, the majority of applications with most active users appeared to be something else rather than explicit game applications. Among the top 35 applications, five clearly present themselves as games: *Word Challenge* (#7), *Quizzes* (#15), *Tower Bloxx* (#16), *Crazy Taxi* (#27) and *Zombies* (#35). A number of other applications are included into the Games application directory, including *Pokey!* ('Adopt an adorable, interactive 3D puppy who

Figure 7.4 The most popular 'posters' shared in funspace (Facebook.com, 26 August 2008).

lives on your profile, plays with you, carries bones to your friends and gives you tons of love!'[32]) and *YoVille* ('YoVille is a world where you can buy new clothes for your player, purchase items for your apartment, go to work and meet new friends'[33]). Rather than 'games' in the classic sense, these applications could be described as 'software toys', which is the concept game designer Will Wright and his company have decided to use to describe popular creations of theirs, such as *SimCity* (Maxis 1989) and *The Sims* (Maxis 2000).[34]

What *Facebook* game applications contribute to the design space of classic board-games or video games is their close integration within the shared social context of online service. *Word Challenge* allows easily inviting or challenging other users that are automatically drawn from the users' *Facebook* friends lists. Another game, *Zombies* is a good example of a simple, 'first generation' *Facebook* game application that does not provide much in terms of actual game play, but rewards by high ranking those users who actively distribute invitations to the game through 'biting their friends'.[35] Active recruiters – who also act as viral agents spreading the game – will soon have impressive titles in their profile page, plus their own 'army'. The improved power points can then be used in challenging other players to fights, which do not require any skilful game play but are rather automatically played out by the game system. Automated battles in themselves are nothing new, many strategy and some role-playing games sort out battle outcomes this way. *Zombies*, *Vampires*, *Slayers* and similar first generation games are mostly remembered for their effective distribution mechanisms, which has also been one of the major sources of 'application spam', resulting in *Facebook* administration stepping in, and forbidding game applications from granting points as a result of sending out invites.[36] There has been some valid criticism among gamers as to whether this style of 'social game' constitutes a proper game at all, being little more than a thinly disguised façade for a viral distribution and marketing application.

Tower Bloxx (developed by Digital Chocolate) is an example of a *Facebook* game that includes a more pronounced skill element; *Tower Bloxx* has originated as a mobile game that has later been converted into a *Facebook* application. Originally a single player game, in *Tower Bloxx*, the player's aim is to build as high and as stable building as possible by dropping building blocks on top of each other. In the *Facebook* version, the basic game play remains the same, but the game scores gained by one's friends are integrated in the screen. The dashed bars are a visible incentive to compete against the best scores gained by those in one's social network. The role of the social networking service is in this case to stand as an audience and also as a competitive setting for a single player experience.

Sampling the *Facebook* application scene a year later in August 2009, the trendy character of these services had started to become clear. *Slide FunSpace*, for example, had already in a year lost much of its active user base, going down to the place 20 in the popularity charts. The most popular applications remained to be both social and playful, but now with a more focused twist: *Causes*[37] (23.5 million active monthly users in August 2009) attempted to avoid the feelings of frivolity by focusing on actions that contribute to positive, real world causes like environmental or health issues. The second most popular *Facebook* application in August 2009 was *LivingSocial* (23.2 million active monthly users), which facilitates sharing of information about books, movies, music albums and other such items of interest, thereby promoting such content as well as the taste and personality of the application user. The number of explicit game applications had increased within a year, as more developers entered this market. Of the 40 most popular applications in August 2009, 18 were categorized as games, with farming game applications, such as *FarmVille, Farm Town* and *Barn Buddy* (18, 15 and 2 million active monthly users, respectively), becoming more popular.[38]

In the sampling carried out in May 2010, the ruling application was *FarmVille*, the social farming game developed by Zynga. Even while the popularity of *FarmVille* had at this point already passed its highest peak (of more than 83 million users in April 2010), it remained still the most popular application in *Facebook* with more than 76 million monthly players. A sense of achievement and sociability come together in this kind of virtual farm-building and caring simulation, where it is possible to help one's *Facebook* friends to take care of their virtual plants and animals. In the August 2009 sample, the most popular *Facebook* applications continued to include several that focus on playful communication and information sharing: *LivingSocial, Hug Me, Food Fling!, How Well Do You Know Me?, Give Hearts, Hugged* and others. In the sample of May 2010, social gaming had grown to take a more prominent share of *Facebook* application space: 23 applications of top 40 were categorized as games, whereas the emphasis in the communication tools and toys category had moved towards mobile applications, with *Facebook for iPhone* and *Facebook for Blackberry Smartphones* applications rising high among the most popular applications.

Designed and programmed mostly on Flash, the game applications of *Facebook* run poorly or not at all within the browsers of contemporary mobile phones. The mobile use is, however, not restricted to handsets – laptop computers have gained in popularity, and some analysts have claimed that laptop sales eclipsed those of desktop computers during 2008.[39] Thus, the 'mobile Internet user' is increasingly typically one that is using Wi-Fi, 3G or some other wireless network to get online from one's laptop. One factor driving this is the ongoing societal and global development where work is becoming increasingly mobile and information based; for example, already in year 2002, 45 percent of the Finnish workforce could be categorized as 'mobile workers' (Gareis *et al.* 2005, 54) . This development will create new kinds of challenges for staying in touch with one's colleagues, as well as with family and friends. Even while working, one might not be in the office. Playing a session of *Tower Bloxx* while waiting for a transfer at an airport might be yet another way of keeping oneself visible in the social map.

The evolution of 'social games' in various platforms is very fast, and my observations appear to verify the industry claims of the lifespan of an average *Facebook* game application being 2 or 3 months.[40] The most favourite *Facebook* games in 2009 had been designed to make much more comprehensive use of social contexts than a game published a year or two ago, and this trend continued in games that were popular in 2010. An example is *Mafia Wars* by Zynga, which exhibited powerful growth during spring and summer of 2009 (see Figure 7.5).[41] In this

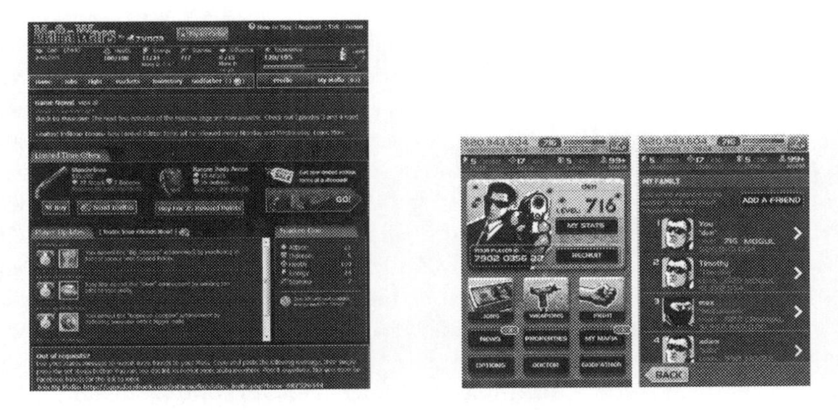

Figure 7.5 Mafia Wars in *Facebook* and on the iPhone.

game, the social presence of other players is tightly interwoven with the theme and metaphor of playing a criminal who is part of a Mafia family. There are mutual in-game rewards gained from actively recruiting, exchanging and linking with players that are part of one's *Facebook* friend network. Having a large Mafia family is an asset that makes a player more likely to succeed in battles that take place in the game, while crime themed status messages and recruiting invitations from one's friends are securing the viral marketing and spread of the application. The effective exploitation of such mechanisms helped to grow Zynga as the most successful social game company; the originality of game design was hardly the key factor: Zynga has been repeatedly accused of copying the competitors' game concepts.[42] There has been also controversy regarding the unethical or clearly fraudulent advertisement schemes applied by Zynga and its partners.[43] Despite the controversy, Zynga has continued to grow and expand its range of offerings. Cross-platform game concepts are one element in this. For example, in addition to the browser application of *Mafia Wars*, Zynga has also released an iPhone version which has a more streamlined interface and simpler game play, and which also implements some sound effects. However, a symptom of proprietary social media networks, at the time of writing, is that it is still impossible to link the *Mafia Wars* account from the *Facebook* game application to that of the iPhone version. Looking at the potentials of social, networked and mobile game forms, there is thus a clear need for open identity and API standards that would generate more extensive social visibility and value for playful activities, regardless of the technology used for communicating with one's friends.

Two core applications, or functionalities, which *Facebook* supports for mobile phone use are status updates and photo uploads. Both of these serve communicative purposes that are important for travelling users: the regular status updates and a stream of camera-phone photos helps in keeping in touch with the people in one's social network. For example, sampling the status update stream from my own friends for a few random days, I can find a mixed collection of messages, some related to their ongoing work, some to personal matters or feelings, some joking or ambiguous in style. Some carry little mobile phone icons next to them, marking them as having been posted from the mobile interface. One person appears to be away from the office on a work matter, another one has posted his status updates from a hospital bed. Announcements about events taking place within some virtual game world are dispersed in the feed among links and discussions that relate to media contents, professional events and changes in the relationships between users that I know and some that are unknown. Play, work and other areas of social life appear as irrevocably tangled up together.

Even while *Facebook* games or other rich media Internet contents are still predominantly accessed through the browser of a personal computer, 'laptop-based mobility' and the mobile Internet accessible through handsets are gaining more prominence. It appears that the borderline between playful and communicational practices of the PC-centric Internet and mobile phone cultures are becoming thinner. Still in 2003, when Eija-Liisa Kasesniemi studied the second generation GSM phone cultures in Finland, her informants could make a clear distinction between what kind of communicational affordances are related to personal computers in contrast to mobile phones:

> *Researcher*: If you had to give up either SMS or email, which one would you drop?
>
> *Kati* [a 14-year-old informant]: Probably email since the computer is so, I mean you can't carry it with you and things like that. [. . .]
>
> *Researcher*: Is IRC more fun than sending SMS?
>
> *Kati*: Mm, there's the thing that you can only do it in one place, sitting in front of your monitor. You can send SMS pretty much anywhere.
>
> (Kasesniemi 2003)

Today, these kind of distinctions appear to be dissolving, as e-mail is increasingly also accessed with mobile phones. As the mobile broadband connections become more common, they facilitate the use of instant messenger applications in phones as well as in laptop computers, whereby the role of the 'application-specific use context' does not remain the same any more. The mobile users of services like *Facebook* or *Flickr* are already blurring this division. Also several micro-blogging services have been created, the most popular of them *Twitter*, which is accessible from personal computers and mobile phones, as well. Playfulness and gaming impulse has permeated this communicational space, and there exist several games that are implemented for Twitter, most of them simple trivia or mathematical competitions that are easy to participate by sending answers via Tweet replies.[44] What remains important, however, in all of these different messaging applications, is the social context. After logging into the service, the same network of contacts and friends is available, irrespective of the device or technology used for getting online and into the communicational space.

Joinson (2008) has studied *Facebook* users' key motivations for accessing the service, and 'keeping in touch' was mentioned as the most important one; the other key motivations were also people-oriented – the desire to go 'virtual people-watching' (social surveillance), re-acquire lost contacts, and just the general need to communicate. Mobile access to *Facebook* is one particular way by which the contact with one's social network can be extended and maintained, and a use mode with rich potential for playful interactions with different location and situation-specific elements. The stream of photos and status updates from mobile users forms one thread in the mediated acts that together contribute to one's social presence. Again, we can find multiple ways in how 'play' and 'games' figure in these hybrid and cross-platform exchanges. The development of how location information has been implemented in *Facebook* is illustrative of the same motivations. At the time of this writing, *Facebook* is still to integrate location information directly into their core functions like status updates, but there are several third-party applications that integrate with *Facebook* and have already started to move the service in this direction. One of the most interesting ones from the contextual play perspective is called *Foursquare*.

Foursquare is a location-based social service that is built around a ludic core – by 'checking-in' various locations you visit, you can become recognized as the 'Mayor' of that particular place. The users earn points from their *Foursquare* activity, and can progress in the game's hierarchy by gaining badges like 'Newbie', 'Adventurer' or 'Superstar'. The game supports links to other social services, making it possible to release news from every check-in, badge or mayor status change to one's *Facebook* or Twitter friends. Many of the places *Foursquare* recognizes are shops, restaurants or other commercial establishments and it is easy to see the commercial potential of social network applications that are operating on such information. Unsurprisingly, the game application also includes an 'add tip/to do' functionality: the users can leave their tips on what food to order in a certain restaurant, for example. The venues on their part can decide to provide discounts and prizes to the customers who are loyally checking in at their location. The boundary between a social game and a playful marketing application effectively dissolves in *Foursquare*. The value of game play, socially shared ludic information and the motivation of applying 'virtual surveillance' into one's social network start to come together with blatant consumerism in a manner that also evokes certain questions and concerns that are discussed in my conclusions.

One of the key elements in most dictionary definitions of 'play' is the free and pleasurable character of playful activities. In an article, McClard and Anderson (2008) have paid attention to the dynamic and social character of identity construction that takes place in *Facebook*: rather than focusing on setting up one's 'profile' as a static page. As in some other services, the application-based nature of *Facebook* allows a representation of identity that is fluid. 'One's 'image' is created by what one does, who one does it with and how it is done; it is constantly in flux', they write. 'On Facebook life is a game', they claim and continue:

> 'Although participants can open chat windows or belong to special interest groups of a more serious nature, the daily drivers of Facebook exchanges are games and quizzes. As technology mediates more and more of our social exchanges, the forms of our interaction change. Gaming – light, breezy and

fun interactions with friends near and far – keeps ties alive without being burdensome.'

(McLard and Anderson 2008)

The role of games in a social networking service like *Facebook* should not be exaggerated – not everyone is using such applications, many focus upon what their friends, co-workers and family are doing. Yet, the role of contextual play is not only restricted to how many people are engaged in game play. As I have argued elsewhere (Stenros *et al.* 2007), on the one hand, there is a visible trend of increasing *ludus* in contemporary society – meaning here both the increasingly visible playful attitude and adoption of game-like practices. A good example is in media, notably in phenomenon like reality television and game shows that are dominating today's broadcast media. On the other hand, it is important to recognize that there still exists some 'serious' media contents that relate to those areas of human life that are not primarily playful. The effects of pervasive playful contexts are difficult to block, however. While the life of dedicated *Facebook* users is not defined by only play and games, it certainly is valid to say more generally that the *paidia* style of playful, social interaction is dominating much of what is going on in this service. It is a character of *Facebook* as media that even the most serious themes of discussion are contextualized in an environment that is saturated by an endless stream of media, games and news-related items, thereby becoming part of a playfully eclectic tapestry.

Considering the different ways of engaging with *Facebook*, it might even be that the *ludus* impulse of (competitive, rule-bound) game play goes against the impulse to participate in the more freeform, playful exchanges taking place through status updates, photo comments and the various 'poking', 'hugging', 'kissing' or 'gift giving' activities that are displayed on *Facebook* user's profiles and application 'walls'. In his analysis of *Facebook* users, Joinson (2008: 1034) notes: 'Interestingly, an increased score on the content gratification scale was negatively related to the number of 'friends' reported to be linked to one's profile.' Thus, the interest in dedicated gaming applications does not appear necessarily to go together with the social interests of *Facebook*. A dedicated *Facebook* gamer might even set up several accounts or user profiles just for the need of having 'alts' (alternative gaming accounts) in order to optimize strategies of gaming success, rather than for any personal interest in 'friends' that are listed as contacts in such a profile[45]. Here we appear to have a case where emphatically social *paidia* is differentiated from *ludus* that is not focused on social networks to a similar degree. However, more research would be required before drawing any more far-reaching conclusions on this.

Towards conclusions

In this chapter, I have examined the rapidly changing landscape of mobile Internet usage and playful behaviours in online social networking services. As 'mobile Internet' is no longer synonymous with dedicated mobile phone services, the character of 'mobility' itself is undergoing transformations. The combination of social networking with playful, or game-like uses and behaviours emerges as an important contemporary form of online communication, mixing and muddling up the boundaries between work and play, as well as leisurely and utilitarian interests. While still a few years back it was typical to focus on various operator-provided utility services like online banking or news services while discussing the mobile Internet, today's landscape appears more centred on users themselves. Perhaps, the much searched-after 'killer app' for mobile Internet is finally found – the people.

Mobile Internet also increasingly appears to be a hybrid one. This is hybridity in terms of technologies, as laptop computers, handheld PC devices and smartphones are utilized in an *ad hoc* manner, making use of whatever network and interface is available. Hybridity is also social, as the users are enrolling in various online services and locate different subsets of contacts in each of them. In the ensuing mélange of contacts and communication, the social spheres of colleagues, personal friends and family are starting to intertwine in an increasingly complex manner. Finally, this hybridity is also existential, as the physical presence (or absence) is augmented by various 'photo streams', 'status updates' and other online acts and representations that together constitute the contemporary presence of an individual in a social context.

It is also easy to criticize the ongoing development. As soon as there is a network-enabled device in most pockets as well as in schoolbags or briefcases, the possibility of 'logging in' might soon be substituted by a social obligation to do so. For some active Internet user demographics, such a situation appears to be already their living reality. The privacy concerns aside, the constant compulsion to 'update oneself' in the online social sphere might also be a symptom of some underlying frailty in contemporary society. The bipolar tension between the Net and the self, as discussed in the context of growing stress on patterns of social communication by Castells (2000) provides one interpretative direction. Management of the various threads that constitute one's identity in a network era is becoming more laborious. One's extended social networks are also busily producing a never-ending stream of contextual information that is in danger of becoming yet another form of 'infoglut'. That our lexicon now includes concepts like 'invite spam' or 'application overload' tells that it already might have become one.

Drawing upon Michel Foucault, Deleuze (1998) has written about the 'society of control', where the confinement of space and time, which was typical to the disciplinary societies in the eighteenth and nineteenth centuries, has been replaced by new forms of internalized, economical and ultra-rapid forms of 'free-floating control'. It is not clear at this point yet what the exact role of mobile and social, contextually enabled applications and services will be when approached from this perspective. The connections these systems allow are always dual: they liberate to explore and express, and are therefore potentially empowering. On the other hand, the constant contact with social networking services is also enabling new, unceasing techniques of control. The examples of commercial contextual play services such as *Foursquare* discussed above hint towards a future, where rather than being deprived by our privacy by some shadowy 'Big Brother', we end up disclosing details of our private life voluntarily, in exchange for perceived commercial and social benefits. My overall conclusions at this point are, nevertheless, predominantly positive. It is inspiring to see the ways in which various 'communication tools' or 'social utilities' are being repurposed by their users to become playing fields. There is certain Dadaist or anarchist – and certainly also infantile – pleasure involved in following how one's colleagues throw sheep at each other, or buy and sell each other as pets. The participation in free flowing energies of play can, nevertheless, also easily turn into compulsions of connection. In this view, contextual play is currently a loaded field, strained between multiple possible directions of future evolution.

Notes

1 IAB (2008) 'Mobile internet penetration hits 12.9 percent in the UK', Internet Advertising Bureau, July 11, 2008. http://www.iabuk.net/en/1/newmobileinternetresearch110808.mxs>. J. D. O'Grady (2008) 'Google; ATandT shocked by iPhone usage', ZDNet, 15 February 2008. http://blogs.zdnet.com/Apple/?p=1316>. (Retrieved 4 August 2009).

2 ITU (2009) *Measuring the Information Society: The ICT Development Index*. Geneva: International Telecommunication Union. http://www.itu.int/ITUD/ict/publications/idi/2009/material/IDI2009_w5.pdf (Retrieved 4 August 2009). ITU (2010) 'ITU Sees 5 Billion Mobile Subscriptions Globally in 2010', ITU Newsroom. Geneva: International Telecommunication Union. http://www.itu.int/net/pressoffice/press_releases/2010/06.aspx (Retrieved 6 May 2010).

3 Relevant research projects include Mogame – The Wireless Gaming Solutions of the Future (2003–2004), Mobile Content Communities (MC2; 2003–2005), Integrated Project on Pervasive Gaming (IPerG; 2004–2008), Transformation of Digital Play (TDP; 2008–2010).

4 Anthony Giddens has summarised work of psychologists such as Erik Eriksson and D. W. Winnicott on the evolution of 'basic trust' and points out how, from a sociological and philosophical perspective, these links to other people are 'connected in an essential way to the interpersonal organisation of time and space' (Giddens 1991). My approach is based on earlier published work (e.g. Ermi and Mäyrä 2005; Mäyrä 2007, 2008).

5 See the Flickr web site: http://www.flickr.com (Retrieved 22 August 2008).

6 See the Eye.fi web site: http://www.eye.fi (Retrieved 22 August 2008).

7 See J. J. Garrett (2005) 'An Interview with Flickr's Eric Costello', *Adaptive Path*, August 4 2005. http://www.adaptivepath.com/ideas/essays/archives/000519.php (Retrieved 15 August 2008).

8 D. S. Butterfield *et al.* (2006) 'US Patent Application 20060242139: Interestingness ranking of media objects'. http://appft1.uspto.gov/netacgi/nph-Parser?Sect1=PTO1andSect2=HITOFFandd=PG01andp=1andu=/netahtml/PTO/srchnum.htmlandr=1andf=Gandl=50ands1=%2220060242139%22.PGNR.andOS=DN/20060242139andRS=DN/20060242139 (Retrieved 18 August 2008).

9 Official Flickr Blog, '4,000,000,000', 12 October 2009. http://blog.flickr.net/2009/10/12/4000000000/ (Retrieved 11 May 2010).

10 W. Hein, blog note 'Deconstructing Flickr's 'Interestingness!'' http://wes2.wordpress.com/2006/05/12/deconstructing-flickrs-interestingness (Retrieved 18 August 2008).

11 The interestingness criterions have been discussed in online how-to-guides that collect tips for optimizing one's chances of success in Flickr; see e.g. 'What It Takes to Get Your Photo on the Flickr Explore Page', *PhotoPreneur.com*. http://blogs.photopreneur.com/what-it-takes-to-get-your-photo-on-the-flickr-explore-page (Retrieved 6 May 2010).

12 'Flickr Sudoku', *Becky's Web.* http://www.beckysweb.co.uk/sudoku/flickrsudoku.asp (Retrieved 18 August 2008).
13 'Fastr – A Flickr Game', *RandomChaos.com.* http://randomchaos.com/games/fastr/ (Retrieved 18 August 2008).
14 G. Robinson, 'Guess the Google', http://grant.robinson.name/projects/guess-the-google/ (Retrieved August 18 2008).
15 'ESP Game', *Gwap.com,* School of Computer Science, Carnegie Mellon University. http://www.gwap.com/gwap/gamesPreview/espgame/ (Retrieved 4 August 2009).
16 P. Fox, 'PhotoMunchrs'. http://imagine-it.org/flickr/PhotoMunchrs.html (Retrieved 18 August 2008).
17 See, for example, S. P. Aune, 'Flickr Toolbox – 100+ Tools for Flickr Addicts', *Mashable.com,* August 4, 2007. http://mashable.com/2007/08/04/flickr-toolbox/ (Retrieved 4 August 2008).
18 See the Facebook web site: http://www.facebook.com (Retrieved 4 August 2009).
19 See 'Company Timeline' in the Facebook web site: http://www.facebook.com/press/info.php?execbios=#/press/info.php?timeline (Retrieved 4 August 2009).
20 See the factsheet and statistics pages available at the Facebook web site: http://www.facebook.com/press/info.php?statisticsandhttp://www.facebook.com/press/info.php?factsheet (Retrieved 7 May 2010). In May 2010, the web site analytics firm Hitwise reported Facebook as the most visited web site in the United States, followed by Google and Yahoo (Yahoo mail and Yahoo main site). http://www.hitwise.com/us/datacenter/main/dashboard-10133.html (Retrieved 7 May 2010).
21 See the Facebook statistics page (the reference above). Also: D. Beaver (2007) 'Facebook Photos Infrastructure'. A blog note. http://blog.facebook.com/blog.php?post=2406207130 (Retrieved 4 August 2009). The article 'Internet 2009 in Numbers' by Pingdom web site reported the monthly photo upload number as 2,5 billion. http://royal.pingdom.com/2010/01/22/internet-2009-in-numbers/ (Retrieved 7 May 2010).
22 See Facebook Mobile web page. http://www.facebook.com/apps/application.php?id=2915120374andb>. See also M. Slee (2007) 'Facebook Your Phone'. A Facebook note, 10 January 2007. http://blog.facebook.com/blog.php?post=2228532130>. (Retrieved 4 August 2009.) Rather than a single application, 'Facebook mobile' comprises of ability to upload photos and notes directly from one's mobile phone to the service, and the of the special interface athttp://m.facebook.comthat is customised for small-screen browser use.
23 See the blog post 'Facebook for iPhone is Here' by Joe Hewitt, posted 15 August 2007. http://blog.facebook.com/blog.php?post=5353402130 (Retrieved 7 May 2010).
24 See the LinkedIn web site: http://www.linkedin.com (Retrieved 4 August 2009).
25 A. Ustinova, 'Developers compete at Facebook conference', *San Francisco Chronicle,* 23 June 2007. http://www.sfgate.com/cgi-bin/article.cgi?f=/c/a/2008/07/23/BU7C11TAES.DTL (Retrieved 4 August 2009).
26 D. Parrack, 'Facebook use drops – Could application spam be the cause? *Blorge,* February 21 2008. http://tech.blorge.com/Structure: %20/2008/02/21/facebook-use-drops-could-application-spam-be-the-cause (Retrieved 4 August 2009). *Time* has also discussed the phenomenon, under the heading of 'application overload'; see A. Hamilton, 'Suffering From Facebook Fatigue?' *Time,* April 16 2008. http://www.time.com/time/business/article/0,8599,1731516,00.html (Retrieved 4 August 2009). The Facebook response is reported in blog post 'Application Spam' by Paul C. Jeffries, posted 20 February 2008. http://blog.facebook.com/blog.php?post=10199482130 (Retrieved 7 May 2010).
27 See Aki Järvinen's blog post 'The Near Future of Viral Design in Social Games' in Gamasutra, posted 7 April 2010. http://www.gamasutra.com/blogs/AkiJarvinen/20100407/4577/The_Near_Future_of_Viral_Design_in_Social_Games.php (Retrieved 7 May 2010).

28 Different sources give somewhat varying estimates for the average time investment of a Facebook user. The official Facebook statistics page reports in May 2010 that 'Average user has 130 friends' and that 'People spend over 500 billion minutes per month on Facebook'. Facebook statistics page. http://www.facebook.com/press/info.php?statistics (Retrieved 7 May 2010). That would roughly translate into 37 minutes per user in a 30-day month. The analytics company Nielsen released figures in February 2010 that estimated an average Facebook user to spend more than seven hours per month in Facebook. This overshadows time spent in any other monitored online service, but on the other hand translates to only 14 minutes per day. See Ben Parr, 'Facebook Is the Web's Ultimate Timesink', posted February 16 2010. http://mashable.com/2010/02/16/facebook-nielsen-stats/ (Retrieved 7 May 2010). The preciseness of such time usage evaluations should be taken with a grain of salt.

29 S. D. Farnham, 'The Facebook Application Ecosystem: Why Some Thrive, and Most Don't', O'Reilly Media webcast, 13 May 2008. http://www.oreillynet.com/pub/e/968 (Retrieved 4 August 2009).

30 The 'Most Active Users' listing in Facebook Application Directory. http://www.new.facebook.com/apps/ (Retrieved 27 August 2008). This source has been later replaced by 'Featured by Facebook' listing that does not provide the same information. Application user popularity analyses for August 2009 have thus been based on data provided at Appdata web site: http://www.appdata.com (Retrieved 4 August 2009, and again 10 May 2010).

31 Facebook Application Directory. http://www.new.facebook.com/apps/ (Retrieved 4 August 2009). Slide FunSpace web page. http://www.facebook.com/apps/application.php?id=2378983609 (Retrieved 4 August 2009).

32 FooPets web page. http://www.facebook.com/apps/application.php?id=10062052582 (Retrieved 4 August 2009).

33 YoVille web page. http://www.facebook.com/apps/application.php?id=21526880407 (Retrieved 4 August 2009).

34 The 'software toy' concept was publicised by Maxis *Software Toys Catalog*; quoted in T. Friedman, 'Making Sense of Software: Computer Games and Interactive Textuality', updated version. http://www.duke.edu/~tlove/simcity.htm (Retrieved 4 August 2009).

35 Zombies web page. http://www.facebook.com/apps/application.php?id=2341504841 (Retrieved 4 August 2009).

36 Jeffries (2008) 'New Design, New Powers, New Responsibilities'. A blog note, 21 July 2008. http://developers.facebook.com/news.php?blog=1andstory=137 (Retrieved 4 August 2009).

37 Causes web page. http://apps.facebook.com/apps/application.php?id=2318966938 (Retrieved 7 August 2009).

38 *FarmVille*, *Farm Town* and *Barn Buddy* web pages. http://apps.facebook.com/apps/application.php?id=102452128776>,http://apps.facebook.com/apps/application.php?id=56748925791>,http://apps.facebook.com/apps/application.php?id=60884004973 (Retrieved 7 August 2009).

39 Hruska (2008) '2008 could be the year laptop sales eclipse desktops in US', *Ars Technica*, 3 January 2008. http://arstechnica.com/apple/news/2008/01/2008-could-be-the-year-laptop-sales-eclipse-desktops-in-us.ars (Retrieved 4 August 2009). K. Thompson (2008) 'Laptop Sales Surpass Desktop Sales in 2008'. A blog note, 29 December 2008. http://rentourlaptops.blogspot.com/2008/12/laptop-sales-surpass-desktop-sales-in.html (Retrieved 7 August 2009).

40 Jon Swartz, 'For social networks, it's game on', *USA Today*, 15 October 2009. http://www.usatoday.com/tech/gaming/2009-10-15-games-hit-social-networks_N.htm (Retrieved 15 October 2009).

41 See Mafia Wars application page. http://www.facebook.com/apps/application.php?id=10979261223>. Mafia Wars statistics. http://statistics.allfacebook.com/applications/single/mafia-wars/10979261223/ (Retrieved 12 October 2009).

42 See Michael Arrington, 'Zynga Settles Mob Wars Litigation As It Settles In To Playdom Fight', TechCruch, 13 September 2009. Online. http://techcrunch.com/2009/09/13/zynga-settles-mob-wars-litigation-as-it-settles-in-to-playdom-war/ (Retrieved 10 May 2010).
43 See Michael Arrington, 'Scamville: The Social Gaming Ecosystem Of Hell', TechCrunch, 31 October 2009. http://techcrunch.com/2009/10/31/scamville-the-social-gaming-ecosystem-of-hell/ (Retreived 10 May 2010).
44 See Twitter web site: http://twitter.com (Retrieved 4 August 2009). About Twitter games, see the article '6 Twitter Games To Make Tweeting Fun', posted 28 March 2009, in Mashable. http://mashable.com/2009/03/28/twitter-games/ (Retrieved 7 May 2010).
45 We have found anecdotal evidence of such practices while interviewing Facebook users in the research projects carried out in University of Tampere Game Research Lab. It is however hard to establish how common or rare such 'hardcore' gaming practices are among Facebook gamers.

References

Caillois, R. (2001) *Man, Play and Games (Trans. M. Barash)*, Urbana (Ill), University of Illinois Press.

Castells, M. (2000) *The Rise of the Network Society*, Oxford, Blackwell.

Deleuze, Gilles (1998) 'Society of control', *L'autre journal.*Online at:http://www.nadir.org/nadir/archiv/netzkritik/societyofcontrol.html

Ellison, N. B., C. Steinfield and C. Lampe (2007) The benefits of Facebook friends: social capital and college students' use of online social network sites. *Journal of Computer-Mediated Communication*, 12(4): 1143–68.

Ermi, L. and F. Mäyrä (2007) Fundamental Components of the Gameplay Experience: Analysing Immersion. In S. de Castell and J. Jenson (eds), *Worlds in Play: International Perspectives on Digital Games Research.* New York, Peter Lang: 37–54.

Frasca, G. (2003) Simulation Versus Narrative: Introduction to Ludology. In M. J. P. Wolf and B. Perron (eds), *The Video Game Theory Reader.* London, Routledge: 221–35.

Gareis, K., L. Stefan and A. Mentrup (2005) Mapping the Mobile Eworkforce in Europe. In J. H. E. Andriessen and M. Vartianinen (eds), *Mobile Virtual Work: A New Paradigm.* Berlin, Springer: 45–69.

Giddens, A. (1991) *Modernity and Self-Identity Self and Society in the Late Modern Age*, Cambridge, Polity Press.

Guarneri, R., A. M. Sollund, D. Marston, E. Fossback, B. Berntsen, G. Nygreen, G. Gylterud, R. Bars and A. Kerdraon (2004) Report of State of the Art in Personalisation. Common Framework Eperspace – IST Integrated Project. Retrieved from http://www.ist-eperspace.org/deliverables/D5.1.pdf.

Joinson, A. N. (2008) Looking at, Looking up or keeping up with people? Motives and use of Facebook. *Proceedings of the 26th Annual SIGCHI Conference on Human Factors in Computing Systems*, Florence, Italy.

Kasesniemi, E.-L. (2003) *Mobile Messages: Young People and a New Communication Culture*, Tampere, Tampere University Press.

Katz, J. E. and S. K. Acord (2008) Mobile games and entertainment. In J. Katz (ed.), *Handbook of Mobile Communication Studies.* Cambridge, MIT Press: 403–418.

Lampe, C., N. B. Ellison and C. Steinfield (2007) A familiar Face(Book): profile elements as signals in an online social network. *Proceedings of the SIGCHI Conference on Human Factors in Computing Systems*, San Jose, U.S.A.

Lerman, K. and L. Jones (2006) Social Browsing on Flickr. Retrieved 15 August, 2008, from http://arxiv.org/abs/cs/0612047.

Mäyrä, F. (2007) The contextual game: on the socio-cultural contexts for meaning in digital play. *Proceedings of DiGRA*, Tokyo, Japan.

Mäyrä, F. (2008) *Introduction to Game Studies: Games in Culture*, New York, Sage Publications.

McLard, A. and K. Anderson (2008) Focus on Facebook: who are we anyway? *Anthropology News*, 9(3): 10–12.

Rourke, L., T. Anderson and W. Archer (1999) Assessing social presence in asynchronous text-based computer conferencing. *The Journal of Distance Education/Revue de l'Éducation à Distance*, 14(2): 50–71.

Stenros, J., M. Montola and F. Mäyrä (2007) Pervasive games in Ludic Society. *Proceedings of the 2007 conference on Future Play*, Toronto, Canada.

Suchman, L. (2007) *Human-Machine Reconfigurations: Plans and Situated Actions*. 2nd edn, Cambridge, Cambridge University Press.

Tester, J. (2006) All the World's a Game: The Future of Context-Aware Gaming. *Technology Horizons Program*. Palo Alto, USA.

8 The whereabouts of play, or how the magic circle helps create social identities in virtual worlds

Thiago Falcão and José Carlos Ribeiro

This chapter considers the process of online social identity construction and how it is presented in virtual worlds – environments that although carrying a solid heritage of the pure online interaction tools, such as chat rooms, are still organized around a very specific logic, portraying objectives and structural repetition dynamics inherent to the game as a system. We posit that the presence of what is called the 'magic circle' by Salen and Zimmerman (2004) exerts a mediating power, not only by circling, limiting the game environment, but specially by transforming social identities formations processes, attaching it not only to the dynamics built in by the contact with technology – in the form of computer-mediated communication – but also in particular to the dynamics imprinted by the game structure present in such worlds.

The whereabouts of play

There is a certain agreement in the study of various aspects of media when it comes to considering the spheres of work and play. As Yee (2006) has coherently pointed out, these activities remain conceptualized as separate poles of the same dichotomy. Such framing has received wide support, along the course of history, figuring also in the ideas of some of the classic game theorists: Huizinga (1955) and Caillois (2001), for instance, believed that in order to play a game, the individual has to consciously step outside 'normal life' (Huizinga 1955) and voluntarily engage in an activity considered 'not serious' – suppressing both time and space.

This place in space and time in which the play activity happens received a thorough theoretical treatment – and a proper terminology – when Salen and Zimmerman (2004) published their treatise on the general analysis and development of the constitutive processes of the game. Salen and Zimmerman (2004) inspired by a passage in Huizinga's classic *Homo Ludens* utilize the term the 'magic circle' to identify and describe where the essence of play is located. According to the authors:

> Although the magic circle is merely one of the examples in Huizinga's list of 'play-grounds,' the term is used here as shorthand for the idea of a special place in time and space created by a game. The fact that the magic circle is

> just that – a circle – is an important feature of this concept. As a closed circle, the space it circumscribes is enclosed and separate from the real world. As a marker of time, the magic circle is like a clock: it simultaneously represents a path with a beginning and end, but one without beginning and end. The magic circle inscribes a space that is repeatable, a space both limited and limitless. In short, a finite space with infinite possibility.
>
> (Salen and Zimmerman 2004: 96)

Although Salen and Zimmerman's notion is based in one of the most classic treatises on the relationship between culture and the ludic expression, it has been extensively debated in recent years (such as, see Juul 2008; Pargman and Jakobsson 2008; Crawford 2009).

However, we should consider the idea of the transposition of realities supported by Salen and Zimmerman (2004) when asking what psychological attitudes would be necessary from a player the moment she engages in a game. Such ideas – if generally taken – may be considered a direct reference to the understanding of the real–virtual relationship as a dichotomy. We would like to point out here that such a dichotomous view has been widely challenged by the recent cyberculture theories (see Van Kranenburg 2008; Tuters and Varnelis 2006). This notion, then, takes us to this point: is it really valid to acknowledge the existence of such a barrier between realities, or reality and game, that needs to be disrupted?

Perhaps so, if we consider video games in which narratives are more salient; games that proclaim the necessity of dealing with particular narrative progression, implying absorption of the player into the narrative world, as would be highlighted by Murray (1997). When addressing other contemporary game categories, though, it becomes somewhat difficult to uphold such separation. (1) Pervasive, (2) alternate reality games (ARGs) and (3) massively multiplayer online (MMOs) games are some such examples – examples that work in the sense that they try to effectively blur the borders between 'normal life' and the game.

It is important to point out that our aim is not to derogate Salen and Zimmerman's notion but to reconsider it through the premise that some ludic forms have assumed – through time and due to their contact with digital and network technologies – complex structures for which both Huizinga's (1955) and Caillois' (2001) approaches seem to lack specificities – and the formalist notion of Salen and Zimmerman seems to impose a very dichotomous view. Therefore, we question not the totality of the notion of the magic circle, but rather its application to these borderline cases, in which it is clear that there is a much larger dialogue between the game structure and 'normal life'.

Thus, the point would not be to consider the magic circle as something that necessarily encapsulates the player, suppressing space–time and projecting her into an alternative zone. Instead, we acknowledge the existence of the magic circle – alongside its inherence to the game structure – but as a mediation element, which facilitates the player dialogue to both the game space and reality outside of it. Such mediation may be presented as a fluid form – drawing blurred borders, in the sense that they cannot be plainly identified, which allows fiction and reality to meet. Alternatively they can be understood as a harder, more defined, solid form, which really enables the sense of displacement (space–time suppression) through an immersive process.

Thus, we understand the magic circle as a cognitive structure whose action depends on an undetermined number of variables. This assumption offers a less limiting and wider perception of the classic understanding of the magic circle theoretical concept. Our proposal, then, is that the magic circle does not separate effectively the game world from reality, rather it acts as a mediation tool assisting the player in how they deal with the different sides of the universe – and not with *two universes*.

Such proposition finds support in the ideas of some game studies theorists, like Juul (2005), who points out that, to Huizinga (1955), the space in which the game develops is as separate as other daily spaces, as court houses, churches and classrooms – places that are integrated to the flow of life, but that ensue on a different behaviour – which explains the fact that Caillois (2001) refers to these places as 'separate'.

As Juul (2008) points out: 'For Huizinga, the space of game-playing is but one type of space governed by special rules, and as with other types of space, the space of game-playing is social in origin. People make special spaces, be they courthouses, religious spaces or game spaces'. This proposition takes us to a more specific understanding: although we may refer to games as objects, there is another aspect that should be considered: the interaction between games and people. The space–time construction, inherent in playing a game, demands interaction between people and object – or objects, as a matter of fact. Hence, according to Juul's (2008) proposition the magic circle is not only formed by the structure of the rules, but requires the players to uphold the illusion of the world, the borders of the magic circle are, therefore, negotiated and defined by the players.

Thus constituted in the moment the *object* game becomes the *activity* game, the magic circle may be understood as a mediation structure due to the fact that it does not act like a space limiter or a mechanism of transport to another dimension, but is merely the point where the 'normal life' (Huizinga 1955) meets this 'separate place'. Perhaps an analogy would help us to illustrate our point: when we enter a Catholic church, say to attend a Mass, or if we are in a courthouse, say as an observer, the circumstances of the context transport us socially and psychologically to a delimited space; conducted by a number of laws and upholding, a number of principles, both the former and the latter supported by the structure of tradition.

The laws and principles that maintain these spaces are nothing but rules that manage an object – both a court audience or a Catholic Mass – and the people are nothing but individuals behaving like expected – according to these social rules. None of the situations, however, keeps us from dealing with common life situations concerning other aspects of our lives, which may be different from, but intersect with, these 'consecrated spaces' (Huizinga 1955) – because also in these spaces, it is possible for us to talk about a certain number of 'indulgences', like going out of a ceremony to answer the cell phone, for example.

We therefore see the magic circle as configured as a mediator, exactly because it is a channel through which the separate worlds establish contact. In the common analogy of 'stepping through the looking glass', absorbed into culture due to Lewis Carrol and his *Through the Looking-Glass* (1944), the magic circle exerts the function of the mirror – considering that the world on the other side is nothing more than an extension of the world where we live.

The game structure, even though it has solid rules and codes of conduct, possesses this aforementioned indulgent essence explored by the analogy between consecrated spaces and game spaces. This indulgent quality appears precisely in the way that players explore, appropriate and adapt to the rules. 'Players do not simply adopt the rules of the game as given but regularly create their own achievement paths and make sense of the frames of play in ways not always prescribed by the designers' (Taylor 2007: 113). This adaptation varies according to the essence of the game – in a soccer match between two teams of five people, for example, it is not appropriate to stop the other nine players in order to answer the cell phone, or to check meteorology; in the game play of an ARG, as the relationship between player and both space and time is eminently different from the relationship experienced by a sport, like soccer, such indulgences are acceptable. As Taylor (2007: 113) elaborates:

> Games are typically thought of as closed systems of play in which formal rules allow players to operate within a 'magic circle' outside the cares of everyday life and the world. This rhetoric often evokes a sense that the player steps through a kind of looking glass and enters a pure game space. From Monopoly to Final Fantasy, commercial games in particular are often seen as structures conceived by a designer and then used by players in accordance with given rules and guidelines. Players, however, have a history of pushing against these boundaries, whether through feedback processes that change the game over time or, as Mizuko Ito suggests, via their engagement with games within an extended set of linked media practices and social identities – a subject that this chapter explores as well. However absorbing the game experience proves itself to be, player culture has never existed in a completely rarified space: we can see all kinds of players – multi-user, first-person shooter, console, simulation, classic – pushing back at and tweaking the structures of play they encounter.

In relation to the magic circle, we should notice that the mediation occurs in two distinct dimensions: the first (i) is related to the game as object – symbolic structure commercially built up and consumed – and concerns the technical assets involved in its creation: the rule and the narrative structure, elements that congregate the essence of the object aspect of the game; the second dimension (ii) is related to the way the game shows itself in the moment of the game play – it concerns the game as activity, and, as pointed out by Juul (2005), in the moment that the structure composed by rules and fiction becomes available to potential players; these are able to start a process of adaptation that becomes, later, a process of appropriation. These two dimensions of the game, which are an important part of the understanding of the consecrated place and therefore part of the understanding of the magic circle, are considered in more detail in the next section of this chapter.

Of living in a virtual world

'Virtual worlds' are computer-mediated multi-user environments spatially navigable and closely mechanically related to multi-user domains (MUDs). They contain potent graphics, complex mythologies and the capacity to harbour millions of simultaneous players on their systems.

To present the argument more clearly, we need to find our way through some related definitions and thoughts, such as, video games, virtual communities and virtual worlds, which then provide a solid context to identify our objects, especially when addressing the construction of social identity and its relationship to computer-mediated communication.

Objectively, a virtual world is a simulated environment based on interaction via computers, in which users 'inhabit' these spaces through their avatars. The inhabitance metaphor here is supported by a notion that Danish researcher Klastrup (2003) discusses in her doctoral thesis the idea of *worldness* – a notion that is reviewed further in this chapter. Such inhabitance is generally mediated through (most commonly) a human(oid) figure representation drawn using 2D or 3D perspectives.

> A virtual world is a persistent online representation which contains the possibility of synchronous communication between users and between user and world within the framework of a space designed as a navigable universe. 'Virtual worlds' are worlds, you can move in, through persistent representation(s) of the user, in contrast to the imagined worlds of non-digital fictions, which are worlds presented as inhabited, but are not actually inhabitable. Virtual worlds are different from other forms of virtual environments in that they cannot be imagined in their spatial totality.
>
> (Klastrup 2003: 27)

Network technologies went through a series of transformations that were mainly started off in the 1960s by non-predicted interaction between computer users. This endless chain of appropriations, i.e. still ongoing nowadays, harbour, in its essence, the first traces implying the expansion of virtual worlds. Since the first communal experiments, bulletin board systems and the Usenet, the myriad of chat room types – web chat rooms, graphic chat rooms, IRC and so forth – to MUDs and social networking sites, what we can see is that gradually the environments become more and more sophisticated and complicated. This occurs both within its technical side and on its sociability processes, with complex group dynamics (as pointed to by several of authors, such as Turkle 1995; Taylor 2007, and Alves 2005, among others).

The idea of inhabitance pointed out by Klastrup (2003) is specially related to the sense of presence – strongly supported by a persistent *locus* – in a virtual world. Klastrup's thesis is centred on the effort of the conception of a poetics of these environments, which is clearly devoted to find out what transforms any given virtual world into an *experienced* virtual world. The author, thus, pursues the essence of this experience between user/player and the world as a symbolic structure. This essence of a virtual world would then translate the idea of *worldness*, within any given virtual world.

Therefore, it is only coherent to consider that the way the worlds are perceived by the users/players becomes a fundamental piece of their composition, since it is through the processes of (re)-reading and (re)-configuration of the narrative elements – both provided by the possible use of technical devices in the environments, and individualized demands from the users (the former considered as technical variables, while the latter, as psychosocial variables) – that the virtual world will finally acquire the expected immersive ambience.

Brief digression throughout immersion

In order to properly weave the notion of the magic circle and the idea of *worldness* into the online social identity construction process, we need to explain the role immersion plays, how it is presented and how it affects the processes of contact between user/player and virtual world. Video games carry out a lot of information, composed by rules and fiction (Juul 2005), which acts as an interactive force in the process of production of meaning, experienced by the player in the moment she 'lives' the designed story, the cybertext (Aarseth 1997).

What transforms the video game experience are the immersive processes inherent them, as Ferreira and Falcão (2009) have pointed – while traditional media, like films and television, claim a passive immersion, where the reader merely absorbs what is transmitted and the ideal of interaction is condensed to the related hermeneutics, video games transpose this process to different ground, where immersion happens in a much more palpable way.

Returning, then, to the main argument of this chapter, we believe that the degree to which the magic circle establishes borders from the world outside is specifically linked to player's cognitive game immersion. The borders of the magic circle, then, redefine themselves and solidifies, and the more the immersive processes develop, the more we gain the illusion of stepping onto a new reality, the more like going 'through the looking glass'.

Klastrup (2003) who upholds the idea of hybridism when it comes to the nature of virtual worlds has a complementary view to the proposed model – and that suits the argument of this chapter. For the Danish researcher, immersive processes in virtual worlds work only as one of many elements of composition of her poetics, which is represented by (i), a model of production of meaning based on a multi-user produced textuality and (ii), an analytical matrix intending to study the scope of interaction, agents, interaction forms and interaction-in-time – the latter enabling the emergence of a story (an event that may be told, narrated, but that does not necessarily relate to the narrative core of the virtual world).

Klastrup (2003) believes that the two above-cited subjects are essential for the user/player to develop a relationship to the world, persuading the player to eventually return to the world. Such forces would, then, help create a specific essence of the world. These ideas are addressed by the notion of *worldness*, as well as helping the individual to fabricate experiences that may satisfy her and the community in which she is positioned.

Given, then, this particular issue in the immersive processes related to the virtual world experience, in which the suspension of disbelief would be reinforced by the need of interacting and spatial navigation, we seek to further analyze Klastrup's (2003) notions. Klastrup expands the discussed idea by mentioning Marie-Laure Ryan's work on narrative as virtual reality, in which she classifies immersion as being 'the experience through which a fictional world acquires the presence of an autonomous, language-independent reality populated with live human beings' (Ryan 2001). Intending, therefore, to defend her argument against the commonsense assumption that immersion generally refers to a process of passive seduction by the medium's language, especially regarding the ones bearing the traditional narrative model, Klastrup suggests that perhaps the ideas of presence and engagement should be brought up, for they are more fit to represent the experience of 'being' in a virtual world.

Marie-Laure Ryan believes that the combination of immersion and interactions would explain the experience of reality to a user/player. To the researcher,

> '[T]o apprehend a world as real is to feel surrounded by it, to be able to interact physically with it, and to have the power to modify this environment. The conjunction of immersion and interactivity leads to an effect known as telepresence.
>
> (Ryan 1994, online)

Therefore, by introducing the idea of (tele)presence, Klastrup (2003) brings together the above-cited notions and the characteristics of a virtual world, creating an understanding of the relationship between users/players and the world, this then helps us to understand the occurrence of a particular point in the process of construction of an online social identity. As Klastrup (2003: 295) writes:

> We can view immersion as the creation of belief in the world, consciously pretending that all that happens in the virtual world is 'real' or 'true', while automatically interpreting events according to their 'in-game' meaning (. . .). (Tele)Presence, then, is following the feeling of 'being there', experiencing that you are your avatar, and other people are their avatars through the interaction with other players, the simulation of the world and the successfulness of immersion. Engagement in the world is what emerges through the experience of the world in time; other players start to matter to you and you begin to experience events of a tellable nature which make you stick to the world because of your interest in completing the story on the level of individual characters.

It is noteworthy that, although Klastrup (2003) evokes the narrative aspects of virtual worlds to address the experience, her analysis does not favour the study of the aspects of the game as a system, as described by Salen and Zimmerman (2004), aspects hereby represented by the notion of the magic circle.

The power of mediation of the magic circle

According to what was posited, then, the immersive process, alongside the experience of world addressed by the notion of *worldness*, experienced by participants in a virtual world presents a number of characteristics that denote a repertoire of non-usual social practices that differ from the practices adopted in daily offline relationships. Even those that could be identified as present in the offline aspects of the world – underlining the fact that, although this chapter focuses on the online aspects, we do not support the dichotomous real world – virtual world view – seems to go through slight transformations, on the context of being adapted to fit the supposed ethos of the virtual worlds.

Such premises raise questions we might consider: (i) what differences are related to the process of immersion when we consider other interactional environments not necessarily related to the aspects of game? (ii) What are the effects in the process of the construction of identity – in its social-representational aspects – derived from these differences? We hypothesize here that the particular traits invested in the magic circle promote the development of a myriad of interchanging social behaviours that, in time, help create the circumstances which, then, will produce a number of wavering social identities.

To enhance our argument, it is necessary that initially we explore the social arena of such virtual worlds, keeping, as reference, their regulatory mechanisms. Having as base the studies of Lakatos and Marconi (1999) about how interaction may produce social processes, Brazilian researcher Recuero (2005), in her analysis of the formation of online social networks, highlights the presence of three important mechanisms that control the social dynamics shaped by the interpersonal contacts in which the users are involved: (a) cooperative processes, (b) competitive processes and (c) conflictive processes.

Such categorization seems to fit our attempted analysis of the relational aspects present in the virtual worlds. The first, cooperative processes, are related to the pursuit of a common goal for both individuals and groups. The cooperation in the virtual environments may be attested through the various events, since the one involved in helping a new user – by pointing easier ways of overcoming challenges or survival tricks of the specific environment – to the more sophisticated ones, derived from a friendship centred ethics, built up over the idea of camaraderie that reigns over a significant cut of players.

The competitive processes represent the effort of individuals and groups to attain their goals in game worlds where resources are often limited – be they material or symbolic. Therefore, the users/players weave a competitive process among them, trying to reach a desired state of welfare, by being rich or powerful, or even unique, but keeping in mind that such welfare brings within the possibilities of being noticed as a notorious character, which is translated in social benefits to the player. In the environments of virtual worlds, this process is commonly explored, since it stimulates the desire of the user to work towards the gradually better social situations, seeking status among the other participants, be it related to her (i) technical prowess (being an indispensable player, for raid groups or other combat situations, e.g.) or to her (ii) social abilities (by convincing other players of using a specific strategy or by assuming the leadership of groups).

In turn, the conflictive processes would radicalize the dispute with the others, be it an individual or another group/faction/nation, and so on. The context is modified to better fit the atmosphere of hostility, and the associated desire is related to the elimination of the opponent. Every action is directed to surpass the other, through suppression and humiliation. This mechanism reigns over yet another significant part of the social dynamics found in the virtual worlds. It promotes the creation of groups, alliances, material stockpiling, and especially, it upholds the idea of winners and losers, fostering an eternal atmosphere of war between the participants.

Employing an interactionist perspective, we may notice that the scope of literature widens, as we evoke the work of Goffman (1961, 1974, 1990) and Gergen (1990), for instance. Such researchers believe that the process of identity formation is derived, in its essence, from the situations we experience in the most diverse social contexts as well as from our ability to represent such experiences. Therefore, resulting from the social-representational aspects built up from the appropriation of the particular traits of the various social and communicative practices from a specific environment.

When it comes to the relationship between user/player and the environment, game inherent content exerts a force that mildly transform social processes and behaviours. The main issue given such assumptions is whether such occurrence reveals only a limited number of character circumscribed aspects experienced in the game, hence transforming the in-game experience, without any interference in the process of identity construction outside the rules of the magic circle, or if it shows experiential façades that could be assembled within the set of elements that format the configuration of identity of the user/player.

According to our view, the latter assumption figures as the more plausible since the presence of the magic circle fabricates effectively either the temporary suspension of the commonly used social references – highlighting, thus, its encapsulating function – or the reshaping, regarding the importance and the utilities, of these references – highlighting, thus, the function of mediation element – without blocking the social learning inherent to the experiences of contact with such structures.

The experience of the magic circle for the users of virtual worlds, be it a restriction or mediation experience, brings forth the emergence of a particular social situation that, though temporary, may serve as useful to the exploration of existential and cognitive territories, augmenting, then, the myriad of experiences that constitute identity.

It is still valuable to realize that according to several researchers (e.g., Turkle 1997; Ribeiro 2003, among others), a deeper experience – that is developed in any given circumstances and due to a diversity of reasons – may occur in the contact with the virtual world; the implications of this specific phenomena are circumscribed not into the online representation, not in the number of roles played by a user of virtual worlds, but it may be transposed, influencing the structure of the user's personality. Such assumptions evoke the work of Yee (2007) about behavioural modifications – some users may experiment with these online social identities with such intensity that some of the online traits could be gradually embodied, literally, to the offline social context. In a more extreme context, such views could uphold a perception of equivalence between the two relational spheres that should not be viewed just as a casual occurrence, but as the result of a complex process of successive interchangeable, temporary identifications, provided and adjusted to the various demands and variables present in the experience of these social contexts.

Final thoughts

According to the argument presented here in this chapter, we suggest that there are particular issues in the process of construction of social identities of users in virtual worlds. This suggestion is based on the fact that in such environments, besides the mediation processes related to the computer in the construction of identity, as highlighted by Ribeiro (2003), another element figures as a factor of uniqueness.

The innate presence of ludic systems that brings forth the presence of solid goals exerts a strong influence in the way the social roles and relationships are drawn all the way through the world experience lived by the user/player. During the immersive process, and considering the possibilities involved in the idea of inhabitance translated by the notion of *worldness*, the user/player is inserted into a tool that may suppress time and space and transform the virtual environment in an extension of the real, in which she would establish relationships through the avatar.

Such hypothetical suppression, resulting in the insertion of the individual into the magic circle, mediates the relationship of the users, weaving the process of construction of social identity into the narrative and ludic fabric, associating it to technical, perceptual, cognitive and existential factors.

It is curious to realize that despite the supposed freedom inherent in the process of identity formation associated with computer-mediated communication, such freedom is still conditioned by the contexts planned by the team that designed the environment. Therefore, future research needs to focus on a more in-depth exploration of these environments, in order to map social dynamics and processes, and understand the power this relatively new form of entertainment holds over gamers.

References

Aarseth, E. (1997) *Cybertext: Perspectives on Ergodic Literature*, London, John Hopkins University Press.

Alves, L. (2005) *Game over Jogos Eletrônico E Violência*, São Paulo, Futura.

Caillois, R. (2001 [1958]) *Man, Play and Games (Trans. M. Barash)*, Urbana, University of Illonois.

Carroll, L. (1944) *Throught the Looking Glass*, London, Penguin.

Crawford, G. (2009) Forget the Magic Circle (or Towards a Sociology of Video Games) – Keynote. *Proceedings of the Conference: Under The Mask 2*, Luton.

Ferreira, E. M. and T. Falcão (2009) Through the Looking Glass: Weavings between the Magic Circle and the Immersive Process in Video Games. *Proceedings of DiGRA*, London.

Gergen, K. J. (1990) *The Saturated Self: Dilemmas of Identity in Contemporary Life*, New York, Basic Books.

Goffman, E. (1961) *Encounters: Two Studies in the Sociology of Interaction*, New York, The Bobbs-Merrill Company.

Goffman, E. (1974) *Frame Analysis: An Essay on the Organization of Experience*, New York, Harper and Row.

Goffman, E. (1990 [1969]) *The Presentation of Self in Everyday Life*, New York, Harper and Row.

Huizinga, J. (1955 [1938]) *Homo Ludens: A Study of the Play-Element in Culture*, Boston, Beacon.

Juul, J. (2005) *Half-Real: Video Games Between Real Rules and Fictional Worlds*, Cambridge, MIT Press.

Juul, J. (2008) The Magic Circle and the Puzzle Piece. *Proceedings of the Philosophy of Computer Games: DIGAREC Series 1*. Postdam.

Klastrup, L. (2003) *Towards a Poetics of Virtual Worlds*, Copenhagen, IT University of Denmark (Unpublished PhD Thesis).

Lakatos, E. and M. Marconi (1999) *Sociologia Geral*, São Paulo, Atlas.

Lummis, M. and Vanderlip, D. (2005) *World of Warcraft: Official Stretgey Guide*, BradyGames.

Murray, J. H. (1997) *Hamlet on the Holodeck: The Future of Narrative in Cyberspace*, Cambridge, MIT Press.

Pargman, D. and P. Jakobsson (2008) Do you believe in magic? Computer games in everyday life. *European Journal of Cultural Studies*, 11(2): 225–43.

Recuero, R. (2005) Comunidades Virtuais Em Redes Sociais Na Internet: Uma Proposta De Estudo'. *E-Compós*, 04. Retrieved from http: //www.compos.org.br.

Ribeiro, J. C. (2003) *Um Olhar Sobre a Sociabilidade No Ciberespaço: Aspectos Sócio-Comunicativos Dos Contatos Interpessoais Efetivados Em Uma Plataforma Interacional on-Line*, Salvador, Federal University of Bahia (Unpublished PhD Thesis).

Ryan, Marie-Laure (1994) 'Immersion vs. Interactivity: Virtual Reality and Literary Theory', *Postmodern Culture*, 5(5), Online at: http://www.humanities.uci.edu/mposter/syllabi/readings/ryan.html

Ryan, Marie-Laure (2001) 'Beyond myth and metaphor: Narrative in digital media', *Poetics Today* 23(4): 581–609.

Salen, K. and E. Zimmerman (2004) *Rules of Play: Game Design Fundamentals*, London, MIT Press.

Taylor, T. L. (2007) Pushing the borders: player participation and game culture. In J. Karaganis. (ed.) *Structures of Participation in Digital Culture*. New York, SSRC: 112–130.

Turkle, S. (1995) *Life on the Screen: Identity in the Age of the Internet*, New York, Simon and Schuster.

Tuters, M. and K. Varnelis (2006) Beyond locative media: giving shape to the internet of things. *Leonardo*, 39(4): 357–63.

Van Kranenburg, R. (2008) *The Internet of Things: A Critique of Ambient Technology Adn the All-Seeing Network of Rfid*, Amsterdam, Network Notebooks.

Yee, N. (2006) The Labor of Fun: How Video Games Blur the Boundaries of Work and Play. *Games and Culture*, 1(1): 68–71.

Yee, N. (2007) *The Poteus Effect: Behavioural Modification Via Transformations of the Digital-Self Representation*, Palo Alto, University of Stanford (Unpublished PhD Thesis).

9 Framing the game

Four game-related approaches to Goffman's frames

René Glas, Kristine Jørgensen, Torill Mortensen and Luca Rossi

Framing was a buzzword heard at games research conferences about the time when this chapter was written. After Salen and Zimmermann (2004) adapted the 'magic circle' from Huizinga's (1955) pioneering work, seeing games as activities that take place within a 'frame' (whatever you conceptualize that to be) that separates the game from the rest of the world has been central to discussions focusing on how we should understand games and the activities that take place within them. However, the magic circle has shown to be fragile, and there have been several debates about its applicability. Games cannot be defined by a simple distinction between inside versus outside game spaces, and realizing this, game scholars have searched elsewhere for tools to describe the distinctive modes of play. Erving Goffman's *frame analysis* looks to be a useful instrument in that respect.

Games in general, and massively multi-player online gaming environments (MMO) in particular, consist of a range of different situations and modes that players need to understand how to interpret when playing. MMOs are especially interesting in this respect since they are social environments that encourage player interaction and where players themselves may define many play activities. Created and maintained in order to support group efforts and social interactions, the games themselves become the backdrop for the creation of social systems. In this chapter, we discuss some examples of player initiated activities in the MMO *World of Warcraft* (*WoW*) that add new frames to existing game framework analysis. In discussing role-play in *WoW*, we study how players sustain a certain state of mind by adding a new frame of reference to the existing game system. In the second example, we discuss strategy guides and walkthroughs as paratexts that change the experience of the game by putting players in a mode that de-emphasizes the fictional world and transforms the traversal of the game world. A very different approach takes on the distinction between work and play, and discusses how using *WoW* as an arena for networking with fellow researchers may depend on what perception the participants have of the situation. The last example discusses the phenomenon gold-farming and how it breaks the frame of the game by letting out-of-game resources account for progression in the game.

Framing games in virtual worlds

Virtual worlds are defined as persistent online environments, open for instant and ongoing communication and interaction among players, where events take place independent of the individual player. Klastrup defines virtual worlds in this manner: 'First and foremost, online worlds are characterized by the fact that they are realized, alterable and permanent fictional universes in which real people can explore the world from within and with each other' (Klastrup forthcoming). In this definition, Klastrup points towards the permanence of online worlds, but most important, she underlines the fact that the participants are real people who can cooperate in these environments.

Since its earliest incarnations in late 1970s and early 1980s (the often fantasy culture-oriented multi-user dungeons or MUDs), virtual worlds have evolved into roughly two types: there are game-oriented and social interaction-oriented virtual worlds or, Klastrup puts it, *game worlds* and *social worlds* (Klastrup forthcoming). It remains difficult to distinguish between the two types, particularly when they both often contain both a fantasy setting and a strong social structure. As Reid noted in her work on MUD communities, categorizing a virtual world is dependent on the styles of interaction they encourage, and not on the way they are programmed (Reid 1999). Nevertheless, the focus for this chapter is on virtual worlds.

WoW, our main focus here, is not described as a game-oriented virtual world; *WoW* is marketed and sold as a game. And successfully so: since its release in 2004, it has become one of the most commercially successful titles available on the PC/Mac platform. Commercial interests aside, it is important not to confuse a game world like *WoW* with what typically constitutes a game. Using a wide array of scholarly and design-oriented previous definitions, Juul (2005: 36) has distilled what he calls the classic game model:

> A rule-based system with a variable and quantifiable outcome, where different outcomes are assigned different values, the player exerts effort in order to influence the outcome, the player feels emotionally attached to the outcome, and the consequences of the activity are negotiable.

Such definitions of a game work fine for most games but encounter difficulties when talking about MMOs. These do not feature a fixed quantifiable outcome. Instead, MMOs are 'structured like serial narratives that grow and evolve from session to session' as Salen and Zimmermann put it, adding that 'sometimes they end; sometimes they do not' (Salen and Zimmerman 2004: 81). Juul also recognizes MMOs as being an exception to the rule as, through the open-ended nature of the MMO, 'the player never reaches a final outcome but only a temporal one when logging out of the game' (Juul 2003: 43). While smaller quantifiable goals do exist in an MMO-like *WoW* (finishing quests is good example), a key to understanding MMOs as game worlds is that they offer players the possibility to set their own goals, which may or may not be game-oriented; successfully running a guild, killing other players' characters to harass them, role-playing a marriage or exploring the environment. This way, the MMO is 'a larger system that facilitates game play within it, giving rise to a series of outcomes that build on each other over time' (Salen and Zimmerman 2004: 82).

The magic frame

What is characteristic for playing games is that the player defines the specific situation as a game situation. Often the game rules themselves help the players to define a situation as a game. In the definition of games above, Juul suggests that games are specific limited systems in which certain rules and actions apply. It is therefore implied that games take place within a specific frame of reference that define them as games. With reference to Huizinga (1955), Salen and Zimmermann call this frame of reference *the magic circle* (Salen and Zimmermann 2004: 94–95). The magic circle separates the game-internal from the game-external, and is therefore a central frame of reference for understanding what it means to play a game. The magic circle may be conceptual, like role-playing games where the players must establish and uphold the boundary through communication and imagination, or it may be visual and defined, like the chalk border of hopscotch or the geographical or exploratory borders of a virtual world. The magic circle has received a lot of criticism in-game studies (Consalvo 2009; Juul 2008), and according to Juul, this is due to a misunderstanding of the original concept and a rushed application of existing theories onto video games (2008: 56). This chapter supports Juul's objections, and in particular we believe that Salen and Zimmermann's interpretation is too narrow and based on a misconception of Huizinga's original mentioning of the magic circle. He wrote:

> All play moves and has its being within a play-ground marked off beforehand either materially or ideally, deliberately or as a matter of course. [. . .] The arena, the card-table, the magic circle, the temple, the stage, the screen, the tennis court, the court of justice, etc., are all in form and function playgrounds, i.e., forbidden spots, isolated, hedged round, hallowed, within which special rules obtain.
>
> (Huizinga 1955: 10)

In the quote, there is no definition of the magic circle, the concept is only mentioned as an example of a range of different and defined situations. When mentioning games and the magic circle as examples that are equal to the court of justice and the screen, Huizinga clearly sees this demarcation as a more general principle than focusing on games alone. As a general principle, the idea that games are defined by such a boundary comes close to how Goffman understands situational *frames* (Fine 1983). Instead of talking about games as defined by one frame known as the magic circle, we should rather discuss the different frames that games may be delimited by. This opens up the idea that games may be framed in different ways depending on, among other things, the social and game-contextual situations, and supports the open-ended and emergent modes of play that characterize MMO game play.

Applying a frame in the Goffman sense of the concept means to direct the gaze or attention to certain aspects of the subject of study. Framing is related to how human beings interpret social situations, and how one distinguishes a situation as different from other situations based on conventions for the specific situation. In the same way as a picture frame marks what constitutes an artwork, social framing defines what should count as part of a given situation. More importantly for this chapter, within a certain frame, we can expect to find that activities that are 'out of frame' will take place. Goffman addresses this in *Frame Analysis* (Goffman 1974: 201):

> Given a spate of activity that is framed in a particular way and that provides an official main focus of attention for ratified participants, it seems inevitable that other modes and lines of activity (including communication narrowly defined) will simultaneously occur in the same locale, segregated from what officially dominates, and will be treated, when treated at all, as something apart. In other words, participants pursue a line of activity – a story line – across a range of events that are treated as out of frame, subordinated in this particular way to what has come to be defined as the main action.

This means that the many different frames through which we view, in this case, *WoW* are not surprising. All arenas for human interaction are also arenas of multiple, occasionally conflicting but mostly peacefully co-existing frames. The use of frames as a tool for understanding the many layers of interaction in *WoW* is still useful, as it creates a common lens for our criticism, which lets us understand, present and analyze the many different practices observed within the online world of *WoW*. Using Goffman's frames also allows us to study the play activities in a complex environment such as MMOs without reducing it to a simple 'inside versus outside' argument in which one sacrifices a functional perspective of game play for a formalist (Consalvo 2009).

In the following text, we will discuss four examples of how players of *WoW* may use the game frame as a point of departure of creating additional frames that increases the value of the games in terms of social interaction and play experience.

Framing *WoW* with role-play

A lot of people use MMO environments for role-play where they envision backgrounds and personalities for their characters, and even make their characters pursue goals and storylines related to fate, future and romance. Role-play can be seen as a *mindset*, or 'a manner of playing that can be adapted to playing any kind of game' (Heliö 2004: 70). It is a mode added by players in order to enhance the play experience. Although role-play can be a solitary experience, its potential is only fully realized when it becomes a social event, where players interact and respond to each other's characters. As a mindset, and not part of the game design, role-play is generated and sanctioned by players, and misunderstandings and disagreements regarding what counts as role-play is likely to occur. When role-play is added to a game environment, the frames players relate to become more complex than the magic circle suggests. Players must be able to distinguish not only between game-relevant and game-irrelevant communication – they must also understand whether a person is role-playing or not.

The classical pen-and-paper role-playing game (such as *Dungeons and Dragons*) relies on face-to-face communication and the player's imagination to uphold the status of the activity as a game. The different frames of reference may therefore be difficult to spot for a novice, as they come into being through verbal conversation between the players. Players need to constantly be aware of the distinction between two frames that position the player either as a character in the fictional world or as a player who refers to the game world as fiction (Fine 1983; Thorhauge 2007). *In-character* (*IC*) communication frames the player as an individual in the game world, and refers to communication between characters in the fictional universe. It is typically represented as direct speech or dialogue from fictional characters. *Out-of-character* (*OOC*) communication, on the other hand, is player communication taking place outside the fictional game universe (Konzak 2007). The typical OOC communication happens when a player asks the game-master or fellow players for the clarification on a game-related issue, or when game rules are discussed. Players must understand when one or the other frame of communication is at use (Fine 1983). In order to identify which frame a player is talking from at a given moment, players search for clues in the play context or what a player is saying. As an illustration, an empirical example of a table-top game with a young, inexperienced role-player is useful. When his character started to act insane and uncontrollable, making it troublesome for the group to continue their quest, the character was tied up and left in the back of the group's wagon by one of the remaining characters. This happened from an IC point of view, but the young player found the treatment unfair and took it for an OOC insult, and left the game. The player confused IC reprimands for OOC hostility, which is a typical example of how difficult it may be for inexperienced players to understand the relationship between the two frames.

When studying MMOs as a specific case, we find that communication is framed by *chat channels* that allow written communication between players. These are bound to specific frames of reference, and may have global or local range with respect to the game world. In *WoW*, the 'Say' channel broadcasts what is written to characters within 'listening' range. The 'Guild' channel, on the other hand, is dedicated to communication between fellow members of a certain guild, regardless of whether they are at the same or different locations in the game world. Other channels are dedicated to meta-gaming[1] or to general communication. The presence of other players demands shared frames of reference, and the chat channels assist players not only in being able to interact, but also in being able to make contracts and define the meaning and content of the different frames (Mortensen 2007).

When adding role-play to an MMO, we get a complex combination of frames. As mentioned, role-play is often a source to misunderstandings and disagreements, since people frame their experiences and interpretations differently when role-playing. In MMO-role-play, certain chat channels are normally dedicated to OOC communication, and a certain role-play community will have additional rules. While some role-players may frown upon any kind of OOC communication in the 'Say' channel, other role-players tend to use this channel for all kinds of communication with player characters within listening range, marking their OOC communication with brackets. Some role-players will ignore communication that runs the risk of ruining their sense of immersion in the game world, and bracketed text is thus regarded as irrelevant (Goffman 1961). Role-play guilds in *WoW* may use the guild chat channel as an IC channel, in which they role-play with their fellow guild members while their character is doing unrelated things such as questing or grinding, thereby making the guild channel their primary means of role-play regardless of location or presence to other characters. Supporters of this 'remote role-play' practice may interpret IC guild channels as an in-game magical feature that allows characters to communicate via telepathy, while others mentally transfer their character to an imagined guild hall, even though the character is situated at a specific location in the *WoW* geography (Montola 2007). This practice seems disturbing for some role-players, either because they find the telepathy explanation weak or because they do not like the idea that their character is in an imagined guildhall when it actually does something else in the game world.

Looking beyond the relationship of chat channels to the game world, we may also trace other kinds of framings. On dedicated role-play servers in *WoW*, only a limited number of players actually role-play, and role-players have different ways of dealing with non-role-players. Some stop OOC when addressed by someone who is clearly not role-playing, while others stay IC and ignore them or treat them as madmen or disturbed characters. Frames are not always shared, and this does not only apply for role-players compared with non-role-players. Also role-players may have different role-playing styles and may treat the game world in different ways. While certain role-players focus on acting out a character personality, others are occupied with creating narrative storylines and events (Harviainen 2003). While the former might stay IC more or less all playtime, the latter may focus their role-play to certain pre-planned situations. Also, while some players would treat the game world as a random virtual platform for role-play without paying attention to the lore of the game universe or how the game mechanics guide behaviour and character relations, others would be careful to immerse into the game universe.

To illustrate how different styles may lead to misunderstandings even when both parties are IC, we give an example from one of *WoW*'s role-playing servers: a low-level character claimed to be a 'Scourge champion' who easily could slay anyone. A group of higher-level characters accused the so-called champion of being a liar. They challenged him and brought him to high-level Plaguelands, where he was killed by level 57 monsters. The low-level player got angry and stepped OOC to accuse the high-level players of meta-gaming since they apparently judged him on the basis of level. The high-level players, however, claimed that their motivation was IC since the 'champion's' armour was too modest for a high-ranking hero. This example illustrates two groups that framed role-play differently in the context of *WoW*: the low-level character believed game mechanics (character level) should have no relevance for role-play, but the high-level characters disagreed, making an example of why they thought role-play should relate to game mechanics. Also, since the high-level group wanted to make an example of what they believed were 'correct' role-playing, they indeed had an OOC motivation for doing what they did, and that they operated within (at least) two frames at the same time.

When included in MMOs that do not have any mechanics specifically intended to support it, role-play adds a new frame of communication to the existing frames of multi-player computer gaming. Thus framing becomes increasingly complex, and as the new frame of role-play adds a new experiential and immersive layer to the game, the frames are woven into each other. These frames are not necessarily intuitive but depend on the conventions of individual role-playing communities and the players' personal attitudes towards role-play. Different interpretations of what constitutes 'good role-playing', the possibility to create new channels for communication, together with aspects of meta-gaming and OOC communication, form the potential for the emergence of additional frames in this context.

Climbing the levels – fast!

As Fine would see it, *WoW* as a game world is a combination of a fictional world and a game that suggests at least three basic frameworks of meaning in which players move. There always is the 'primary framework' of the real world, there is the framework of the game with its rule systems through which players manipulate their in-game characters, and there is the framework of the fictional world in which they *are* these characters (Fine 1983). During play, players constantly switch between these frameworks depending on their level of engrossment. The voluntary nature of frameworks and the 'fun' factor contribute highly to the increasing or decreasing of the likelihood of frame-shifting (Fine 1983).

When a game world prevents or hinders engrossment due to design flaws or constraints, players will try to find their own ways to become caught up in the game again. Strategy guides, for instance, provide helpful tips for players, who are stuck, lost, bored or in other ways unsatisfied with the game world as it stands. As Consalvo (2007) discusses in her book on the practices of cheating, strategy guides can be seen paratexts accompanying a game. As defined by literary theorist Genette, a paratext gives meaning to all the information accompanying the main text of a media object, like the preface, table of contents and index of a book. They form 'thresholds of interpretation', existing outside of the main texts while, nevertheless, containing the potential to control its reading (Genette 1997). In terms of framing, a different reading of a text (in this case the game world) as a result of paratext might also signal a shift in the way engrossment is experienced.

In this section, a type of strategy guide called a walkthrough will feature as an example of paratextual influences on the game experience. Before focusing on two distinct walkthroughs, a short description of *WoW*'s design, which has lead to players using these strategy guides, is in place.

In *WoW*, the rules, structures and dominant strategies as implemented by design in many ways try to control or at least guide a player in their activities. Players are given clear goals, like fulfilling a quest or slaying a monster, which yield rewards like improvements in gear or weaponry and provide experience points, the main 'currency' needed to gain levels and therefore progress further into the game. Just getting a character up to the highest level, the point at which most of the social activities take place, can take an average player weeks, even months to achieve. Not all players can or wish to invest so much time in levelling characters to be able to participate in group play activities. For some, the levelling process is not even fun but a chore, making it hard to stay engrossed in the game experience.

Walkthroughs provide players dissatisfied with slow levelling a step-by-step approach to speed up the levelling process, thus promising improved enjoyment and engrossment. Two walkthroughs are discussed and compared here, the first an officially licensed one published by commercial strategy guide publisher Bradygames (Lummis and Vanderlip 2005), the second an amateur-created guide sold online by its creator (Joana 2007). The first is aimed at newcomers to *WoW*, offering a broad range of tips and tricks, the second is made for seasoned players and focuses on powerlevelling and speedrunning, practices of getting one's character through the game as fast as possible, with the minimum input and the maximum result. Both focus on doing quests, which in *WoW* function both as main storytelling device and a means to provide players with goals and reward them with experience points.

Creating a step-by-step guide out of months, even years worth of game content, including an inverted-tree branching structure of thousands of quests, is not just a daunting but an impossible task. Walkthroughs therefore have to limit themselves to a certain approach to play. Under the title 'Your first day', the official walkthrough describes the steps a player has to take with a minimum of references of the level of the game, addressing the player as if he was existing in the fictional world. This style of translating a game into paratextual form is aiming for a narrative telling of events through *WoW*'s fiction, 'encouraging the reader to envision their own progress on the same journey' (Consalvo 2007), in an effort to keep engrossment within the framework of the fictional world intact.

The exact opposite, the powerlevelling guide exchanges the most engaging and logical succession of quests story-wise for the most *useful* succession in terms of instrumental progression. Powerlevelling and its competition-based relative speedrunning is all about route planning (going through a game as fast as possible by planning ahead), sequence breaking (reordering or skipping sections of the game to ensure this route is actually the best route) and tricks (of which some can be exploitations or 'exploits' of game design flaws) to achieve such breaks (Carless 2004). The walkthrough offers players the opportunity to copy its writer's record-breaking speedrunning endeavours for their own *WoW* experience.

When comparing the guides, we see a strong de-emphasizing of *WoW's* fictional world in the powerlevelling guide. It wastes no time on fancy descriptions of characters to meet or places to go but sticks to the instrumental goals of the quests. Consalvo, who signalled similar differences between descriptions within the official strategy guide for the game *Myst*, calls this process a 'de-*Myst*-ification' of the game's challenges (Consalvo 2007). Levelling a character by following speedrunning tactics though not only demystifies the game's fiction but also transforms the way the virtual space is traversed. To improve speed by avoiding unnecessary travel, groups of quest are bundled together when their goals are roughly in the same area. Any coherency between quests on a fictional level – going where the story goes – is replaced with a coherency of quests on a spatial level – going where the other quests goals are. Linking quests together like this makes reading the quest descriptions or following quest series in the right order – the basic narrative tools of the game – unnecessary for progress. They become 'narrative obstacles' (Kücklich 2004: 3) which hinder speedy levelling and can be ignored altogether by following a pre-planned route.

Another almost unavoidable consequence of play using a strictly goal-oriented walkthrough like Joana's Guide is the devaluation of socially oriented play. Joana's Guide is hesitant with group play during the levelling process. PvP or player-versus-player content is generally seen as entirely counterproductive to powerlevelling. No experience points are gained through PvP, stopping the levelling process in its tracks, and should therefore be avoided. Most group quests are skipped or, when they are worth good experience points, advised to do only when a group can be found fast, right there and then (going back later to finish a group quest breaks the planned route sequence). Instances, uniquely generated areas aimed at dedicated group play, are worth doing according to the guide, especially for casual levellers, as they tend to earn a player good experience points. If you are serious about powerlevelling or speedrunning though, they too are best avoided: as the guide puts it: 'I personally skip instances while racing to 70, cause I don't want to take the chance that I will have a bad group' (Joana 2007: 38): with 'bad' pointing at insufficient knowledge about the rules and game play mechanics or sheer lack of skill. Whether these other players might be fun company, or good role-players, does not matter. As Taylor (2006: 190) pointed out in her work on other highly goal-driven play practices, a strong focus on the instrumental side of *WoW* can dwarf all other aspects of both individual and group play. The emergent nature of interaction with other players, who may not be as dedicated or evenly skilled as one would want to, is simply too hazardous for the dedicated speedrunner and powerleveller to get involved with.

To conclude this section, we can say that as thresholds of interpretation, walkthroughs, like the ones discussed in this chapter, not only potentially transform the way *WoW* is read and played, some, like the one from the official strategy guide, try to keep players engrossed in the framework of the fictional world, addressing them as characters within it. Others, like the powerlevelling guide, downplay or downright ignore the fictional, giving the experience of playing *WoW* an entirely different meaning in which the player is just that, a player. In such a case, it leaves the *WoW* player with a minimalist, goal-oriented experience of individualized progression instead of the feeling of narrative-driven discovery resulting from emergence in a shared environment.

Blending frames: academia inside games

Two frames that appear to be mutually exclusive are the frames of gaming and working. In 2005, a group of game researchers and their friends realized they were all playing *WoW*. From thinking 'this was a nice coincidence' to the thought that 'who knows, we might be a kick-ass guild' was a short step, and soon the Horde-side[2] European game-researcher's guild was born called 'The Truants'. The guild was formed in 2006 and still exists in 2009, including as members the authors of this chapter. This was not a 'kick-ass' guild though. It turns out that researchers are not always efficient gamers, and that a group mainly consisting of mature players with jobs, families and careers means the majority will wander off for long periods of time while they deal with real life; all behaviour which is not ideal for power-gaming

A lot did, however, work. One of the first guild generated projects, which started in 2006, was an anthology published in 2008. In 2007, many of the game researchers in this group met in Gothenburg, Sweden at a game conference, and in 2008, several of the members of the guild participated in and organized a 3-day game research symposium in Umeå, Sweden. This small researchers' community meets in a digital space, and at the same time research activities are maintained and administrated through an e-mail list. The original weblog ended up in a vacuum, as the wiki and, most importantly the e-mail list turned out to be better and more efficient for the purpose of mixing work and play and work again – as all play leads back to work in this context.

Let us examine the two frames 'play' and 'work'. We have already seen several definitions of play, particularly in the understanding of gaming, and several of these are quite compatible with both frames. Juul's (2005) definition of *game* is, for instance, fully compatible with *work*, as both games and work are circumscribed by rules, values, effort, outcomes and consequences. According to such a definition of games, there should be no difference between an arena of productive labour and a game. Still, we are quite aware of a distinction which positions work and play at seemingly opposite ends of a scale, and the key to the contrasting opinions of the two activities lie in Huizinga's statement that 'play is superfluous' (1955). Play is, in this definition, free, done at leisure and is never a chore. When play becomes a chore it stops being play and becomes extremely work-like. One of the more frequent statements from players of *WoW* before they leave or as a reason for their leaving the game is 'it started to feel like work'. When the game has become routine and it starts to feel restricting rather than rewarding players lose interest, as their frame of understanding the game has moved from the frame of play to the frame of work.

On the other hand there is pleasure experienced from work, and activities that makes work play-like. The research of Csikszentmihalyi (2008, 2002) discusses how people find pleasure in work situations, and the concept of *flow* is his response to how people can enjoy situations that are otherwise characterized by chores dominated by monotonous tasks.

> In normal everyday existence, we are the prey of thoughts and worries intruding unwanted in consciousness. Because most jobs, and home life in general, lack the pressing demands of flow experiences, concentration is rarely so intense that preoccupations and anxieties can be automatically ruled out.

> Consequently the ordinary state of mind involves unexpected and frequent episodes of entropy interfering with the smooth run of psychic energy. This is one reason why flow improves the quality of experience: the clearly structured demands of the activity impose order, and exclude the interference of disorder in consciousness.
>
> (Csikszentmihalyi 2002: 58)

Csikszentmihalyi's model of the flow state in work, where there is a balance between challenge and mastery, has been adopted as a model for how to balance games in order to keep the interest of players. His discussions of flow show how work can be viewed as play. This describes how 'work' and 'play' are mainly a matter of framing, that there is no obvious distinction between these two fields except the perception of the viewers or participants. The *autotelic* experience can be presented as a matter of framing. By reinterpreting and hence, reframing, the experience, the boring routine can shift position in the hierarchy of foci in the mind of the worker. An autotelic or optimal experience is an end in itself (Csikszentmihalyi 2002), not an imposed task the participant needs to finish.

But what came first, the game or the flow? The description of how workers achieve a flow state fits surprisingly well with Huizinga's descriptions of play. Play often 'breaks out' in situations of strict routine, either in a 'disruptive' or a 'constructive' manner. Disruptive play can be if two people in a bucket brigade decide to pour the water over each other rather than send the bucket on to the next. Constructive play can be when the chain of hands and buckets falls into a rhythm guided by a word-game or a singing game, turning the movement and the togetherness into an opportunity for community-strengthening play. The second is the *flow* experience as described by Csikszentmihalyi, but both examples are play. In these examples, there is a transition from one state to another, from a regular work experience to an optimal experience.

If the difference between work and play is merely a matter of reframing, there should be examples of game experiences turning into work, and more productive results from play. What we instead have is a sociality where playing is ideal for the productive situation. Work dreams of being play, but play only dreams of being itself. Let us take an example from a situation that is extremely work-heavy, but still playful, Jane McGonigal's study of *I Love Bees* (McGonigal 2008: 199).

I Love Bees is an interactive fiction, straddling a wide range of digital and analogue platforms, with a playful component. It was a promotional event for *Halo 2* commissioned by Microsoft. McGonigal was the lead community designer, and the event or alternate reality game (ARG) relied heavily on what is called CI: Collective Intelligence. McGonigal describes the event and the huge amount of research and analysis that went into the problem solving by the participants. During the event participants solving the puzzles self-organized across large distances, had long complex discussions online, and launched several different theories, based on very vague hints. While this is a wonderful example of creativity and community design, it is also a very powerful reminder of the amount of energy a motivated but loosely organized group can muster in order to solve a task. It describes the power of collective intelligence, and of the motivating power of playfulness. When teachers and coaches use games in learning and training, it is this motivation to work they try to recapture and apply to more mundane asks.

As we all know, games tend to resist this. It is possible to apply the 'power of fun' for a short period and make otherwise complicated and uninteresting experiments interesting and challenging through a game-like approach. But in the long run, the effect tends to wear off. Huizinga claims that play wants to be free, Csikszentmihalyi claims that flow comes with an absolute focus at the task at hand, from doing something for its own sake (2002). The Bee-game was played for the love of playing, but if it was organized in order to teach people how to cooperate they would miss out on the autotelic experience, as the final knowledge would become the goal.

What does that do to the distinction between *work* and *play* as frames? While at the one hand the distinction is vital to the perception of a situation, and that the situation changes in a fundamental manner if the frame changes, at the other hand the examples of *I Love Bees* coupled with an understanding of the autotelic experience shows how vital the different frames are. Too many demands, too many tasks that must be finished in order to reach a goal, and the process stops being the goal, and so, flow dies.

'The Truants' were in an enviable position as gamers and researchers. Research is perhaps the most classic example of applied CI, as each individual researcher builds on the knowledge of others in order to solve a riddle – whether that riddle is the inner universe of humans, or the secrets of the distant stars. Also, research is a task where the autotelic experience hovers close to the surface. Gathering a group of researchers in a game where all cooperation is voluntary and done for the hell of it is a perfect setting for work in an autotelic manner. But it did not stay successful permanently. From being framed as playful and flowing on enthusiasm, it turned into work, and as such was easily exchanged with other work-like tasks.

Games going real

Dealing with MMOs makes it harder to be able to define a unique frame of the game itself. MMOs seem to exceed themselves reaching aspects of society that have been traditionally considered far from gaming practices. The social definition of games and gamings, when it comes to MMOs, seems to be larger everyday, trying to include all the new territories that constitute the contemporary online gaming experience. This characteristic that forces the social definition of games to its broadest limits is undoubtedly one of the most innovative aspects of the genre: MMOs, as Castronova puts it, are important 'because events inside them can have effects outside them' (Castronova 2005: 4).

The expansion of what can be defined as in-game activities, or game-related, activities is a consequence of the new status that players have today, especially when it comes to MMOs. Gaming experience can no longer be closed around what happens between the login and the logoff procedure. A wider lens to describe these games is, if not necessary, surely possible.

Networked players, as a specific kind of networked audience, play a game in which they are, at the same time, players and creators of tools and providers of information for other players. Every contemporary game comes followed by a thousand related products, both official and unofficial: online forums, fans sites, guides, various game-support tools and so on. While this happens almost for every game that is launched today, this seems to be a core part of the MMOs experience where the presence of the community of players is a key part of the game. If many of these products, guides, walkthroughs and strategy books, can be considered as meta-texts, able to enrich and even to change radically (as highlighted earlier in this chapter) the game experience, there are several objects here with an unclear status. How can we consider the online services that guilds use to coordinate their in-game activity? Are those part of the game? Can they be framed as part of the game experience? These tools surely play an important role in how guilds manage complex situations that happen during the game, like the preparation of 25 persons, raid or a role-play event. Nevertheless, calendar tools have been, for a long time, external resources, not part of the game–client itself, not officially supported, or maintained by Blizzard. Only recently, with the *WoW* expansion *The Wrath of the Lich King*, the calendar service moved from being an external, web-based, support application to be part of the official game client. The evolution of *WoW* is full of examples like this, where innovations are first generated and proposed by players and later introduced into the game itself: from the voice chat system to the UI improvements, from the guild vault to the calendar service, the full list would be very long. Every new version of the game could be described in terms of a partially user-generated version.

If MMOs make it harder to draw a border around gaming activities, they make it even harder to be sure about what games consequences can be, or about what the final outcome of thousands of gaming actions could be. MMOs, being social events that involve millions of networked people around the world, give the opportunity to describe them as a complex emerging phenomenon with many unexpected and unpredictable outcomes in different fields.

It is possible to speak about emerging phenomena when a number of relatively simple interactions produce, on a macro level, perceivable qualities of radical novelty (Corning 2002). The way that MMOs are interacting with the legal, or the economic system, since they reached a major diffusion, can surely be read in this way, framing the game itself as a massive phenomenon, able to impact on the global scale on different macro systems. This is exactly the frame we are using in this last part of the chapter: the frame of emergence, trying to describe what happens when the game goes 'real' and millions of gaming actions can produce a new reality able to impact the real world.

One key example is goldselling. Goldselling is the practice of selling game currency to MMO players, which are therefore defined as goldbuyers. The reason why people should spend 'real' money in order to get some amount of 'virtual' currency is quite easy to understand. Usually MMOs are designed to reward, and the more you play the more you get, both in terms of game money and other in-game achievements. Within this specific situation a player could easily be tempted to pay (more) money in order to speed up his levelling process or to buy a specific item they really want. The player could obviously collect the same amount of money by themselves but it would cost a lot of time, mostly spent in activities that are often very boring and repetitive. Buying gold, as the virtual currency is called in *WoW*, leaves the boring part of the game to someone else and lets the player focus only on the fun part. Unlike speedrunning, goldbuying is not based on the will to exploit deep game dynamics in order to achieve something in a shorter amount of time, it is mainly based on the economic evaluation of how much a player is ready to pay in order to skip a boring activity. When the amount of real money the player is able to pay is enough, the space for a new business model is created. Evidence suggests that there is a large demand for this kind of business and goldselling is today a huge phenomenon for every major MMO out there. In 2008, the monetary impact of this phenomenon in the real world economy has been estimated at more than 10 billion USD (Heeks 2008).

According to the economic dimension of the phenomenon goldselling seems to work quite well. Nevertheless, the simple existence of such a practice should force game scholars to enlarge game definitions. Players acting as goldbuyer are, in fact, pushing the borders of the game out of the game itself or, at least, out of the game as it was thought by the developers. However, according to Terms of Service and to the End User Licence Agreement (EULA) of *WoW*, goldselling and goldbuying are explicitly prohibited practices that damage the game by adding real money dependent variables which cannot be easily kept under control. In order to stop the goldselling practice, as long as the real-money trade practice – the sale of in-game goods for real money – the EULA of *Word of Warcraft* states:

> All title, ownership rights and intellectual property rights in and to the Game and all copies thereof (including without limitation any titles, computer code, themes, objects, characters, character names, stories, dialog, catch phrases, locations, concepts, artwork, character inventories, structural or landscape designs, animations, sounds, musical compositions and recordings, audio–visual effects, storylines, character likenesses, methods of operation, moral rights, and any related documentation) are owned or licensed by Blizzard.
>
> (World of Warcraft EULA, §4 Ownership).

Since the EULA is a legal agreement between the player and the game company the meaning of this paragraph is that by buying the game–client players are allowed to play the game but they will never be entitled to own anything in the game. If players cannot own objects or anything else in the game, as the EULA states, it appears evident that from a strictly legal point of view, goldselling and goldbuying are not only illegal but also impossible: how can you sell something you do not own?

This is a paradox that *WoW* forces us to face: something that works on the economic level as a huge and growing market of services appears to be illegal (and technically impossible) from the legal point of view. In spite of the legality of the practices, the game-related economy keeps working day after day toward a *de facto* acknowledgement: in 2008, the Chinese government approved a bill to impose taxes made by selling digital money.

This brief excursus on the reality of *WoW* economy was aimed only at stressing how far the border of gaming can be pushed when we move into the realm of MMOs. Highly networked players can shift easily between in and out of the game using out-of-the game resources to improve their in-game experience and vice-versa. This process will produce an emergent reality made up by thousand of millions of everyday simple 'gaming actions' able to reverberate in many different aspects of the society. Every analysis will therefore require a different perspective in order to investigate specific phenomena; again the observer's perspective will be crucial to frame that highly interconnected emergent social phenomenon also known as game.

So what? Some conclusions

The four examples discussed above show very different player activities, which on the surface only seem to be connected by the fact that they take place within a specific MMO. However, as we have tried to demonstrate, by using Goffman's frame analysis, these cases are examples of different approaches to gaming that all happen within the very frame of this game. When adapting to different play activities such as role-play, goldselling, speedrunning and using the game as a platform for networking, the players do not leave the original frame of the game behind, but add new frames to the existing framework. They are still playing within a frame that defines the game as a specific kind of mindset in which actions have a different status than in the rest of the world. By defining their own goals and challenges within that game frame, players are able to mentally remain within the dominant frame, while also adding their own interpretations of what constitutes an interesting game activity. In this sense, adding new frames is an *emergent* activity; something that was not designed into the game, but that comes into being when players are invited to play with the system long enough. It is therefore the structure of MMOs that facilitate this playing with and adding of new frames. This should, however, not be surprising, as the MMOs indeed are called 'worlds', which implies that they are arenas for social interaction and thus only bound by human imagination. When adding frames to MMO activities like this, we may ask ourselves whether we are using Goffman's theory too broadly? Can everything be called 'framing', or does this concept lose its meaning when applied to so many different play activities? We believe that it is possible to use frame analysis if we are very clear about limiting the use of frames to clear cases as specific examples of frames. In order to make our application of the theory as specific as possible, we use the illustration below in order to specify how the frames apply to specific aspects of the game.

The model in Figure 9.1 is an illustration of how the four examples in this chapter relate to the game world of *WoW*. The game world itself is seen as a frame of reference separated from the real world, and for MMO players, it is the common frame of reference regardless of their approach to the game. Within the game world, however, there are different contexts in which the four examples of this

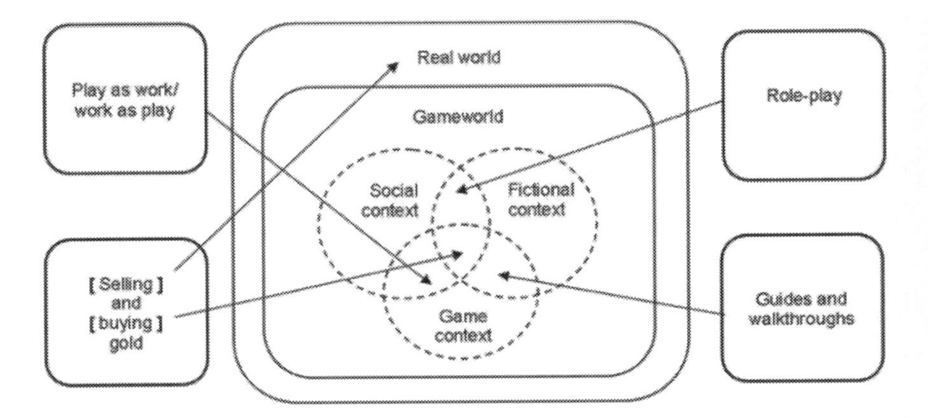

Figure 9.1 An illustrative model.

chapter are situated (Fine 1983). Parallel to Fine's three frames operating within fantasy role-playing games (1983: 194), and Juul's motivational frames for all game actions (2008: 61), we call these the in-game *social context*, the *fictional context* of the universe presented in the game, and the *game context*, relating to the rule system and design of the game. In this model, these are seen as overlapping contexts that colour the play experience of MMOs. Individual players may decide to focus on one or more of these contexts, and the model illustrates where the four *examples* in this chapter are situated with respect to this.

The different contexts for play demonstrates that the game world of MMOs is a flexible construct in which new play experiences may emerge depending on player attitude and framing. This has been illustrated by adding each of the four framing situations discussed in this chapter as satellites that are not originally part of the game context but can be added at the player's will. As we argue, role-play is both fictionally and socially contextualized within the game. Here players enter a frame in which they are supposed to play the role of an individual living in this world, an activity that only makes sense when interacting with other players. Using the game as a frame for professional and social networking and thus blending play and work emerges from the social and game contexts. As an environment for networking, *WoW* becomes a social arena connected to work, at the same time, the enjoyment of playing the game and mastering the system is a game-centric activity. The use of walkthroughs and guides concerns the game context since it is motivated by a wish to progress faster in the game than traditional game exploration allows, and the fictional context, since it tends to de-emphasize the fictional world of the game. The activity of selling and buying gold has a dual position, since selling gold is a real-world activity that takes place outside the game context, while buying gold is an activity with direct consequences not only for the player's performance within the game world. Buying gold also affects the social opportunities for the player, and de-emphasizes the fictional world of the game. In this sense, these four gaming activities are already covered by the frame of the game, although they represent very specific emergent features.

Notes

1 Originally a term from mathematical game theory, meta-gaming is generally understood as the use of game-external information to advance in the game. In this specific example, meta-gaming would include asking other players how to solve a quest or about how a certain mechanic works. In role-playing contexts, an example of meta-gaming is when a player makes the character take action and make decisions based on game-technical information that the character cannot be aware of. See also discussion on meta-gaming on ENWorld forum at http://www.enworld.org/forum/showthread.php?t=131782and page=1andpp=15 [Retrieved 03.07.08].

2 The 'Horde' being one of two warring factions in the game World of Warcraft, the other being the 'Alliance'.

References

Carless, S. (2004) *Gaming Hacks: 100 Industrial-Strength Tips and Tools*, Sebastopol, O'Reilly Media.

Castronova, E. (2005) *Synthetic Worlds: The Business and Culture of Online Games*, Chicago, University of Chicago Press.

Consalvo, M. (2007) *Cheating: Gaining Advantage in Videogames*, Cambridge, MIT Press.

Consalvo, M. (2009) There is no magic circle. *Games and Culture*, 4(4): 408–17.

Corning, P. A. (2002) The re-emergence of 'Emergence': a venerable concept in search of a theory. *Complexity*, 7(6): 18–30.

Csikszentmihalyi, M. (2002 [1992]) *Flow: The Classic Work on How to Acheive Happiness*, London, Rider/Random House.

Csikszentmihalyi, M. (2008 [1990]) *Flow: The Psychology of Optimal Experience*, New York, Harper and Row.

Fine, G. A. (1983) *Share Fantasy: Role-Playing Games as Social Worlds*, Chicago, University of Chicago Press.

Genette, G. (1997) *Paratexts: Thresholds of Interpretation*, London, Cambridge University Press.

Goffman, E. (1959) *The Presentation of Self in Everyday Life*, New York, Doubleday Anchor Books.

Goffman, E. (1961) *Encounters: Two Studies in the Sociology of Interaction*, New York, The Bobbs-Merrill Company.

Goffman, E. (1974) *Frame Analysis: An Essay on the Organization of Experience*, New York, Harper and Row.

Harviainen, J. T. (2003) The multi-tier game immersion theory. In M. Gade, L. Thorup and M. Sander (eds), *As Larp Grows Up: Theory and Methods in Larp*. Frederiksberg, Projektgruppen KP03: 4–8.

Heeks, R. (2008) Current Analysis and Future Research Agenda on 'Gold Farming': Real-World Production in Developing Countries for the Virtual Economies of Online Games, Manchester.

Heliö, S. (2004) Role-playing: a narrative nxperience and a mindset. In M. Montola and J. Stenros (eds), *Beyond Role and Play: Tools, Toys and Theory for Harnessing the Imagination*. Helsinki, Ropecon RY: 65–74.

Huizinga, J. (1955 [1938]) *Homo Ludens: A Study of the Play-Element in Culture*, Boston, Beacon.

Joana (2007) Joana's 1–70 Horde Leveling Guide. Retrieved 30 November 2009, from http: //www.joanasworld.com.

Juul, J. (2003) The Game, the Player, the World: Looking for a Heart of Gameness. *Proceedings of DiGRA and Level Up 2003*, Utrecht, The Netherlands.

Juul, J. (2005) *Half-Real: Video Games Between Real Rules and Fictional Worlds*, Cambridge, MIT Press.

Juul, J. (2008) The Magic Circle and the Puzzle Piece. *Proceedings of the Philosophy of Computer Games: DIGAREC Series 1*. Postdam.

Klastrup, L. (Forthcoming) Understanding online (Game)Worlds. In J. Hunsinger, L. Klastrup and M. Allen (eds), *International Handbook of Internet Research*. Berlin, Springer Verlag.

Konzak, L. (2007) Larp Experience Design. In J. Donnis, M. Gade and L. Thorup (eds), *Lifelike*. Copenhagen, Projektguppen KP07: 82–92.

Kücklich, J. (2004) Other Playings: Cheating in Computer Games. *The Other Players Conference*, Copenhagen, Denmark.

McGonigal, J. (2008) Why I love bees: a case study in collective intelligence gaming. In K. Salen (ed.), *The Ecology of Games: Connecting Youth, Games and Learning*. Cambridge, MIT Press: 199–228.

Montola, M. (2007) Breaking the invisible rules: borderline role-playing. In J. Donnis, M. Gade and L. Thorup (eds), *Lifelike*. Copenhagen, Projektguppen KP07: 82–92.

Mortensen, T. (2007) Me the other. In P. Harrigan and N. Wardrip-Fruin (eds), *Second Person, Roleplaying and Story in Games and Playable Media*. Cambridge, MIT Press: 297–307.

Reid, E. (1999) Hierarchy and power: social control in cyberspace. In M. A. Smith and P. Kollock (eds), *Communities in Cyberspace*. London, Routledge: 107–33.

Salen, K. and E. Zimmerman (2004) *Rules of Play: Game Design Fundamentals*, London, MIT Press.

Taylor, T. L. (2006) Does Wow change everything? How a Pvp server, multinational player base, and surveillance mod scene caused me pause. *Games and Culture*, 1(4): 318–37.

Thorhauge, A. M. (2007) *Computerspillet Som Kommunikationsform, Spil Og Spillere I Et Medievidenskabeligt Perspektiv*, Copenhagen, University of Copenhagen (Unpublished PhD Thesis).

Ye, J. (2008) Real Taxes for Real Money Made by Online Game Players. Retrieved 25 May 2010, from http://blogs.wsj.com/chinarealtime/2008/10/31/real-taxes-for-real-money-made-by-online-game-players/.

Part III

Communities and Communication

10 Identity-as-place

The construction of game refugees and fictive ethnicities

Celia Pearce and Artemesia[1, 2]

Prologue: an imaginary homeland

My Homeland Uru

From my beautiful homeland
From my beloved homeland
I hear the Bahro cry
and Kadish's wife sing her song of despair

And a refrain is sung by a sister who lives far from her homeland
And the memories make her cry
The song that she sings springs from her pain and her own tears
And we can hear her cry

Your homeland strikes your soul when you are gone
Your homeland sighs when you are not there
The memories live and flow through my blood
I carry her inside me, yes its true

The refrains continue, as does the melancholy
And the song that keeps repeating,
Flows in my blood, ever stronger
On its way to my heart

I sing of my homeland, beautiful and loved
I suffer the pain that is in her soul
Although I am far away, I can feel her
And one day I'll return

I know it

The homeland to which the author of this poem is referring is an entirely fictional and entirely virtual place. The poem was written by Raena,[3] a player from the massively multiplayer online game *Uru: Ages Beyond Myst*,[4] about 3 months after the game closed in 2004. In the months and years that followed, hundreds, possibly thousands of *Uru* players immigrated into other online games and virtual worlds, many self-identifying as 'Uru refugees' and referring to the game as their 'homeland.' This chapter presents results from a multi-sited ethnographic study (Marcus

1995) of players from *Uru*. *Uru* opened as a beta test in 2003 and closed less than 6 months later, leaving about 10,000 players self-termed 'refugees'. Many of these players migrated to other online worlds, carrying their *Uru* identities with them and forming collective 'fictive ethnicities' within games and virtual environments such as *Second Life* and *There.com*. These players described themselves in terms of a collective identity, referring to *Uru* as 'our homeland' and to themselves as 'refugees'. They also created artefacts and environments in these virtual worlds either directly derived from or inspired by *Uru*, thus using place as a means to express their collective identity. Throughout these worlds, they became known as 'the Uru people', or simply, 'Uruvians'. Players who migrated into *There.com* were initially discriminated against; however, over time, due to their group size (450 at its peak), as well as their demographics (50 percent female and largely 'Baby Boomers' with disposable time and income), they began to assert significant influence on the *There.com* culture at-large. Because of the shared experience of *Uru*'s closure, characterized by players as being highly traumatic, they also had a strong commitment to community and became actively engaged in community-building activities in their new homes. Through a dynamic process of iterative social feedback and 'transculturation' (Ortiz 1947), they eventually transformed from 'Uruvians' to hybrid 'Uru–Thereians', integrating both virtual places into their collective identity.

This chapter explores the connection of identity to virtual place, referencing in particular anthropology, humanist and socio-geography and Internet studies to look at the construction and performance of 'fictive ethnicity' tied to a specific, though virtual and fictional, locality. I will argue that in the current historical moment, in which connections between identity, community and place are being supplanted by the generic placelessness and identilessness of 'global markets', the tendency of players in the Uru Diaspora to construct a shared, place-based identity may reflect a larger need by individuals to associate themselves with affinity groups and reclaim a sense of connection between place, community and identity.

Uru history, context and players

To lay the groundwork for understanding the dynamics of this relationship, it will be helpful to understand the nature and the history of *Uru* and its players. *Uru: Ages Beyond Myst* was one of the final games in a successful game franchise that began with *Myst* in 1993 (Figure 10.1). *Myst* was the first blockbuster CD-ROM game, and continued to hold its chart-topping status, until it was outstripped by *The Sims* in 2001. *Myst* was classified by many as the first video game that was truly a work of art (Rothstein 1994; Carroll 1994). It broke new ground in digital interactive narrative, established a new genre – the adventure puzzle game – raised the bar for both visual and audio design, and broadened the audience of games to women and older players (Miller 1997). The majority of *Uru* players were *Myst* fans that had already been 'living' in the Myst world for a decade when the multiplayer game launched.

This history is important for two reasons: one, it helps us to understand who these players were: the *Uru* demographic is unusual for an MMO, both in terms

Figure 10.1 *Myst* drew players into a fully realized fictional world.[5]

of age and gender. The majority of players in the group who were the focus of this study were 'Baby Boomers' and exactly half were women. This is a very unconventional demographic for an online game, most of which attract predominately college-age and young males (Yee 2001; Castranova 2001; Seay *et al.* 2004). And two, because *Uru* was part of a decade-old franchise, players already had an affinity with the fictional world in which the game took place before even setting foot within *Uru* itself. There was an entire community of people with whom they could share what had previously been a solitary experience, which was something of a revelation to most players, who, in interviews, commonly identified themselves as 'loners'. This background helps explain how players could form such a strong affinity with a virtual place that they had only inhabited for less than 6 months.

The *Myst* world and the *Uru* narrative

The *Myst* world concerns a fictional, likely ancient, vaguely steampunk culture that draws from a number of traditional influences, including the old and new testaments, and the visual iconography and culture of Native Americans, to create a complex and extensible universe. Its cosmology revolves around a humanoid race, the D'ni, who possess the godlike ability to write entire worlds into being, but who are embattled by their own hubris. These worlds are represented by books, each of which contains an entire sub-world and its inhabitants. These books, called Ages, serve as the primary transportation device in the game, and within each of their resident worlds are complex puzzles that can take hours or even days to solve. In addition to this elegant conceit for the creation of an extensible world, some have posited that *Myst* also provides a compelling metaphor for game design and computer programming.

Like other successful imaginary worlds, such as Tolkien's *Lord of the Rings*, *Star Trek* and *Star Wars*, the complexity of the *Myst* universe, story and game have inspired a thriving fan culture. *Myst* and *Uru* fans have created D'ni dictionaries, elaborately conceived and crafted game walkthroughs, a magazine, and all manner of fan fiction and art in homage to their beloved fictional world. In spite of its success, the *Myst* world is one of the few video game franchises that has never been extended to any media beyond written or recorded books. At various points, everything from a feature film, to a television series, to theme park on an actual physical island, were proposed, but none of these projects ever came to fruition.

Uru, while not the last game in the series, formed a kind of bookend to *Myst* for a number of reasons. First, it was the first and only multiplayer game in the series. There was a single-player mode, but the puzzles were designed to be solved in groups. Second, *Uru* was real-time 3D. This means it was both navigable and inhabitable in a way that most (though not all) of the earlier games, which were essentially nonlinear slideshows, were not. Third, *Uru* was the first and only game in the series to feature a player avatar (Figure 10.2). Prior games were played in first person with an undefined player role, referred to obliquely as 'The Stranger'. My research showed that the ability to visualize oneself as a unique and personalized character in the *Myst* world introduced both an experience of proprioception,

Figure 10.2 The introduction of a player avatar in Uru enhanced players' sense of presence.

enhancing players' sense of embodiment in the world, and also a sense of unique identity. This sense of identity was further enhanced by the presentation of this avatar to others. As MacKinnon has astutely put it, in cyberspace 'I am perceived, therefore, I am' (MacKinnon 1995). I refer to this phenomenon as 'seeing and being seen'. Players' sense of presence was enhanced not only by seeing themselves, but also by being seen by others (Pearce 2006). The introduction of the avatar is the pivotal contributing factor to the emergence of the *Uru*'s diasporic cultures. Without a player avatar, there would be no affordance in the game for constructing either individual or group identity, and without this ability, players would not have been able to construct the 'fictive ethnicity' that this chapter describes.

Finally, there is the matter of the *Uru* narrative itself. The narrative of *Myst* centred on Atrus and Catherine and their errant sons Sirrus and Achenar. The central character of *Uru* is their younger sister Yeesha, who has learned and exceeded her father's craft of Age-writing. The hub of the game is the abandoned city of D'ni Ae'gura, housed in an underground desert cavern, and settled by the D'ni *when their own world was destroyed*. In other worlds, the D'ni themselves were refugees. The word 'Uru' is attributed to have two meanings in the D'ni language: one, 'a large gathering or community;' the other, a more informal definition, 'you are you'. Players gather in the 'cavern' as 'Explorers', but the implication is that they play themselves. They arrive in D'in Ae'gura long after it has been deserted, guided by cryptic clues left by Yeesha.

The chief conflict in the game narrative was the restoration of the city (Figure 10.4). Yeesha was pitted against the D'ni Restoration Council (DRC), a fictional group of humans, played by Cyan employees (known colloquially as 'Cyanists'), who were the official caretakers of the cavern and had the authority to restore Ages so they were 'safe' for explorers. As each new Age was restored, it would be opened up for players, hence the game was envisioned as having an episodic structure in which new Ages would be released at regular intervals. However, Yeesha's messages suggested a darker side to the D'ni, whose pride was ultimately their downfall, and who appeared to have enslaved a group of 'beast-people,' called Bahro, who Yeesha calls 'The Least'. The City and some of the Ages were peppered with their markings, pictograms that told their story. The implication is that the Bahro were the 'indigenous' peoples of the cavern, which had, in essence, been colonized by the D'ni. Thus, a controversy was set in motion among players who sided with the DRC and wished to restore the D'ni culture, and players who sided with Yeesha and wished to free the Bahro. In practice, there were also players who preferred to remain outside this conflict, which created still a third factionless faction.

Uru is unlike any other MMO, in that it has no points, no levels, no player statistics, no competition, and no killing. The mechanic is based on collaborative puzzle solving and scavenger hunt style exploration, similar to that of other *Myst* games. The game play is for the most part nonlinear and un-timed, promoting open-ended exploration. Ages can be solved in a single play-through or over a series of sessions.

Becoming refugees

Uru was released as a CD-ROM in the Fall of 2003, but the full multiplayer game was never opened to the public. Instead, the game was released as a beta test, known as 'Prologue', to which players had to request an invitation. Eventually, due in part to a 'clerical error', all the players who had requested an invitation received one, and at its peak, *Uru Prologue* hosted 10,000 players. The following February, with only a few days warning, publisher Ubisoft and developer Cyan announced that *Uru Prologue* would be closing and that the full public release, *Uru Live*, was being cancelled. Players gathering in-world for the server shutdown is remembered by players as 'Black Monday' – or 'Black Tuesday', depending on which time zone players were in at the time of the closure (Figure 10.3).

Players in the study described their reaction to this announcement and its aftermath as 'post traumatic stress', many using those precise terms, and most of them were surprised by the strength of their emotional response. As mentioned earlier,

Figure 10.3 The final screen of *Uru Prologue*.

many identified in interviews as 'loners', prompting me to refer to them in publications as 'a community of loners'. For most, *Uru* was their first exposure to an online game, and they found themselves surprised at how quickly and easily they bonded with their fellow explorers. In a relatively short period, players had formed a cohesive community, especially within the 'hoods' (the *Uru* equivalent of MMO guilds), with The Gathering of Uru (TGU), the main focus of this study comprising the largest of these.

Migration

As soon as players heard about the closure, they began to make preparations. TGU leaders had created a forum, which included a live text chat area, to help players keep in contact with each other. Players began to scout for new games and virtual worlds as candidates for migration. They would report their findings to the forums, complete with screenshots and detailed descriptions of each world's characteristics. These comparative discussions are fascinating because they reveal a great deal about what software features players valued, and what points of contention might arise. There was no formal decision made to move to *There.com*. As with virtually every other aspect of *Uru* diasporic culture, the 'decision' was made in an emergent, bottom–up fashion. While players scattered to a number of worlds, a feedback loop began in which the more players who signed up for *There.com*, the more players who followed. When Leesa and Lynn, the group's mayor and deputy mayor, adopted *There.com* as their new home, though they were explicitly told that everyone could settle where they wished, the majority chose to follow their leaders to the new settlement. In order to facilitate this migration pattern, players adopted (again emergently) a practice of 'trans-ludic' identities: when they started their accounts in *There.com* and other virtual worlds, they maintained the same names and approximated the same appearances as their avatars from *Uru*. This is the common practice among all players who identify themselves as refugees or immigrants: they often carry their online identities with them between worlds, thus facilitating continuity among group members in different virtual worlds (Figure 10.4).

Uruvians, who migrated into *There.com* would often refer to themselves in the collective 'we'. In early interviews with members of the TGU, they would highlight the differences between themselves and other gamers, and speak about their shared values. 'We are explorers', they would say, or 'we are puzzle solvers'. They were also very clear that they did not like violent games, which was one of the reasons they chose not to go to some of the MMOs that were considered, such as *Ryzom*. Part of what they sought in a new home was a place with pleasing scenery that they could explore together, one of their favourite group activities in *Uru*. They also wanted an environment that was easy to learn and use and support their main objective, which was being together.

Initially, the *Uru* refugees were ostracized by native Thereians for a number of reasons. First, the group size, which, when joined by not-TGU Uruvian refugees, swelled to around 450 at its peak, was perceived as a threat to the small and

Figure 10.4 Uru refugees in There.com wearing player-created ethnic garb.

nascent Thereian culture. Second, Thereians feared that the incoming migrants would try to take over *There.com* and make it into *Uru*. In fact, Uruvians did try early on to recreate *Uru* in one area of the world, but due to the constraints of *There.com*'s building apparatus, they were unable to realize this plan except through emergent and incremental means.

Third, and perhaps most interesting, was the question of limited natural resources. In virtual worlds, as Castranova has pointed out, what we typically think of as 'resources' are theoretically unlimited: an infinite number of digital copies can be made of anything, with no scarcity (Castranova 2001). However, virtual worlds do contain one very valuable and very limited 'natural' resource for which players often compete: server processing. The great irony of massively multiplayer environments is that the more massively they are populated, the more the experience is degraded by the limits of server processing, creating lag, loss of fidelity, and in some instances, complete server breakdown. In *There.com*, high player traffic causes avatars to revert to low resolution 'blockheads'. In *Second Life*, if a 'sim' (distinct geographic area) is overcrowded, participants will be expelled into another area of the world. In *Uru*, the lag problem was so severe that when the city was highly populated, it became almost impossible to move, often resulting in software crashes.

Thus, the presence of a large group that insisted on travelling in packs generated enough lag to degrade the experience of their neighbours. The TGU group was forced to move no less than seven times. In each place they settled, they were driven away by neighbouring communities who resented the excessive lag they were causing. They were also 'griefed' in a variety of ways, including the posting of obscene signs adjacent to their settlements. Despite utopian discourses, the politics of difference still follows us into virtual worlds.

In contrast with Thereians' reactions, There, Inc., then owners of *There.com*, wanted to accommodate this large and lucrative community. They had actually been forewarned about their arrival by *Uru* community managers and saw this as an opportunity to quickly absorb a large influx of paying subscribers, which were sorely needed by the struggling start-up. In an attempt to hold onto this new subscriber base, they eventually identified a small island, far removed from other areas, where Uruvians could settle with minimal impact on neighbouring communities (Figure 10.5).

Transculturation

Although the initial acclimation to *There.com* culture was challenging, over time, *Uru* refugees underwent a process of 'transculturation' (Ortiz 1947). This included embracing their new home and integrating its constituent identity with the group identity they had brought with them from *Uru*. They also began creating *Uru*-derived and *Uru*-inspired artefacts and environments, thus using the space of *There*.

Figure 10.5 TGU's island settlement in There.com. The flag at left bears TGU's logo.

com itself as an expression and marker of group identity. They began to create *Uru* places in *There.com*, much the way a new immigrant group creates a Little Italy or a Chinatown in a new city. Many of their creations also become popular with the Thereian population at-large, so a sort of emergent cultural exchange occurred through the transaction of player-created artefacts. Today, you can see *Uru*- and *Myst*-inspired artefacts, buildings, and landscape elements throughout *There.com*, many of which have been placed by non-*Uru* players who are entirely unaware of their history of cultural significance. In addition to expanding their *Uru* community settlements, which contributed significant real estate fees, they also became involved in other aspects of *There.com* culture and governance. Members of the *Uru* community began to hold posts on the Member Advisory Board, a representative body nominated by players and appointed by the world's owners. Uruvian immigrants also launched and continued to manage The University of There, a large campus that hosts free educational activities open to all Thereians. At one point, when *There.com* was threatened with closure, rather than abandon ship as many players, *Uru* refugees stood their ground and helped assure that their new home remained open.

As this new hybrid Uru–Thereian culture began to emerge, their own reflection on this transculturation process was highly sophisticated, and players were able to integrate it into their discourses of refugee identity. Cola, a disabled player who later passed away due to a congenital illness, gave voice to this transition in a now-famous letter to her community as follows:

> The merging of the soul of the Urufugee into the citizen of There is happening. It wasn't without its tantrums of not wanting to merge, not wanting to believe Uru was gone and the guilty feelings of actually enjoying something other than Uru. But time does tell and there will always be the memories of D'ni and having been together there. Perhaps we could have a dual citizenship; Uruvian and Thereian. I have ***Myst*** being in D'ni, my soul, heart and being were **Riven** from D'ni, I am an **Exile** from the place where I want to be yet **Uru** has been put to bed. But perhaps the ending has not yet been written.

The closing language is typical of Uruvians, who exalt in wordplay and puns; here she has integrated the names of some of the Myst games into a sentence expressing her feelings about the transculturation process. Note the last line, which is quoted from *Uru*, and is often invoked as an expression of the continual longing that, even while embracing this new community, the group may still someday return to its 'homeland.'

Self-determination

In the meanwhile, other Uruvians continued to pursue this aim in various ways. One TGU member, Erik, had set to work building a copy of the *Uru* neighbourhood (the gathering place for each of the 'hoods') in Adobe Atmosphere, a 3D authoring environment. Players from other groups created *Uru* mods, including replicas of the game and entirely new Ages, using popular game engines such as *Doom*. Another group of player/hackers, communicating via forums and other forms of mediation, began to reverse-engineer the *Uru* servers. Initially a 'black ops' hacking enterprise, they reached out to Cyan and eventually received the developer's blessing and support to allow players to run their own *Uru* servers. Finally, Uruvians could return to their homeland once again.

Return(s) to the homeland

The launch of the player-run *Until Uru* server system had the empowering effect of providing members of the Uru Diaspora with a sense of self-determination, a symbolic gesture that was appreciated by all. And while many Uruvians did return to the player-run servers, some, still too traumatised by the closure, opted not to. Others felt that since they had already played all the Ages, there was little point in returning to an *Uru* frozen in time that provided no new content (although the hackers were working on that as well). Furthermore, and unexpectedly, players who had migrated into other games, including another group of Uruvians who had immigrated into and built an *Uru* island in *Second Life*, did not abandon their new settlements. Rather, they added the new instantiation of their homeland to their repertoire. This is one way in which immigration in virtual worlds differs significantly from real-world immigration. Players can 'live' in multiple worlds simultaneously. This is less common traditionally; MMO players tend to move from one game to another, rather than pay for multiple accounts. But with older players, who have more income and free time, as well as strong community ties, maintaining multiple game accounts allow for a more dynamic and flexible relationship to the virtual worlds they inhabit. Players also used a variety of multi-modal and extra-virtual communication strategies such as voice-over-IP (for worlds that had text chat only or for which voice was technically problematic), forums, and web sites. These practices suggest that the oft-lauded 'magic circle' (see Salen and Zimmerman 2004) is more porous than might have been previously believed. Players seem to move with fluidity and frequency between worlds and communication modes, and it is becoming less unusual for players in certain demographics to maintain multiple accounts in multiple games and virtual worlds concurrently.

The year 2006 saw a new chapter in the Uru saga. A new *Until Uru* test shard (server) was set up by a commercial company interested in re-launching *Uru*. I was enlisted as a consultant to help recruit 3000 players (the goal that was set by the company) to make a business case for the re-launch. By February of 2007, 3 years after its closure, *Uru* re-opened, in its originally envisioned form, complete with episodic Age openings, as *Myst Online: Uru Live* (*MO: UL*), under the auspices of Turner Broadcasting's GameTap 'classic' game portal. Sadly, the publishers felt that the player-run shard system would compete with the new, commercial version the *Until Uru* system was shut down, precipitating a second-wave migration among non-US players and those who did not or could not get GameTap, for whatever reason.

Even with the official homeland restored, complete with new episodic content from Cyan, players did not, as it was assumed they would, abandon their new homes. Now, culturally acclimated to their new communities, where they were exercising significant social, economic and political influence, and having established and built their own extensive homesteads, players had placed too high an investment in their new settlements to abandon them. Instead, they opted to, as they had done with *Until Uru*, split their time between their new homes and their 'homeland.' Players engaged in 'world-hopping,' where they would participate in an activity in one world, then go over to the next in the same evening. Even more interesting, they began recruiting new *Uru* players from within these communities. Of their own accord, players in *There.com* placed promotional signage for the game in *Uru*- and *Myst*-themed areas, and even built an '*Uru* Visitors Center' providing instructions on how to get into the game. They also conducted guided tours of *Uru* for *There.com* residents. *Second Life* players actually partnered with GameTap and re-opened their *Uru* Island, which had closed a few months before for lack of funds. The new instantiation of *Uru* in *Second Life* included signage and links to a special site that offered a special GameTap promotion for aspiring *Uru* players in *Second Life* (Figure 10.6). This novel extension of inter-game immigration, in which players actually pointed to games from inside other games and virtual worlds suggests a significant shift in play habits away from the traditional 'one game at a time' mode of MMO subscribership.

Figure 10.6 Uru Re-creation in *Second Life* to promote 2007 re-release of the game.

For a brief time, players had, literally and figuratively, the 'best of both worlds.' They could at once inhabit their homeland, complete with the long hoped-for new content, and at the same time maintain their *Uru* settlements in other worlds such as *There.com* and *Second Life*. The trans-ludic synergies brought more new migration in both directions, precipitating a new 'Second Wave' of *Uru* immigrants into *There.com*, as well as a new generation of Uruvians recruited from other worlds.

The Uru story does not end here, however. April 2007 saw yet a third closure, albeit with more forewarning and grace than the initial closure. Uru players were forced into a third wave of migration, with first and second-wave immigrants acting as emissaries to those newly traumatised by the loss of the game. Again, Uruvians in *There.com* as well as *Second Life* staged tours and events welcoming their refugee brethren. But as always, 'the ending has not yet been written'. Turner subsequently released the game's rights back to Cyan, which announced its intention to create *Myst Online: Restoration Experiment* (MO: RE), set of tools that will allow them to turn over much of the game's content creation to players, although this has yet to come to fruition as of this writing.

The *Uru* story suggests that, in graphical virtual worlds, place and identity can be tightly knit together. In the case of the Uru Diaspora, this plays out in two ways: where players are 'from' becomes a key marker of identity. In addition, in worlds like *Second Life* and *There.com*, spaces that players create through 'productive play' practices also become an expression and marker of both individual and group identity (Pearce and Artemesia 2006). In a sense, the place itself becomes an 'avatar,' a spatial embodiment of players' individual and collective identities (Figure 10.7).

Figure 10.7 The Uru 'hood' fountain, an important cultural artifact as instantiated in (clockwise from upper left): *Uru*, *There.com*, *Atmosphere*, and *Second Life*.

Methodology

This research can be characterized as 'multi-sited cyberethnography,' combining methods from Margus (1995), Mason (1996) and Hine (2000). Margus suggests 'multi-sited ethnography' as a way to address the waning instances of hermetically sealed real-life cultures, and proposes a method of 'following' various aspects of culture from one field site to another (Margus 1995). Hine (2000) builds on Mason's notion of 'virtual ethnography' (1996) in describing an immersive approach to engaging with the lived experience of online cultures on their own terms.

This primary research activity was an 18-month ethnographic field study across multiple sites that took place from April of 2004 to September of 2006, as well as subsequent field research. The subject group consisted of the largest of the *Uru* 'hoods' (similar to a guild in other games), comprising approximately 300 players, most of whom migrated into *There.com*, as well as a secondary group of 200 players in *Second Life*. The research question focused on the specific ways in which software design in online games influences social patterns in the construction of emergent cultures.

The theoretical framework combined several different perspectives. As the group was 50 percent female, an atypical ratio for an MMO community, aspects of feminist ethnography came into play, including: exploring the relationship between ethnographic accounts and fictional or literary approaches to writing (especially relevant in a 'real' study of a 'fictional' world); foregrounding the role of gender in online culture; privileging the authority of the subjects and interrogating the subjectivity and authority of the researcher (Visweswaran 1994).

I found Turner and Schechner's ethnography of performance and ritual (1985) of particular utility in the study of online games, which can be interpreted as a kind of ritualized co-performance; I have also theorized that role-playing is a natural extension of the rituals that Turner and Schechner describe (Fron *et al.* 2007, Pearce and Artemesia 2009). I was also influenced by Goffman's notions of the performance of everyday life (Goffman 1959) and frame analysis (Goffman 1974) and Denzin's subsequent observations of the increasing performativity in everyday life (2003). In the context of games and virtual worlds, where the researcher must engage in the role-playing activity, I framed ethnographic research itself as both a performance practice and a game. Clifford (1986) highlights this performative aspect in his writings on ethnographic allegory and Janesick (2000) describes a choreographic approach to participant-observation using a repertoire of improvisational techniques. An improvisational approach is key to studying culture, which requires, as Strathern (2004: 5) puts it 'collect[ing] information on unpredictable outcomes'. Because play is inherently labile and improvisational, its study requires a flexible, responsive approach. Through this process, my role developed such that I eventually came to be recognized as the group's ethnographer, a kind of insider–outsider status, which positioned me as their advocate and spokesperson to the outside world.

The writing used both 'thick description' (Geertz 1973) and polyphonic voices at a number of levels. Following Wolf (1992), Powdermaker (1966), Smith-Bowen

(1964), I explored issues of subjectivity, as well as utilizing polyphonic texts (Fisher 1990; Miles and Huberman 1994; Helmreich 1998), privileging direct quotes (such as the poems above), and soliciting feedback from players via a 'participant blog'. These efforts were all aimed towards taking a collaborative approach to the research and its presentation, acknowledging 'native authority' while at the same time finding my own voice in my unique role as the group's ethnographer.[6]

The construction of 'refugee' and 'homeland': discourses of displacement

Uru is not the first instance we have of game refugees, although this is an under-studied aspect of MMO culture. While game refugees can result from closure, such as was the case with *Meridian 59* (the first Tolkien-inspired graphical MMO) *Ultima Online* and *Asheron's Call*, they can just as often be the outcome of changes in a game, a lapse in popularity, or the opening of a new game in the same genre. A change in rules of *Star Wars Galaxies* that made the game easier to play, in effect negating the accomplishments (and status) of more experienced players, precipitated a mass exodus in 2005 (Kohler 2005). When I first began my *Uru* research in 2004, there was a group of 800 self-identified *Sims Online* Refugees in *There.com*. These players had independently adopted the practice of 'trans-ludic' identities, recreating the same names and appearance as their avatars from *The Sims Online*. The launch of *World of Warcraft* in 2004 precipitated a mass migration of players from *EverQuest*, which had previously been the most popular US-based MMO.

For Uruvians, the construction of 'refugee' and 'homeland' as markers of group identity arose entirely out of the play practices and discourse of the players themselves, clearly influenced by the connotations evoked by these terms in the culture at-large. In prior work, I have described the 'social construction of identity' and pointed out that game communities are involved in a dynamic interplay between the individual and the group identity, and that the construction of individual identity is an intersubjective process (Pearce and Artemesia 2007). Nowhere can this be seen in greater relief than the Uru Diaspora. There was no one person who deemed the group 'refugees'; rather, players began individually adopting this terminology as soon as the game closed, and began to spontaneously construct and perform, in an emergent fashion, what I term a 'fictive ethnicity'.

It is interesting to note that neither the concepts of 'refugee' nor of 'homeland', appeared within the *Uru* community, while the original game was in operation. Axel describes 'homeland' as the 'defining locality', the 'place of origin', also the defining characteristic of diaspora; yet he also points out that diaspora is defined less by context but more by '*loss* of context' (2004: 28). Given that discourses of 'homeland' emerged among *Uru* players only when this 'defining locality' was lost, it is fair to say that the very notion of homeland was precipitated by its absence. We might think of the word 'homeland' as a replacement for what is absent, a kind of phantom limb for what has been lost. Even the construct of 'Uruvian' only emerged after the players were foisted from their 'homeland.' As they established communities in other virtual worlds, it was this difference of having come from another place – a place that was destroyed in much the way original D'ni homeland was destroyed – that provided the scaffolding for this fictive ethnicity. Had *Uru* never closed, there would have been no need to formulate neither this fictive ethnicity nor the trans-ludic identities that supported it. Furthermore, the bond among Uru refugees was based on their shared history, memory and trauma of the loss of their homeland, and, to a lesser extent, a history that marked them as different from those around them.

As Gupta and Ferguson (1992: 11) point out, 'remembered places have often served as symbolic anchors of community for dispersed people . . . 'Homeland' in this way remains one of the most powerful unifying symbols for mobile and displaced peoples. . . .'

Homeland

'Attachment to homeland can be intense . . .' asserts leading humanist-geographer Tuan (1974: 149). '. . . home is the focal point of a cosmic structure . . . Should destruction occur we may reasonably conclude that the people would be thoroughly demoralized, since the ruin of their settlement implies the ruin of their cosmos' 'Yet . . .' he continues,

> 'human beings have strong recuperative powers. Cosmic view can be adjusted to suit new circumstances. With the destruction of one "centre of the world", another can be built next to it, or in another location altogether, and in turn becomes "the centre of the world". Thus 'centre' is a 'concept in mythic thought rather than a deeply felt value bound to unique events in locality'.
>
> (Tuan 1974: 149–150).

Another way to state this might be that 'homeland' is a state of mind.

The notion of 'homeland' is invariably tied to community. Homeland is seldom a singular association, is never simply ones 'house' of origin, nor merely a location. Rather, it is a web of shared meanings, histories and narratives that individuals associate with both a specific place and with a shared identity. Homeland is at once experiential and symbolic and, as Jackson (1998) might assert, inherently intersubjective. A community's 'homeland' may have a different significance or no significance whatever to persons outside the group.

'Attachment of a deep though subconscious sort may come simply with familiarity and ease, with the assurance of nurture and security, with the memory of sounds and smells, of communal activities and homely pleasures accumulated over time' (Tuan 1974: 159). These associations are deeply tied to a sense of nostalgia, which is often embellished or modified in the continual retelling and reconstruction of narratives of homeland.

Placeness

Uru can be characterized as what Ito (1999) calls a 'network locality', a sense of place achieved over the network, or to put it another way, a sense of 'placeness'. Ito uses this term to refer primarily to text-based community sites, but the *Uru* experience brings this concept into bold relief by creating a virtual 'place' that is visually rendered, explorable and inhabitable (Pearce and Artemesia 2009), and which players identify as a legitimate locality, though not a 'real' place. Further these players think of this virtual network locality not only as a place they visit, but also, distinctly, in the context of their trans-ludic activities, as the place they are *from*.

So what do we mean by terms such as 'place' and 'placeness'? Tuan (1997: 4) describes 'space' as 'more abstract than 'place.' What begins as undifferentiated space becomes place as we get to know it better and endow it with value.' So in the broadest terms, we might describe space in terms of its qualities – spaciousness, expanse, horizon, vista, and so forth, while place can be defined as a specific location imbued with specific cultural meaning. Tuan (1974) points out that the meaning of a place is often coupled with its narrative, such as Kronberg Castle in Denmark, also known as Helsingør, Shakespeare's Elsinore. These narratives, whether 'real,' 'fictional,' or, as is often the case, somewhere in between, form much of the foundation of what gives a generic space a sense of 'place.'

I have previously posited that the need for such synthetically narrativised environments, such as theme parks, themed malls, and MMOs, arises from the growing sense of placelessness brought on by a highly westernized and genericized global marketplace (Pearce 1997, 2007). Gupta and Ferguson (1992: 10–11) have put it this way:

> The irony of these times, however, is that as actual places and localities become ever more blurred and indeterminate, ideas of culturally and ethnically distinct places become perhaps even more salient.It is here that it becomes most visible how imagined communities (Anderson 1983) come to be attached to imagined places, as displaced peoples cluster around remembered or imagined homelands, places, or communities in a world that seems increasingly to deny such firm territorialized anchors in their actuality.

In their wonderfully astute pre-Internet analysis *Spaces of Identity*, Morely and Robins (1995) argue that the new 'audiovisual geographies' of electronic media have precipitated a global identity crisis, by shifting the focus from 'imagined communities' of nation states to the 'placeless and non-referential sense of identity' of global markets. They point out that the capitalist framing of the global village fails to recognize that, as Rustin (1987: 34) asserts, '. . . collective identities are formed through the common occupancy of space'. They also cite Williams (1983: 39), who argues that in a world of 'false and frenetic nationalisms and of reckless and uncontrollable transnationalism', the struggle for meaningful communities and 'actual social identities' is more and more difficult. '[We] have to explore new forms of variable societies . . .'. What is called for, say Moreley and Robins (1995: 39) is the 'reclamation or reimagination of a sense of referential identity, the revaluation of particular and concrete experience'. In a strange twist, *Uru*, the very embodiment of an "audiovisual geography" has become the site just such a reclamation.

Fictive ethnicities and imaginary communities

The notion of fictive ethnicities, that is, attachments to ethnic identities of nonexistent, fictional cultures are neither new nor even unique to networked communities. We see it instantiated in many forms, most famously among 'trekkies,' Star Trek fans who have adopted the fictive ethnicities of the races within the series, a form of what Jenkins called 'participatory culture' (Jenkins 1992). At Dragon*Con, a multi-themed science fiction and fantasy fan convention held in Atlanta each year, groups of costumed players convene under the auspices of races and ethnic groups from fictional worlds. These can include elves and hobbits, zombies, vampires, and werewolves, pirates and denizens of the Renaissance, Anime and Neko characters, furries, steampunks, and space aliens of various races and origins, such as Martians, Wookies and Klingons. At Dragon*Con 2007, word spread that all attendees were invited to a Klingon wedding, provided they dressed as Klingons. Many fans of Tolkien's *Lord of the Rings* series and its variegated derivatives, from *Dungeons and Dragons* to *World of Warcraft*, dress as and adopt the personae of fictional races at fan events. Part of the pleasure of this type of roleplay is the instant connection with a community of shared values and interests. As we found during our 2007 study of cosplay at Dragon*Con, the costume creates an affinity among roleplayers and costumers 'from' the same fictional world, even if they are perfect strangers (Pearce *et al.* 2008).

It is important to note here the distinction between Anderson's (1983) 'imagined communities,' and what might be better termed 'imaginary communities'. Tuan asserts that 'Culture is driven by imagination and is a product of imagination' (Tuan 1974: xiv). In this sense, all communities are, to some extent, imagined. Anderson's 'imagined communities' arise from the complex interplay between history, geography, power, language, religion (belief) and culture and often explain the formation of nation states or historically rooted communities that do not always exist in reality. The boundaries of nation states, often the artefact of colonial occupation, can define as communities groups that have no sense of natural affinity (1983). In spite of the fact that they 'live' in different 'nations', one could argue that members of the Uru Diaspora have a stronger sense of shared ethnicity, than the residents of, say, 'Iraq', an imagined community by Anderson's definition.

We tend to think of ethnicity as assigned by location and context of birth. But the Uru Diaspora's discretionary ethnicity forces us to interrogate fundamental issues of nation, territory, culture and identity. Gupta and Ferguson point out that there is often a tendency to take for granted the relationships between these elements of culture, as epitomised by the bright colours used to distinguish nation states on a map; however, in practice, the relationship between culture and place is much more complex, especially in the context of post-colonialism, dispersed immigrant and refugee communities, and border cultures that exist at the interstices between territories. They assert that:

> . . . issues of collective identity today do seem to take on a special character, when more and more of us live in what Said (1979: 18) has called 'a generalized condition of homelessness,' a world where identities are increasingly coming to be, if not wholly deterritorialized, at least differently territorialized'.
>
> (Gupta and Fergerson 1992: 9)

The Uru Diaspora is such a 'differently territorialized identity', which in a fictional or virtual world is adopted as a 'homeland', and thus by definition, becomes a marker of identity.

'Distributed diasporas'

While the terms 'refugee' and 'homeland' were part of the terminology for their status invented by players themselves, I introduced the 'diaspora' into the lexicon to describe the trans-ludic *Uru* community. What does it mean to say a distributed community is diasporic? Since they are already 'not in the same place,' geographically dispersed, is not this term somewhat redundant? While it is true that all distributed communities are, in some way, by definition, diasporic, what differs here is that the *Uru* community is 'trans-ludic', meaning that they are not distributed merely in geographical space, but also distributed in virtual space. We might call such a group a 'distributed diaspora' as a way to describe their dispersion in a variety of distributed online environments.

I realize that the term 'diaspora' is highly contested, but since an extended debate on the various interpretations of the term is well beyond both the scope of this chapter and the expertise of this author, it might be useful to at least select one that is fairly well recognized, though debated, as a starting point. Drawing on Safran, Clifford (1994) idea of diaspora as characterized by a history of dispersal, myths/memories of homeland, including the desire for eventual return, ongoing support of the homeland, alienation in the host country, and a collective identity importantly defined by these relationships. Safran (1991) describes a diaspora as a group or community which (1) is dispersed from an original 'centre' to at least two 'peripheral' places; (2) maintains a 'memory, vision or myth about their original homeland'; (3) has a real or imagined sense that they are not fully accepted in their host country; (4) sees the ancestral home as a place of eventual return; (5) is committed to the maintenance and restoration of this homeland; and (6) maintains a group consciousness and solidarity that are 'importantly defined' by this continuing relationships to homeland (Safran 1991). While this definition may be debatable for some communities – the African and Jewish Diasporas, for instance, may not fit all these criteria – there is no question that the Uru Diaspora fulfils these requirements. Nothing in either Safran or Clifford's schema suggests that the homeland must be 'real'; virtual and even imaginary homelands are not disqualified.

Consuming production/play practice as metaphor

I realize that there is risk of falling into some traps here so before closing I wish to address the obvious issue: *Uru* is the product of a capitalist system of media production and consumption. Much of what this chapter describes falls under the rubric of what I call 'productive play', an emerging phenomenon in which play transgresses into creative or work-like activities (Pearce and Artemesia 2006). Virtual worlds such as *There.com* and *Second Life* have a structure that inverts the classic producer/consumer relationship and creates a scenario where players are actually consuming the right to produce. These notions echo de Certeau's (1984) theory of creative consumption. They also reflect Willis' (1978, 2000) related concepts of life as art practices, perhaps best epitomised in his studies of British 'bike boys', use of motorcycle customization as an expression of individual and group identity formation.

All of the player discourses described in this chapter fall within a range of practices around participatory and productive play culture. I also want to make clear that invoking terms such as 'homeland', 'refugee' and 'diaspora' is in no way meant to trivialize the traumas, violence and tragedy that typically accompany real-world refugee narratives.

Conclusion: Uru diaspora as allegory

In his classic essay 'On Ethnographic Allegory', James Clifford treats 'ethnography itself as a performance emplotted by powerful stories' (Clifford 1986). Allegory, says Clifford (1986), is a 'representation that interprets itself'. The narrative of the Uru Diaspora operates as just such an allegory.

It is difficult to ignore the multifaceted complex allegories and ironies that span the 'fiction' of *Uru* and the lived experience of its players. Players engage in a game about restoring the lost culture of a refugee community, only to become refugees themselves. They then set about restoring their own culture, engaging in both a performance and a game that is at once 'real' and 'fictional'. Building off a narrative about creating worlds, they create their own sub-worlds within other, larger virtual worlds. That this narrative sits within the larger context of a 'global village' that is increasingly diluting the relationship between identity and place forces us to take a step back and ask ourselves: what does the story of the Uru Diaspora say about culture at-large?

We inhabit a moment in which abstract and generalized global markets, combined with Western notions of individualism, have preempted the primacy of local communities and the reassurance of a connected identity. Tuan (1974) points out that the reification of individuality can lead to a sense of isolation and vulnerability which, in nonwestern and even earlier western cultures, was mitigated by ties to place and community. As pointed out by some of the authors cited here, the growing sense of placelessness has precipitated a worldwide identity crisis. Without the emotional connection to specific place and specific community that have sustained our ancestors at many scales for many generations, where does that leave us in terms of identity? We can, as is happening throughout the world, take refuge in the imagined regional communities of nationalism, or the transnational identities of religious sectarianism and extremism, or real-world diasporic cultures that may or may not represent a common cultural affinity. Unlike ourselves and our ancestors who were born into their proximal cultural contexts; however, the *Uru* Diaspora has *chosen* its culture, and its members have adopted a fictive ethnicity at their own discretion. In a sense they have created a new form of discretionary ethnicity, one which is based on imagination, shared values and a collective identity that go with a connection to place that is elsewhere fast fading from our cultural landscape.

Notes

1 An earlier version of this chapter was as 'Identity-as-Place: Trans-Ludic Identities in Mediated Play Communities – The Case of the Uru Diaspora' at *Internet 9.0*, Association of Internet Researchers, Copenhagen, October 15–18, 2008.
2 Artemesia is the author's 'research avatar.'
3 All individual player and group names used in this chapter are pseudonyms.
4 For an in-depth study of the Uru Diaspora, including details on methodology, please see Pearce, and Artemesia (2009). *Communities of Play: Emergent Cultures in Multiplayer Games and Virtual Worlds*. Cambridge, MA: The MIT Press.
5 All images by the author, except where otherwise noted.
6 I am greatly indebted to Tom Boellstorff for guiding me through this process. For those interested in game ethnography, I highly recommend his writings on virtual worlds and research methods: Boellstroff (2006/2008).

References

Anderson, B. (1983) *Imagined Communities*, New York, Verso.
Ashe, S. (2003) Exploring Myst's Brave New World. *Wired* Retrieved 24 September 2010, from http://www.wired.com/wired/archive/11.06/play.html.
Axel, B. K. (2004) The context of diaspora. *Cultural Anthropology*, 19(1): 26–60.
Boellstorff, T. (2006) A ludicrous discipline? Ethnography and game studies. *Games and Culture* 1(1): 29–35.
Boellstorff, T. (2008) *Coming of Age in Second Life: An Anthropologist Explores the Virtually Human*, Princeton, Princeton University Press.
Carroll, J. (1994) Guerillas in the Myst. *Wired* Retrieved 24 September 2010, from www.wired.com/wired/archive/2.08/myst_pr.html.
Castranova, E. (2001) *Virtual Worlds: A First-Hand Account of Market and Society on the Cyberian Frontier*, CESifo Working Paper Series, No. 618.
Castranova, E. (2002) *On Virtual Economies*, CESifo Working Paper Series, No. 752.
Clifford, J. (1986) On ethnographic allegory. In J. Clifford and G. E. Marcus (eds). *Writing Culture: The Poetics and Politics of Ethnography*. Berkeley, University of California Press: 98–121.
Clifford, J. (1994) Diasporas. *Cultural Anthropology*, 9(3): 302–38.
de Certeau, M. (1984) *The Practice of Everyday Life*, Berkeley, University of California Press.
Denzin, N. K. (2003) *Performance Ethnography: Critical Pedagogy and the Politics of Cultures*, Thousand Oaks, Sage Publications.
Fisher, M. J. (1990) *Debating Muslims: Cultural Dialogues in Postmodernity and Tradition*, Wisconsin, University of Wisconsin Press.

Fron, J., T. Fullerton, J. F. Morie and C. Pearce (2007) Playing Dress Up: Costumes, Roleplay and Imagination. *Proceedings of the Philosophy of Computer Games*, Modena, Italy.

Geertz, C. (1973) *The Interpretation of Cultures*, New York, Basic Books.

Goffman, E. (1959) *The Presentation of Self in Everyday Life*, New York, Doubleday Anchor Books.

Goffman, E. (1974) *Frame Analysis: An Essay on the Organization of Experience*, New York: Harper and Row.

Gupta, A. and J. Ferguson (1992) Beyond 'culture': space, identity, and the politics of difference. *Cultural Anthropology*, 7(1): 6–23.

Helmreich, S. (1998) *Silicon Second Nature: Culturing Artificial Life in a Digital World*, Bertkeley, University of California Press.

Hine, C. (2000) *Virtual Ethnography*, London, Sage Publications.

Ito, M. (1999) Network Localities: Identity, Place, and Digital Media. *Meeting of the Society for the Social Studies of Science*, Sandiego, USA.

Jackson, M. (1998) *Minima Ethnographica: Intersubjectivity and the Anthropological Project*, Chicago, The Unviersity of Chicago Press.

Janesick, V. J. (2000) The choreography of qualitative research design: minuets, improvisations, and crystallization. In N. K. Denzin and Y. S. Lincoln (eds). *Handbook of Qualitative Research*. Thousand Oaks, Sage Publicatons Inc: 379–99.

Jenkins, H. (1992) *Textual Poachers: Television Fans and Participatory Culture*, New York, Routledge.

Kohler, C. (2005) Star Wars Fans Flee Net Galaxy. *Wired* Retrieved 25 September 2010, from http://www.wired.com/gaming/gamingreviews/news/2005/12/69816.

MacKinnon, R. C. (1995) Searching for the Leviathan in Usenet. In S. Jones (ed.). *Cybersociety: Computer-Mediated Communication and Community*. Thousand Oaks, Sage: 112–37.

Marcus, G. (1995) Ethnography in/of the world system: the emergence of multi-sited ethnography. *Annual Review of Anthorpology*, 24: 95–117.

Mason, B. L. (1996) Moving toward virtual ethnography. *American Folklore Society News*, 25(2): 4–5.

Miles, M. B. and A. M. Huberman (1994) *Qualitative Data Analysis: An Expanded Sourcebook*. 2nd edn, Thousand Oaks, Sage Publications Inc.

Miller, L. (1997) Riven Rapt: How Myst and Its Riveting New Sequel Won Our Hearts and Minds – by Dazzling Our Eyes and Disabling Our Trigger Fingers. *Salon*. Retrieved from http://archive.salon.com/21st/feature/1997/11/cov_06riven.html.

Moreley, D. and K. Robins (1995) *Spaces of Identity: Global Media, Electronic Landscapes and Cultural Boundaries*, London, Routledge.

Ortiz, F. (1947) *Cuban Counterpoint*, New York, Knopf.

Pearce, C. (1997) *The Interactive Book: A Guide to the Interactive Revolution*, Indianapolis, Macmillan.

Pearce, C. (2006) Seeing and Being Seen: Presence and Play in Online Games and Virtual Worlds. *Position Paper for Online, Offline and the Concept of Presence When Games and VR Collide*. Berkeley, California.

Pearce, C. (2007) Narrative environments: from Disneyland to World of Warcraft'. In F. von Borries, S. P. Walz and M. Bottger (eds). *Space, Time, Play: Computer Games, Architecture and Urbanism: The Next Level*. Basel, Birkhauser: 200–05.

Pearce, C. and Artemesia (2006) Productive play: game culture from the bottom up. *Games and Culture*, 1(1): 17–24.

Pearce, C. and Artemesia (2007) Communities of play: the social construction of identity in online game worlds. In P. Harrigan and N. Wardrip-Fruin (eds). *Second Person, Roleplaying and Story in Games and Playable Media.* Cambridge, MIT Press: 311–18.

Pearce, C. and Artemesia (2009) *Communiteis of Play: Emergent Cultures in Multiplayer Games and Virtual Worlds*, Cambridge, MIT Press.

Pearce, C., E. Barba, G. Gunter, R. Schweizer and R. Vichot. (2008) Conventional Dress (Digital Video). Retrieved 25 May 2010, from http://www.conventionaldress.com.

Powdermaker, H. (1966) *Stranger and Friend: The Way of the Anthropologist*, New York, W. W. Norton.

Relph, E. C. (1976) *Place and Placelessness*, London, Routledge.

Rothstein, E. (1994) A New Art Form May Arise from the 'Myst'. *The New York Times* Retrieved 24 September 2010, from http://www.nytimes.com/1994/12/04/arts/a-new-art-form-may-arise-from-the-myst.html.

Rustin, M. (1987) Place and time in socialist theory. *Radical Philosophy*, 147: 30–36.

Safran, W. (1991) Diasporas in ,modern societies: myths of homeland and return. *Diaspora*, 1: 83–99.

Said, E. W. (1979) Zionism from the standpoint of its victims. *Social Text*, 1: 7–58.

Salen, K. and E. Zimmerman (2004) *Rules of Play: Game Design Fundamentals*, London, MIT Press.

Schechner, R. and M. Schuman (1976) *Ritual, Play and Performance: Readings in the Social Sciences/Theater*, New York, Seabury Press.

Seay, A. F., W. J. Jerome, K. S. Lee and R. E. Kraut (2004) Project Massive: A Study of Online Gaming Communities. *CHI '04 extended abstracts on Human factors in computing systems.* Vienna, Austria: 1421–1424.

Smith-Bowen, E., (aka, L Bohannan). (1964) *Return to Laughter*, New York, Anchor Books.

Strathern, M. (2004) *Commons and Borderlands: Working Papers on Interdisciplinarity, Accountability, and the Flow of Knowledge*, Oxford, Sean Kingston.

Taylor, T. L. (1999) Life in virtual worlds: plural existence, multimodalities and other online research challenges. *American Behavioral Scientist*, 43(3): 436–49.

Tuan, Y. F. (1974) *Topophilia*, Englewood Cliffs: Prentice Hall.

Turner, V. and Schechner, M. (1985) *Between Theater and Anthropology*, Pennsylvania: University of Pennsylvania Press.

Williams, R. (1983) *Towards 2000*, London, Chatto and Windus/Hogarth Press.

Willis, P. (1978) *Profane Culture*, London, Routledge.

Willis, P. (2000) *The Ethnographic Imagination*, Cambridge, Polity Press.

Wolf, M. (1992) *A Thrice Told Tale: Feminism, Postmodernism, and Ethnographic Responsibility*, Palo Alto, Stanford University Press.

Yee, N. (2001) The Norrathian Scrolls: A Study of EverQuest. Retrieved 24 September 2010, from http://www.nickyee.com/eqt/home.html.

11 The rise and fall of 'Cardboard Tube Samurai'

Kenneth Burke identifying with the *World of Warcraft*

Christopher A. Paul and Jeffrey Philpott

Guilds[1] in online games often have a tumultuous life of ups and downs. In this chapter, we examine the rise and fall of the Cardboard Tube Samurai (CTS), a *World of Warcraft* (*WoW*) guild, and explain three key phases in the guild's existence using the ideas of Kenneth Burke. We argue that rhetorical theory can offer substantive insights into the events of online games, in this case focusing on the roles of identification, division, and consubstantiality in massively multi-player online games (MMOs) by explaining how a guild can build for 2 years to their greatest triumph then fall apart 2 weeks later.

CTS was a *WoW* guild that grew from a handful of people talking in an online forum to become an aggregation of more than 200 different user accounts and the third ranked raiding guild on the Muradin server.[2] Within weeks of its greatest success, and almost 2 years after it was founded, the guild went from exultation to dissolution. Christopher, as the human mage Alruna, spent literally weeks of his life as part of the guild over its 2-year existence. He went from being one of CTS's first members on *WoW*'s launch day to an officer of the guild-to-guild leader to guildless when CTS dissolved. By analyzing the history of CTS through Christopher's playing experience, conducting interviews of CTS members and analyzing the guild's forums, we saw how the theories of Kenneth Burke helped to explain life within the guild. The rise and fall of CTS is an all too common tale for online gaming guilds that demonstrates the relevance of rhetorical analysis as a tool for understanding key elements of online gaming behaviour.

A leading 20th century rhetorician, Kenneth Burke, argued that identification and consubstantiality were keys to bringing people together to overcome their inherent differences (1950). For Burke, identification was the central appeal of rhetoric; the key reason why people chose to work together and the primary determinant of the success of rhetorical appeals. Because of the innovations and social interaction promoted by games, like *WoW*, scholars can find a fertile set of texts to analyze in online gaming spaces, as MMOs make interaction with others a central component of playing the game (Williams *et al.* 2006). MMOs offer a particularly interesting setting for research, as 'a virtual world can be used to replicate research. It is like a Petri dish' (Castronova 2009: 274). The intensely social aspects of MMO design require different ways of understanding events in video games, in this case opening the door for the application of Burke's theories. Recognizing the role rhetorical analysis can play in explaining the phenomena occurring in online games requires three steps. First, we will explain key dynamics of MMOs, with an emphasis on *WoW*. Second, we will explain the theories and holdings of Burke as it pertains to identification, division, and consubstantiality. We then consider the history of CTS. Finally, we will chart the rise and fall of CTS with a focus on how Burke's theories help explain key events in the guild's history and provide suggestions for how gamers and game designers can benefit from Burkean theories.

Online gaming and world of warcraft

The primary distinction between MMOs and most video games is that the other people playing are an integral part of the game, bringing questions about identification to the table. Unlike in *Madden Football*, where one can play against (or with) other players, but also by themselves against the computer, an MMO is defined in large part by the fact that players cannot play on their own; all players are always part of a persistent world in which others are also playing. One can go off on one's own or play 'solo' without the aid of others, but there is no separate, individualized world to which players can retreat. This fully integrated interaction with others means that the other people playing become a tremendously important element of the game. As Taylor (2006: 9) notes, 'shared action becomes a basis for social interaction, which in turn shapes the play'. Castronova (2005) argues that 'the shared nature of synthetic worlds is a critical part of the technology of place, because perceptions of how things are have to be shared and agreed upon by many people before they acquire the flavour of reality'. Players have to negotiate scarce resources, find ways to work together to defeat foes that cannot be beaten by a lone adventurer, and some play these games to interact with other players. Friendships are formed, marriages have occurred, and venom has been spewed at pixilated others, but this presumption of human interaction ensures that MMO game play can be seen as a fundamentally rhetorical process, one which invites a discussion of how players identify within the game.

Although *WoW* enables people to successfully play by themselves, unlike many other MMOs, there are a number of places where group interaction is necessary. Certain elements of the game, often called 'dungeons' or 'instances,' are reserved for groups of people. Individuals can enter, but, with few exceptions, the challenge of these locations requires a group of people to be successful. Instances generally require groups that range from 5 to 40 people, all of whom must work together to accomplish a common goal, generally to kill computer-controlled monsters, often referred to as 'mobs.' Players can group together at any point in the game, but it is in these instances where the structure of the game encourages group interaction, rather than solo game play (Nethaera 2008). Instances, especially those designated as raid instances, are where the best equipment, or 'loot,' in the game can be acquired and where the most powerful and dangerous monsters can be fought. Raiding is an application of Julian Dibbell's belief that the ordering principle of online games is a player's desire 'to own and not be owned' (Dibbell 2006). Raiders get access to things that enhance their characters, which give them powerful social markers in-game and also the ability to 'own' others (cf Yee 2006; Bazazu

2007). However, succeeding in those dungeons requires assembling a large group of people that meet online at the same time and then getting those people to work together for a common goal. Furthermore, as Torill Mortensen notes, 'in order to keep raiding, you have to believe you do the only thing worthwhile, as raiding excludes most other game activities' (Mortensen 2008: 214). Raiding is the ultimate experience for many *WoW* players, an activity that displaces the many other things they could do in the game or outside of it and a pursuit that could easily exceed the hours most people spend at a traditional job in a week (cf Ducheneaut *et al.* 2006). Across servers, those who raid track their progress based on how many 'bosses,' or very difficult raid mobs, each guild has killed. World and server firsts are tracked by the playing community and the most successful raiders generally enjoy a degree of celebrity status within the game (cf Schramm 2007).

In addition to raids, one of the other key social aggregations in *WoW* is the formation of guilds (cf Williams *et al.* 2006). Guilds are a place for players to gather together and talk within the game, as well as a prime indicator of social status. The primary advantage of joining a guild is that it gives players within the guild a dedicated chat channel in which to talk and enables them to easily track fellow player's comings and goings online. Guilds can be formed for virtually any purpose. Most join together through some overarching common interest, whether it is raiding; a common web site member frequent, like Penny Arcade[3]; a shared sexual preference; or the simple desire to talk with other people in the game. Guilds make it possible to talk with other people while doing things in the game and offer a means by which to find players to help accomplish more challenging tasks. Guilds can range in size from a small group of three or four friends to large entities with hundreds of different members. Most large guilds have their own web site and forums, so that members can communicate and coordinate with each other outside of the game. Player's guild affiliations are also broadcast to others encountered within the game. When approached by another player, the default game interface renders the player's avatar, as well as their character's name and guild name. This makes guild affiliation almost as notable as a character's name, so that a player's guild, and the guild's reputation, is a key part of how individual players are viewed online. A high status, successful guild is a clear social symbol, akin to hanging out with the cool kids in a high school cafeteria, while belonging to a loathed guild can contribute to a player being shunned.

MMOs are predicated on social interaction and are designed to reward players for joining together with others. Raiding and guilds are two means by which people come together, but more important for this chapter, they are sites where *WoW*'s game design structures the identifications that happen within the game. Finding other people who share common interests and working together with them is a key part of *WoW*, which is at the core of how identifications are shaped within the game.

Identification, division, and consubstantiality

Video games in general and online games in particular offer an outstanding set of texts for the application of tools from communication studies. As Mäyrä (2009:

313) notes, 'much of the current academic work [in game studies] needs to rely on approaches and findings provided by and rooted in other academic fields'. Other game studies scholars have applied the elements of communication studies to the study of games, ranging from Mia Consalvo's use of communication theory to analyze elements of noise in *Final Fantasy XI* (Consalvo 2009) to applications of elements of rhetorical theory, like Ian Bogost's (2007) analysis of persuasive games, Gerald Voorhees's rhetorical criticism of the systems of representation in the *Final Fantasy* series (Voorhees 2009), or analysis of the reward system in *WoW* (Paul 2010). Although games studies scholars are beginning to employ tools from communication studies to analyze games, we believe the work of Kenneth Burke offers a particularly useful set of tools to understand how people interact within online games.

Central to Burke's writings on identification is a belief that identification is the means by which appeals unify the interests of different people (1950). Burke held that people are inherently divided from each other, and to deal with the feelings of loss stemming from our inherent divisions, we seek to identify with others. Effectively, identification is about 'finding a shared element between the speaker's point of view and the audience's, or finding the audience's point of view and the speaker's convincing them that they share a common element' (Rosenfeld 1969: 183). However, Burkean scholars have articulated how Burke's concept of identification applies beyond speaking and listening, as 'identification, in short, becomes as much a process and structure as a discrete perlocutionary act' (Jordan 2005: 269).

Identification is more than making a message more persuasive, and can be viewed as 'the dynamic social process by which identities are constructed, through which they guide us, and by which they order our world' (Scott *et al.* 1998: 306). What people identify with shapes how they encounter the world and the structures surrounding appeals how identification can be developed. This can be seen in Richard Crable's analysis of Dwight Eisenhower and how he was well received by the American public because of their feelings in the wake of Second World War. Crable argues that Eisenhower's success was largely dependent on him being able to present himself in a very particular way at a very particular time. This idea of timing and social context is a key part of how identifications are established. What we identify with shapes our world and our world shapes how we identify. Gaming, with overt design goals and clear rule sets, offers an outstanding platform for analyzing how identifications are formed.

An initial relationship between identification and game design can be seen in how people seek to overcome division in *WoW*. One way the impact of design can be analyzed is by studying how 'progression' is conceptualized by end-game characters and how the number of people required for high-level raids shapes the terms for raiding game play. If the most desirable ends could be obtained by a means other than raiding, it is unlikely that as many people would seek to come together with other players to raid. If Blizzard dictated a different number of players were permitted in raid groups, players would adjust to match those expectations.[4] The design of *WoW* by Blizzard programmers encourages people to come together to accomplish things in specific ways, shaping the terms on which players encounter each other.[5]

The notion of progression within *WoW* is straightforward until a character reaches the highest level in the game, as there is a linear movement from level to level, much like in many offline games. At the highest level, the game changes fundamentally, as players no longer have levels to gain and are faced with a choice to quit the game, start a new character, or seek to improve some aspect of their character not related to level. In this case, many players find their achievement in the game marked by 'getting better stuff' (Retteberg 2008: 24) and the best stuff is most likely acquired by working with groups of people, rather than by one's self. Blizzard made a meaningful decision to make the highest level relatively easy to obtain, but by doing so, the introduction of a 'much more intensely social game' that replaces 'earlier stages of the game where a large majority of time is spent alone' (Ducheneaut *et al.* 2006: 308) stands to be far more striking for players. Players have an open choice of whether or not to pursue 'better stuff,' but, should they do so, they are identifying not only with a desire for better equipment for their character but also a requirement to navigate the complicated waters that emerge in a game that suddenly requires social interaction.

Coming together as a guild or raid group offers a reason for unity, but allocation of loot, a scarce resource, can also pull the group apart. Burke argues that the complement to identification is our inherent division from others, writing

> In pure identification there would be no strife. Likewise, there would be no strife in absolute separateness, since opponents can join battle only through a mediatory ground that makes their communication possible, thus providing the first condition necessary for the interchange of blows. But put identification and division ambiguously together, so that you cannot know for certain just where one ends and the other begins, and you have the characteristic invitation to rhetoric.
>
> (Burke 1950: 25)

As players consolidate into guilds, they implicitly divide themselves from others, as 'identification always suggests a "we" and a "they"' (Cheney 1983). Further, within individual guilds, players may have different motivations and interests, subdividing them into smaller groups. As long as players are getting what they seek from a guild, it can maintain stability by reifying processes of identification, but those processes of identification necessarily entail divisions that can endanger the group. Often, for raiding guilds, identification is re-established through the successful act of raiding. The common cause and drive of players gives them something with which to identify. As a result, should a guild base its primary identification on raiding, it is necessary to keep raiding *successfully* to use raiding as a basis for identification. Although multiple identifications are possible, one primary link will generally take primacy over others. Further, Burke argues that a spirit of hierarchy drives humans, which can be seen when guilds encounter difficulty with individual players or in achieving new successes within the game, as there must be overarching, ongoing identifications for players to stay together, rather than naturally splitting apart. When identification is successfully established, it 'allows us to cope with the demands the organization [or guild] places on us and, on the other hand, pushes us to act in the best interests of the organization' (Barker and Tompkins 1994: 225). If players cease to identify with their guild, they may stop acting in the group's best interest or seek to join a group with whom they identify more strongly.

In a Burkean sense, to come together within the game, players need to find consubstantiality in the act of playing together. Players join together and when success is found, the high from the common accomplishment reinforces the bonds among members. As Burke (1950: 21) notes:

> A doctrine of *consubstantiality*, either explicit or implicit, may be necessary to any way of life. For substance, in the old philosophies, was an *act*; and a way of life is an *acting-together*; and in acting together, men [sic] have common sensations, concepts, images, ideas, attitudes that make them *consubstantial.*

Through the actions and practice of raiding, groups in *WoW* are able to overcome the divisions that threaten to pull them apart. Through a process of identification, players connect with each other and become 'substantially one' with players other than themselves. However, this connection only lasts as long as fellow players believe, or are persuaded to believe, that they share common interests. If guilds are unified in a common desire for things, whether those things are new gear, talking with one another, completing challenging tasks, or something altogether different, consubstantiality helps to keep the group together. Yet, when interests diverge or when members feel that they are not able to accomplish what they want within a guild, players will no longer feel have the common bond that can help them feel consubstantial. For many in raid guilds, it is the act of raiding and success in raiding that helps establish and sustain consubstantiality among players.

Given Burke's drive to identify how people are pulled apart and drawn together, *WoW* guilds and raiding offer a new means by which to see the connection between game studies and actions in a group setting. To see those connections at work in one particular case study, it is important to provide the back-story for the CTS.

Cardboard tube samurai: a history

Like many other guilds, CTS started out by forming around a common interest. Originally started by a group of Penny Arcade forum members, to gain an invitation to the guild, all one needed to do was post on a thread on those forums and then send a message to a designated person once in-game. Some members were active posters on the forums, others, like Christopher, occasionally lurked, but all generally sought a group of people who shared a common sense of humour as a starting point towards identifying in *WoW*.

Started on the same day *WoW* launched, CTS had a core of members with two clear things in common: an affinity for Penny Arcade and a desire to play *WoW*.

However, as CTS moved into the game itself, things changed for many people as subgroups developed within the guild and the guild had to adapt. Because people played for different amounts of time and with different aims while within the game, the guild began to separate out by level. Although there are incentives to group with others, the differentiation in level and general dynamics of the game led to people grouping only with those who logged similar amounts of game time, raising issues about what individual members were actually identifying with. The persistent world further complicated working with others, as it was necessary to find common times to play, which is made more difficult by differences in time zones and schedules outside of the game. Although the vast majority of members were from the United States, those outside of the country faced substantial difficulties in finding times to play when those in North American time zones were online. Even the 3 hours difference in United States time zones was a factor prohibiting play with others both early in the morning and late at night, when differences in time zones are most apparent. These factors led to separation within the group, as people found friends who played in a manner like their own. Repetition of these patterns meant that cliques formed within the guild, something that could only really be changed at the point when most members had something else in common: achievement of level 60, the maximum level at the time.

At level 60, many of the structural issues that separated guild members still remained, but the central issue of having a variety of guild mates in the same level range was finally resolved. As a result, more of the group could work together to accomplish tasks in-game. No longer seeking a higher level, players pursued alternate goals, many of which were centred around improving their equipment and upgrading the kinds of things their character could accomplish. Some of these tasks could be accomplished in groups of no more than 15; many could be done in groups as small as 5. Members of CTS pursued these avenues to better their characters, but quickly became tired of the limited options available to them as a small group. However, to accomplish more, CTS needed to add more people, as the number of people required for raiding was 40, more than twice as many as required for previous tasks.

There were two primary ways in which CTS could attract additional people: add more members to the guild or seek alliances with existing guilds so that all could achieve something they could not do on their own. Both approaches have drawbacks. The first option stands the risk of destabilizing the guild by adding too many people who do not have the same interests as original members. Furthermore, there are rarely sufficient numbers of people without an existing guild allegiance to add as many people as CTS needed at that point. Guilds may add people over time, but the idea of more than doubling the membership made adding enough people an impossible option at that point. As such, CTS opted to pursue alliances with other, similarly sized guilds so that all could raid. These endeavours were not without their faults, but the primary issue was the underlying temporariness of the agreements.

Alliances within *WoW* enable guilds to do things that they simply cannot do on their own. These alliances are tenuous, often with specific agreements as to how spots in the raid will be allocated to the various guilds and how the loot that 'drops' will be allocated. Both dynamics can cause tensions if one guild feels another is either getting too large a share of the loot or are not fully pulling their own weight in contributing to the success of the alliances. CTS first formed a series of short-lived agreements with other guilds that barely made it out of the initial phases of negotiation. The first successful alliance the guild formed was composed of two other guilds, Blackwing Mercenaries and For Khaz Modan (FKM), and lasted until anticipation of one guild or another breaking the alliance caused trust among the guilds to disintegrate and led to the departure of Blackwing Mercenaries. FKM and CTS stayed aligned, adding a new third guild, Stonewall Champions (SWC). As all three guilds grew, there was an increased competition for spots within the raid. This put each guild in a position where internal guild memberships were pressuring leadership to find a better option. As members of each guild became increasingly dissatisfied, problems were frequently blamed on the other guilds, effectively creating identification within the guilds by promoting division from the alliance. Increased membership made it apparent that CTS would choose one of the guilds to pursue a two-way alliance with and CTS eventually chose to align with SWC. The news struck FKM unaware and set a curious tone for the CTS/SWC alliance.

The alliance with SWC was founded on interesting ground. From the start, both guilds had elected to cut out a third guild on suspicious terms, which made all aware of what division can do to groups in online games. Further, both guilds were growing rapidly and were internally debating when they would start raiding alone, making the alliance temporary. The guilds also had different aspirations, with CTS seeking to expand the amount of time spent raiding and progress to new, unfaced enemies, and SWC favouring maintenance of the schedule from the previous arrangement. Just a couple of months after the alliance began, it dissolved, as disagreements among the officers of the two guilds led to a testy exchange that split the guilds on particularly bad terms. After this, both guilds raided on their own and achieved substantial success. Although the split was far from amicable, several members of SWC defected to CTS to swell our numbers to nearly 40 raiders per night, which put CTS in the new position of being able to raid alone.

Within days of breaking the alliance with SWC, CTS managed to match the alliance's raiding success. Shortly thereafter, we started actively raiding the next instance up in level of difficulty, Blackwing Lair (BWL). To have a better chance of success in BWL, CTS started actively recruiting additional members. As one of the few guilds looking to add members led to a flood of applications, many of whom were turned down, implicitly solidifying the identifications among those within the guild, but several applicants were added into the guild. The addition of new people enabled CTS to complete BWL in a matter of months, but also changed the makeup as the guild, as it was fundamentally shifted from Penny Arcade fans and others met on the server to those who chose to apply to a raiding guild.

Shortly after completing BWL, Blizzard announced that they were enabling players to transfer off of the server we had started on, Proudmoore, to a new server. In addition to two other raiding guilds, CTS opted to take advantage of the opportunity. Upon transfer, CTS entered a server where they were suddenly a much larger fish in a substantially smaller pond. The most significant change was the raiding rank of the guild relative to our server. On Proudmoore, CTS was fairly low in the rankings, as there were many raiding guilds. On Muradin, even though two of Proudmoore's best guilds came with us, CTS was ranked fourth overall. As the summer of 2006 approached, this higher ranking was quite important, as a number of people either took breaks from the game or had more offline plans that cut into their raiding time. As a result, CTS needed to add even more new members. Initially, adding members on Muradin was quite easy, as many of the transfers to the server came unaffiliated. However, many unaffiliates came that way to flee a bad reputation on their old server and quickly reverted to the actions that produce a questionable reputation,[6] while those of quality were quickly snatched up by guilds on Muradin. CTS invited their fair share of new members, but the extraordinarily small population of Muradin meant that, after the first recruiting burst, new recruits were hard to find.

While on Muradin, CTS encountered substantial successes. After completing BWL, CTS moved on to the next instance, The Temple of Ahn'Qiraj (AQ40). CTS experienced both highs and eventually lows because of their progress in AQ40. Two of the best moments for CTS, passing another guild to obtain third place on the guild rankings and defeating the Twin Emperors, one of the most difficult bosses in the game at the time, were highlights of our experience on Muradin. However, as summer struck and the guild leadership opted to continue focusing on AQ40, people stopped showing up for raids. Finding it increasingly difficult to fill raids, CTS was forced to increase recruiting. However, stuck on a low population server, it was almost impossible to conjure up the people the guild needed to fill raid slots.[7] As a result, raids started getting cancelled for lack of attendance, as simply having 36 or 37 people made it difficult to succeed in challenging encounters that were designed for 40 players. Tension built within the guild and accusations were made as to who was pulling their weight and who was not. Leadership tried to motivate through offering additional bonuses for raiding or threatening punishment for skipping raids to get people to show up, but all efforts were unsuccessful. Members regularly turned out for encounters where they were almost assured of winning battles, but for battles where they were learning how to win, people were scarce. Members called to increase recruitment, but there simply were not enough people available to fill the gaps in the raid group, resulting in substantial disagreements between a handful of established members who wanted to actively raid the most difficult encounters in the game and the guild-at-large. The members who felt the tension most acutely were our guild leaders at the time, Anoria and Antinous.

In late September of 2006, less than 3 weeks after our triumphant victory over the Twin Emperors, Anoria and Antinous chose to leave the guild late one night. Without discussing their decision openly with the guild or establishing a timeline for their resignation, posts were made on the web site of Muradin's top-ranked guild announcing their departure for CTS and application to a new guild. Immediately after their departure, three other members opted to leave CTS. Suddenly, a guild that was already shorthanded lost five of the people they needed to compose a raid. Within 3 weeks of the exodus, after losing additional bodies to attrition, CTS leadership announced that we would no longer make an attempt to raid. The decision was made to finally acknowledge the fact that we no longer had the bodies to compose a 40-person raid and that continuing to schedule time for them was laughable. After the decision came, several others left to join guilds that were actively raiding. A new, splinter guild was formed that was almost entirely composed of ex-CTS members. Finally, less than 10 days after the announcement that we would no longer raid the guild was disbanded. Guild assets were dispersed to former members and CTS was no more. Shortly after disbandment, former members restarted the guild to reserve the name, but CTS was a shell of its former self, with only a couple of people online at any given time, a far cry from the forty-plus in days of yore.

CTS is like many other guilds in *WoW*, but the most notable features of the guild's history are indicative of how groups can come together, thrive and fall apart. With a basic understanding of Burkean concepts and CTS's history, it is possible to investigate why the guild initially came together and fought through the difficult process of establishing a raiding nucleus, but how the shift in purpose of the guild and changes in the context of the game ensured our demise.

Burke and the rise and fall of cts

There are many interesting aspects of *WoW*; however, the one that may be the most interesting in a Burkean sense is the inherent division of players and how players seek to remedy division within the game. There are clear, structural elements in *WoW*'s design that encourage players to band together; but without establishing consubstantiality and maintaining identification, relationships are bound to fall apart. The history of CTS points to three key phases in how a cycle of identification and division demonstrates the importance of consubstantiality in the life of a guild. The key phases for CTS were the early period of the guild, as it was formed and as people levelled to 60; the early development of CTS as a member of a raiding alliance; and the subsequent redefinition of CTS as a raiding guild, which preceded its demise. Burke's theory of identification helps explain why CTS rose and would eventually fall apart, as members lost the basis for identifying with their fellow guild mates and raiding became their sole tie.

The first phase, the early period of CTS, is common to all players in the game, as they are beset with the challenge of levelling their character through a linear progression to level 60. Turning to others for help, or just for someone else with whom to talk while one is playing, offers a foundation for building guilds within the game. Since one of the primary benefits of being in a guild is a dedicated chat channel for members, guilds are generally predicated on some sort of commonality. CTS was no exception to this, as the guild was initially organized around on the Penny Arcade web site and based its name on a recurring character within the oeuvre of Penny Arcade comics. The singular barrier to entry, a post on a forum, offered a uniquely honed subset of *WoW* players to comprise the guild. Posting on a forum thread is unlikely to be a difficult task for the average *WoW* player, but knowing about Penny Arcade, frequenting their forums, having interest in *WoW*, and pursuing a Penny Arcade themed guild defined the early members of CTS. This gave all of them multiple points of common interest, ensuring that members had some traits in common and clear ways to foster identification.

Consubstantiality in the early version of CTS was predicated on the initial connection to Penny Arcade. The common link to the site offered up a wealth of material that provided bases upon which to identify. Members were pre-screened for a particular sense of humour and interest in gaming, which was reinforced by the fact that the guild was not advertised beyond the web site. Guild chat could be vulgar, as members were used to particular communication norms based on reading Penny Arcade and its forums. Interests in gaming beyond *WoW* typified many members, as Penny Arcade is a blog about gaming in general. The need to actually go to the forums of the site, which are not as highly trafficked as the main pages, and post, which requires registration, limited the number and type of person that would be in the guild. This pre-screening process meant that, regardless of their substantive interaction, all members had a clear reason to believe they shared something with every other person in the guild.

An additional way in which formation through Penny Arcade aided consubstantiality in CTS was the connection to a base of source material. As new material was produced on Penny Arcade, players would either talk about them in-game or subtly use them to joke with another player. Words and phrases from the most recent comics were often dropped into casual conversation and it was expected that all members would be able to decode the messages and connect them back to Penny Arcade. The guild used the universal password of 'wang' owing to its prevalence in the discourse on Penny Arcade. The expectation of members having a working knowledge of Penny Arcade demonstrated not only the importance of the site to the guild but also how the site played a role in drawing members together within the game. The game became a means by which Penny Arcade fans could connect with and talk to other Penny Arcade fans. Even though some felt a deeper connection to the site than others, it gave all members something in common above and beyond the fact that they played *Warcraft*. This gave members a way to connect, while also dividing them from others who played the game.

The second phase in the history of CTS, when guild members started reaching the level cap of 60 and subsequently sought raiding alliances, shifted the focus of member's game play. Although members had been stratified by level throughout CTS's existence, players at 60 no longer could level. More than anything, game design forced players to reach beyond the bounds of CTS to connect with other people, while using the guild as a location in which to joke with like-minded individuals ceased being enough to keep the guild together. The structure of *WoW* and the incentives to gather larger groups at 60 meant that guilds generally needed to get bigger to pursue goals in the game, even if that diluted the guild's original basis for identification. As getting new equipment required multiple other people, members often reached beyond the bounds of CTS to accomplish greater tasks. Occasionally, this meant that new people were added to the guild, as people inquired about CTS and subsequently sought to be included because they thought they would appreciate being part of the guild. However, the need to look outside of the guild to obtain the 40 people for raiding altered how the guild was structured. People sought to succeed, and because there was not a critical mass in the guild to succeed in acquiring equipment, people looked outside, to other guilds, to align with them in the search for things that could not be done in CTS as it was configured. Effectively, identification was different for each member, and as more members interested in raiding entered the guild, the original identification with Penny Arcade was largely lost. The name and some of the references remained the same, but each individual altered how identification could be fostered within the guild.

The lack of opportunity to progress as a group of likeminded individuals within a guild placed CTS in a tenuous position. There was a desire to align with others to accomplish more, but that desire was single-minded, with a focus on actively raiding. This changed how the group related, shifting from a focus on Penny Arcade and similar personalities, to one that was based on enabling CTS to raid, which would result in obtaining better equipment, additional notoriety, and new experiences within the game. These dynamics led to our series of short-lived alliances. The introduction of raiding alliances opened up new ground for identification and division, as there would frequently be multiple levels of conversation, as members followed the conversation of the raid group in one chat channel, the guild group in another, and messages to individuals in additional channels. This offered opportunities for CTS to talk openly about the shortcomings, or occasionally strengths of others, amongst themselves, furthering identification with guild mates, while fostering division from the alliance and emphasizing its impermanence.

The search for others in a position similar to ours was successful, but the very nature of the relationships precluded us from actually identifying with those in the other guilds. For CTS, raiding alliances were always temporary conditions, a thing to be endured until we could thrive on our own. Although we were engaging in activities together, offering the potential to develop consubstantiality, the fact that we were separated into different guilds built division into our relationships. In one sense, George Cheney's observation that 'names, labels, and titles become the foci for larger corporate identities; they carry with them other identifying 'baggage' in the form of values, interests, and the like' (Cheney 1983) should have indicated that the insistence on maintaining separate guild identifications was a problem. As we remained separate in at least one crucial way, we built structural divisions in to our relationships. In most cases, there was clear indication that people were a member of their guild first and the alliance second, as our baggage was never overcome with a larger, alliance-wide identity. Further, the common interest in slaying monsters was checked by the zero-sum game that was allocation of the equipment that we won. The tensions among guilds came to the fore whenever adversity was faced, from allocating how equipment and positions in the raid should be divided to the disciplining of players. As Burke (1950: 25) observes:

> Man's [*sic.*] moral growth is organized through properties, properties in goods, in services, in position or status, in citizenship, in reputation, in

identification beyond success in raiding or they will face substantial issues overcoming inherent division, which is a prerequisite for raiding success, in the long-run. As there can be only one top raid guild on a server and the members that constitute a guild have relatively free movement from guild to guild and server to server, an identity based solely on being an 'elite' guild will likely lead to dissolution if adversity is faced. Guild leaders can work with members to develop other bonds and ways of identifying that are less contingent on a sole connection or at the least, recognize that the sole connection can be an issue and work to minimize the problems that arise.

Designers face other issues, as the dissolution of a guild is less detrimental, but the departure of a player from a game, particularly if it marks a large exodus from the game, can be quite problematic. As a result, designers should pay attention to more macro-levels of identification and consubstantiality. This is likely most important at a game's launch, finding ways to bind players to the game quickly and then maintaining a hold on them throughout the game's lifecycle. There are many recent developments in online games that demonstrate the role of identification in-game design, from the multiple difficulty levels now implemented into *WoW* raiding that let more people participate in raids to the difficulty new MMOs have had in retaining subscribers in the face of *WoW*, as a game with many more years of live time simply has a larger, richer world within which to find ways to experience consubstantiality with other players and the game itself. Building in sophisticated communication tools that let people communicate and aggregate easily as well as including a wide variety of activities for players to pursue increase the odds that they will identify with the game and keep playing.

In the case of CTS, Burkean theories help explain why the guild grew, thrived, the fell apart. Burkean analysis is particularly well suited for the analysis of player behaviour in online games, especially in games where players are encouraged to work together. Burke can help explain what brings players together and what can drive them apart. In light of this analysis, we hope that more academics look to their home disciplines as a way of understanding the fascinating dynamics of online games.

Notes

1 Guilds are the most common social aggregation in massively multiplayer online games. In *World of Warcraft*, guilds enable groups of people to share a common chat channel and a guild bank within which to store goods and gold.
2 MMOs like *WoW* generally use separate servers to segregate the game's population into manageable groups. In the case of *WoW* each server has a name that somehow relates to the mythology of the game and many have distinct rule-sets, but, at their base, all are replicas of the imaginary world of Azeroth within which *WoW* is set. The Muradin server is a player-versus environment server named after Muradin Bronzebeard.
3 Penny Arcade is a web comic with a substantial and dedicated following. Measures of its popularity range from their ability to host the largest consumer focused video game conference, PAX, to their ability to raise millions for their charity, Child's Play. The site's forums offer a location for interaction for hundreds of readers and it was on these forums that CTS was originally founded.
4 This was borne out in the expansion to *World of Warcraft*, *World of Warcraft: The Burning*

Crusade, where the largest raid group was dropped from 40 to 25. This had a substantial impact on many guilds and on the raid experience at large as guilds needed to cut members to reduce their numbers for the smaller raid sizes. Some of the reaction from players can be found at: http://elitistjerks.com/f15/t11708-raid_sizes_future_wow_raiding/.

5 Players can choose to identify with something other than raiding or notions of gear progression, as Mortensen (2008) discusses in an analysis of 'deviant' play within *WoW*. However, all of these players are still choosing to identify with certain aspects of the game, like role-playing or exploring.

6 The most notable examples of questionable reputation were characters fleeing a reputation as a 'ninja looter,' a person who would take items that should have been given to other players. Ninja looting is one of the larger social transgressions in MMOs. Further reputation issues include personality issues, poor skill at playing the game, or erratic attendance for raids.

7 Joining a low population server was a boon for CTS, until we needed more people. Muradin was far more stable than our previous server and because we moved there with the numbers we needed, we thrived upon the move. Transferring to Muradin was a good decision, but once recruiting became an issue, the low population server simply did not have enough players for CTS to recover from the flagging identification issues.

8 The departure of our leaders in particular raises questions about with what guild members were identifying. Although we believe the fundamental issue with identification was the lack of numbers for raiding, if even a couple of those members who left shortly after the departure of Antinous and Anoria left because they identified with the leaders, then identification with the leaders was also an issue in the demise of the guild. At the very least, the departure of leadership was a warranting action, excusing the departure of anyone else who sought a more successful raiding experience.

References

Barker, J. R. and P. K. Tompkins (1994) Identification in the self-managing organization: characteristics of target and tenure. *Human Communication Research*, 21(2): 223–40.

Bazazu (2007) Status Symbols and Why You Play Wow', Elitist Jerks. Retrieved 14 September 2009, from http://elitistjerks.com/f15/t17374-status_symbols_why_you_play_wow/.

Bogost, I. (2007) *Persuasive Games: The Expressive Power of Vidogames*, Cambridge, MIT Press.

Boyd, J. (2004) Organizational rhetoric doomed to fail: R.J. Reynolds and the principle of the oxymoron. *Western Journal of Communication*, 68(1): 45–71.

Burke, K. (1950) *A Rhetoric of Motives*, Berkeley, Unviersity of California Press.

Castronova, E. (2005) *Synthetic Worlds: The Business and Culture of Online Games*, Chicago, University of Chicago Press.

Castronova, E. (2009) Synthetic worlds as experimental insturments. In B. Perron and M. J. P. Wolf. (eds) *The Video Game Theory Reader 2*. New York, Routledge.

Cheney, G. (1983) The rhetoric of identification and the study of organizational communication. *Quarterly Journal of Speech*, 69(2): 143–58.

Consalvo, M. (2009) Lag, language and lingo: Theorizing noise in online game spaces. In B. Perron and M. J. P. Wolf. (eds) *The Video Game Theory Reader 2*. New York, Routledge: 295–312.

Crable, R. E. (1977) Ike: identification, argument, and paradoxical appeal. *Quarterly Journal of Speech*, 65(2): 188–95.

Dibbell, J. (2006) *Play Money: Or How I Quit My Day Job and Made Millions Trading Virtual Loot*, New York, Basic Books.

Ducheneaut, N., N. Yee, E. Nickell and R. J. Moore (2006) Building an MMO with mass appeal. *Games and Culture*, 1(4): 281–317.

Jordan, J. (2005) Dell Hymes, Kenneth Burke's 'Identification,' and the birth of sociolinguistics. *Rhetoric Review*, 24(3): 264–79.

Juul, J. (2009) Fear of failing? The many meanings of difficulty in video games. In B. Perron and M. J. P. Wolf. (eds) *The Video Game Theory Reader 2*. New York, Routledge: 237–52.

Mäyrä, F. (2009) Getting into the game: Doing multidisciplinary game studies. In B. Perron and M. J. P. Wolf (eds). *The Video Game Theory Reader 2*. New York, Routledge: 313–29.

Mortensen, T. E. (2008) Humans playing world of warcraft: Or deviant strategies? In H. Corneliussen and J. Walker-Rettberg (eds). *Digital Culture, Play and Identity: A World of Warcraft Reader*. Cambridge, MIT Press: 203–24.

Nethaera (2008) Re: Correct Me If I'm Wrong. Solo Instances? Retrieved 14 September 2009, from http://forums.worldofwarcraft.com/thread.html?topicId=1230503915 andpageNo=2.

Paul, C. A. (2010) Welfare epics? The rhetoric of rewards in world of warcraft. *Games and Culture*, 5(2): 158–76.

Retteberg (2008) Corporate ideology in world of warcraft. In H. G. Corneliussen and J. Walker-Rettberg (eds). *Digital Culture, Play and Identity: A World of Warcraft Reader*. Cambridge, MIT Press: 19–38.

Rosenfeld, L. B. (1969) Set theory: key to the understanding of Kenneth Burke's use of the term 'Identification'. *Western Speech*, 33(3): 175–83.

Schramm, M. (2007) Exclusive Interview: Awake from Nihilum Speaks with Wow Insider', Wow Insider. Retrieved 14 September 2009, from http://www.wowinsider.com/2007/06/19/awake-from-nihilum-speaks-with-wow-insider.

Scott, C. R., S. R. Corman and G. Cheney (1998) Development of a structurational model of identification in the organization. *Communication Theory*, 8(3): 298–336.

Taylor, T. L. (2006) *Play between Worlds: Exploring Online Game Culture.*, Cambridge, MIT Press.

Voorhees, G. (2009) The character of difference: procedurality, rhetoric and roleplaying games. *Game Studies*, 9(2). Retrieved from http: //gamestudies.org/0902/articles/voorhees.

Williams, D., N. Ducheneaut, L. Xiong, Y. Zhang, N. Yee and E. Nickell (2006) From tree house to barracks. *Games and Culture*, 1(4): 338–61.

Yee, N. (2006) Motivations for playing in online games. *CyberPsychology and Behavior*, 9(6): 772–75.

12 Analyzing player communication in multi-player games

Anders Drachen

Communication is a key component of most contemporary multi-player games as portions of the game play are reliant on player communication either directly through verbal channels or indirectly through texted chat. Verbal communication between players of video games is, however, a subject that has received minimal attention from game studies as well as the media and communication research fields, whereas communication in, for instance, online discussion forums and other forms of social environment has been studied in detail (e.g. Smith 2009). Because of the lack of knowledge, it is not clear whether inter-player communication in video games is used mainly for *functional* purposes such as coordination, information sharing or negotiation of collective choices or *strategic* purposes such as ensuring that the players' individual objective game goals are met (Smith 2006). A third possibility is that players mainly communicate about issues that are game-external, thus utilizing the game as a backdrop for *social interaction*. There exists only a handful of publications focusing on this topic, and these are typically focused on the more general functions of language in gaming (e.g. Wright *et al.* 2002; Lazzaro 2004; Manninen 2004; Holmes and Pellegrini 2005), rather than analysis of player communication content. Empirical research is as mentioned previously mostly absent, with the few exceptions including Trappl (1997); Hutchinson (2003); Mateas (2003); Eladhari and Eindley (2004); Castronova (2005); Aylett *et al.* (2006). This is also the case for the most research carried out on massively multi-player online games (MMOs) from a social science, communications and ethnographic perspective.

Communication between members of a group is a topic that has been explored in contexts apart from video games and online social environments, for example, within collaborative design or in organizational and managerial contexts as well as for general communications research and even interactive storytelling (Chen *et al.* 2006; Aha *et al.* 2005). It has also been examined within the development and testing of virtual working environments to overcome geographical distances, in architecture, design and software engineering (Maher and Simoff 2000; Gabriel and Maher 2002; Sudweeks and Simoff 2005).

Given that communication between players in multi-player video games and MMOs is an important factor in the core game play of many of these games, it is perhaps surprising that this part of game design remains relatively unexplored and without substantial innovation in recent years. The topic is nonetheless important, not only to game studies – for example, understanding player communication (and player interaction more generally) aids the understanding of the social function and social rules of gaming. That is, how people play games and how the game-playing activity relates to other social activities in terms of norms governing the social interaction. It also aids game design – understanding what people need to communicate about is key to ensuring that this is facilitated in the design of multi-player and MMOs.

This chapter focuses on how to analyze the details of player communication in multi-player games, be they online or off-line. Communication between players has been studied from a top-down, general perspective in, for instance, sociological and ethnographical research; however, there is very little research available on the details of player communication, or how to evaluate this methodologically. This chapter therefore presents an empirically tested method for evaluating and analyzing player communication, based on communication analysis at the utterance level; the individual sentences of the players.

Results from three different sets of interconnected empirical studies are presented: (a) communication between players of a multi-player tabletop pen and paper (PnP) role-playing games (RPG), (b) player groups playing the computer RPG (CRPG) *Champions of Norrath* on a PlayStation 2 console (Smith 2006), and (c) groups of people playing the CRPG *Neverwinter Nights* on PCs, each with their individual monitors in a typical online/LAN setup. The aim of this study is to provide an analysis of the verbal communication of players in these three different multi-player gaming situations, focusing on the content and structure of communication. How players of these games utilize verbal communication to interact and play will be analyzed. RPGs are the perfect test-bed for a methodology focusing on analyzing player communication, as they include some of the most complex player interactions in contemporary digital games (Tychsen 2006). Working across different game media, or formats (e.g. digital and non-digital), permits the examination of the different roles that verbal language plays in the different game contexts. This provides an opportunity to study the effects on communication of transferring a game form between different media, both between the participants, and the participants and the fictional game world. It also assists in determining that functions of language are general across formats and those which are format-specific.

The chapter is based on results and experiences from long-term research projects on player communication, interaction and experience in a collaboration between several institutions, notably the IT University of Copenhagen, Macquarie University and Herriot-Watt University. The method presented has been developed as a part of these collaborations, and the results presented here are key examples extracted from this work, serving to show how communication analysis of game players can operate in practice. For further information on the previous research and results, see Tychsen *et al.* (2006) and Drachen and Smith (2008).

This chapter discusses perspectives on communication in video games in general and then examines the RPG genre specifically to clarify its internal variations as well as its relationship to other game genres. The methodology developed to capture and analyze players' communication is presented, results outlined and perspectives offered for how these results can serve as a basis for further research.

Theories of player communication

A range of hypotheses can be developed to describe what verbal communication it is expected players will use in video games. According to Salen and Zimmerman (2004: 60), a game is 'a system in which players engage in an artificial conflict, defined by rules, that result in a quantifiable outcome'. This is far from the only definition of what games are, but it highlights a problem that they do not address; what to expect in terms of player interaction, apart from noting that players are in a state of competition, with the aim of achieving the game goals. Assuming this is correct, it follows that the primary driver of player communication is to assist players with achieving these goals. It would therefore be expected that they only communicate to the extent that it assists with this achievement. Therefore, it would be expected that players freely give advice in cooperative games, negotiate appropriate behaviour in semi-cooperate games and remain mostly silent in competitive games – unless something is gained by communicating (Smith 2006). This approach fails to explain why communication between players of competitive games is frequent and commonplace (Holmes and Pellegrini 2005; Smith 2006). As suggested by these authors, the basic motivational driver of games – to win – is influenced by the fact that multi-player games also are social situations/frameworks as well as games. This led Drachen and Smith (2008) to formulate three theories for why players communicate:

The functionalist perspective: Communication between players serves as a tool for coordination, information-sharing and negotiation of appropriate behaviour. For instance, players of *Counter-Strike* will often need to agree on where to make their stand, may share information on enemy sightings and may discuss appropriate in-game behaviour (with both allies and opponents). In this perspective, communication is thought to be relatively tightly focused on the game itself (as opposed to game-external topics).

The strategic perspective: Communication between players serves as a tool for furthering the narrow, goal-oriented interests of the specific player. From this perspective, it would be expected that players do not share information, without getting something in return. In competitive games, players would not be expected to communicate beyond taunts and possible short-term alliances. As with the above perspective, communication would presumably be relatively tightly focused on the game itself.

The socializing perspective: Communication between players is only indirectly related to the game. From this perspective, the game itself may be seen as an activity around which players are social. In this context, it would be expected that player communication be unfocused and resemble conversation in nongaming contexts.

The three perspectives are not be mutually exclusive, but rather function as a means for understanding the variation between statements in one and the same game session; that is, one statement may be categorized as mainly functional while the next may be strategic, and so on.

RPGs across media

RPGs exist across a range of media and form one of the core game genres (King and Borland 2003; Lindley and Eladhari 2005; Tychsen *et al.* 2007a, 2007b), thus providing an ideal opportunity to investigate player communication, and how it is affected by the technology platform of the game in question. RPGs form the recurrent case throughout this chapter, and therefore, a brief introduction to these games and how they operate in terms of communication frameworks is presented.

There is an increasing amount of research publications and other resources focusing on the design of RPGs, and the gaming process, including Murray (1997), Hallford and Hallford (2001), Mackay (2001), Björk *et al.* (2001), Fine (2002), Tosca (2003), Bartle (2003), Lindley (2004), Peinado and Gervás (2004) and Tychsen (2006). Additionally, enthusiasts within the live action RPG (LARP) community and the tabletop RPG community have produced a number of useful publications that provide design- and practical knowledge about how these games operate, such as Padol (1996), Edwards (2001), Henry (2003), Kim (2003) and Bøckman and Hutchison (2005). Across the multitude of RPG forms, a number of key features are shared, including the reliance on rules systems, many basic concepts and the focus on character-centric play and often a storytelling-heavy element (Salen and Zimmerman 2004; Tychsen *et al.* 2006). The key differences between RPGs are related to the media they are expressed through and the hardware platform (PC or console, digital or non-digital) as well as the number of players involved (from one to thousands). The hardware platform/media affects the way that the fictional worlds these games take place in are presented and provide different sets of interaction and communication options to the players (Tychsen 2006).

Verbal communication, in theory, serves different functions across digital, non-digital and physical/embodied RPGs (LARPs). In PnPs, verbal communication is foundational to the game and is used to ensure that the state of the shared, imagined game world is consistent among the participants (Edwards 2001; Kim 2003; Young 2005a, 2005b). Participants need to verbally state what actions their fictional characters perform in the game world, and these normally require approved by a game master (GM), who controls the fictional world the player characters operate within, and update it according to the actions of the players while being mindful of the game storyline (Tychsen 2006). In terms of 'speech act theory' (Searle 1969; Austin 1975), the players of tabletop RPGs are performing a series of 'illocutionary acts' subject to the approval of the GM. Specifically, they are illocutionary acts by proxy since the player character performs them – a distinction that in the current context is irrelevant.

In contrast, the computer-created and -maintained fictional worlds of CRPGs and MMORPGs means that the important job of maintaining and updating the game-world state has been taken over by the game software. Additionally, CRPGs may restrict the communication channels between the players – unless they are physically present in the same room, they can only communicate verbally, not using body language. It is also possible that verbal communication is not even facilitated, and the players therefore forced to rely on texted chat. Finally, video games limit the freedom of the players to make choices for their characters – the game software sets the limits, not the imagination.

Case study: player communication in three RPGs

The experimental work performed to obtain raw data on player communication in the three RPG situations is described in detail by Drachen and Smith (2008); however, here the process is briefly described. The primary concern when running experimental studies of games is that they vary, and so do the players that utilize them. Therefore, three RPGs were chosen that were as similar as possible in terms of the underlying rules systems, narrative and linearity of the plot (the type of story may be an important driver of communication) and ease of play. Furthermore, the participants involved in the experiments represented both genders and a varying level of prior experience with RPGs.

The participants for the experiments were recruited at Danish and Australian universities and gaming communities. More than 50 players were involved in the experiments discussed here, with an age varying from 18 to 54 years, with two-third being male. The experience level of the players varied from non-existent to highly experienced (tested using a short survey, see Tychsen *et al.* 2009). Using RPGs with similar storylines and rules, and players with varying experience levels, provides a limited modus of context control and eases comparison of results across the three formats evaluated (non-digital, console and online/LAN).

Two video games (iCRPGs) and one tabletop game (PnP) were used in the experiments. The CRPGs were the Playstation 2-game *Champions of Norrath* (*Sony Online Entertainment* 2004) and the PC-based *Neverwinter Nights* (*Bioware* 2003) (Figure 12.1); the latter using a custom-made game module with a limited playtime, i.e. the entire storyline could be completed in one experiment. For *Champions of Norrath*, the introduction sequence was used, which presents a mini-narrative that is structurally similar to the *Neverwinter Nights* module used. Both of these multi-player CRPGs are relatively combat-heavy but with substantial amounts of NPC interaction and typical representatives of console RPGs and PC-RPGs, respectively: Both feature action-driven game play, a linear storyline and a generally collaborative environment and similar graphics. *Champions of Norrath*, as a console game, is played on a single monitor, with players sharing the one view of the game world, whereas *Neverwinter Nights* is played by multiple players using their own computers/monitors, either online or using a LAN network (local network). The PnP chosen for the experiments was a game module titled *What a Lovely War!* It utilizes similar rules as the two CRPGs and features a relatively linear, pre-planned storyline, which, however, is highly malleable as is normal for PnP play.

Figure 12.1 Screenshots from Champions of Norrath (top) (image © Sony Online Entertainment) and Neverwinter Nights (bottom) (image © Bioware and Atari).

Several game sessions were run with the three games with the participating players, and data from five of each type are included here. The same players could participate in different games but with at least a week in between game sessions. Group sizes varied from four to five, with one example of a two-player group in one of the *Champions of Norrath* sessions.

Champions of Norrath was played using a Playstation 2 and a 50″ plasma screen, whereas *Neverwinter Nights* was played using individual computers. The groups were at the beginning of each game session introduced to the game and game controls in general terms. The game sessions lasted approximately 45 minutes for *Champions of Norrath*, with the *Neverwinter Nights* groups playing through the custom module in 2–3 hours. The *What a Lovely War!* Groups played for 3–7 hours – of their own choice completing the game module quickly or at a more relaxed pace – highlighting the ability of this game form to adapt to the interests of the players. The PnP sessions were managed by one GM each, all highly experienced. They utilized the game module as a blueprint to run the game.

For all the game sessions, an observer was present at all times, either in the game room or in a neighbouring room, but did not interfere with the players beyond answering questions about game play or controls. The sessions were videotaped using two cameras, and verbal communication was recorded using wireless tabletop microphones.

Rather than transcribing the entire game sessions, three scenes were selected based on their narrative content and temporal placement in the individual games (beginning, middle and end) (Table 12.1).

This selection was made in order to be able to compare player communication as a function of the narrative content of the RPG being played. The completion time for each scene varied between the groups and the three game formats, from a few minutes in *Champions of Norrath* and *Neverwinter Nights* to over an hour in the PnP situation. The time transcribed for each game session was between 45 and 60 minutes, with a scene lasting about 20 minutes, except for *Champions of*

Table 12.1 Sections of the tabletop RPG and two CRPGs that were transcribed

	Scene type	*What a Lovely War!*	*Champions of Norrath*	*Neverwinter Nights*
Section 1	Non-stressful planning and prioritizing	Selecting and donning equipment (shopping)	First extended shopping scene	Module start, early NPC encounters
Section 2	Non-threatening combat scene	Initial fight against aliens	Initial confrontation against goblin invaders	Easy fights and NPC interaction
Section 3	Possibly dangerous (lethal) combat scene	Raid on an alien military base, fighting tougher aliens	First boss fight against goblin overlord and his pets	Final boss fight against major demon, story resolution

Source: Tychsen and Smith (2008).

Norrath, where the average time for each scene across the five groups was 13 minutes. The transcriptions were loaded into nVIVO 7.0 for coding-based analysis and an Active Perl script was used to extract information about communication intensity, sentence lengths, most typical words and so forth.

Capturing and analyzing player communication

A hierarchical coding scheme designed for RPGs is used, by which the verbal components of the communicative actions are categorized in a system of hierarchically structured codes, prior to statistical evaluation and analysis, following, for example Gabriel and Maher (2002). While the coding scheme presented here is developed for the RPG experiments, the process is generally applicable to communication studies of other games.

This approach is commonly referred to as 'protocol analysis' (Ericsson and Simon 1993), and is utilized within communication studies, architecture, design and so forth, with the purpose of analyzing the communication and behaviour of individuals or groups (Cross *et al.* 1996; Vera *et al.* 1998). The analysis is based on a protocol containing a recording of the behaviour and communication of the study subjects, and can be expressed in different ways such as computer logs, sketches, transcriptions, audio recordings and so forth (Akin 1986). The advantage of protocol analysis is that the sample sizes are large and statistical significance is obtainable – datasets can be analyzed in terms of raw numbers of different types of categories, direct comparisons and statistics.

Protocol analysis requires a framework for analysis, for example, based on a hierarchy of 'utterance codes', developed for the specific situation. The approach is usually iterative, based on an initial, theory- or experience-based coding hierarchy, where transcriptions are analyzed using the hierarchy, which is updated and changed as the material is analyzed. With each iteration, the coding hierarchy is refined until a saturation level is reached where no new codes appear, or no new redefinitions of codes are necessary, and the analysis framework is stable. Following this step, the entire set of recorded material is analyzed using the developed hierarchy.

For the RPG experiments, the coding hierarchy was developed deductively from theory and RPG models, and then in an inductive manner from categories that arose from an initial test coding of the transcribed game sessions, as described above (Drachen and Smith 2008). As the system reached maturity, the entire set of recorded data (transcriptions) was coded using the coding hierarchy.

Definitions

In this chapter, standard communications theory terminology is employed; however, a few key terms bear definition, as their meanings can vary in the literature:

Statements and utterances: The basic unit of coding is an *utterance*, defined as consisting of (a) a subject who performs the communication, (b) the content of the communication, and (c) an object(s) to which the communication is addressed (Gabriel and Maher 2002). A *statement* is the entire verbally expressed communication by a participant (such as potentially several utterances, until another participant starts talking). Basing the analysis of player, communication on utterances has the advantages of precision and resolution; however, due to the fine scale that utterances represent, the context of communication can be lost. In essence, there is a difference between looking at each line in a book in turn and the entire volume or individual pages. It is therefore important to consider the context – the broader situation occurring in the games under analysis, when making conclusions based on utterance-level protocol analysis.

In-game and out-of-game: An utterance can be given 'in-game' (IG) or 'out-of-game' (OOG). An utterance that is IG relates to the game or game content – for example, a rules question. An OOG utterance has content that is unrelated to the gaming activity – for example, asking a question about real-world politics.

In-character and out-of-character: an utterance can be 'in character' (IC) or 'out of character' (OOC). IC utterances happen in RPGs when players describe actions from the perspective of their characters (from either a first- or third-person perspective), or speak as their characters would (i.e., enact, embody their characters). Examples include 'I lift the rock' (first-person), 'my character lifts the rock' (third-person character action description (CAD)) or directly 'Let us see if the rock can be lifted' (enactment). An OOC statement could, for example, be related to a question about the rules of the game or interface questions such as 'How much damage does a thrown rock do?'. The three examples are slightly different; however, they involve a player directly describing a character action or communication. OOC utterance could, for example, be rules questions or comments about the interface with the virtual world in a CRPG. An OOC utterance can be both IG and OOG, whereas IC utterances cannot be OOG.

Coding hierarchy

The coding hierarchy is in this case designed to work at the level of utterances; however, protocol analysis does not specify this level of detail must be used – other forms of text segmentation can also be used. In the current work, the aim was to analyze player communication at the detailed level. The codes that were predefined (specifying the speaker and to whom the person is speaking to, whether the person is communicating IG or OOG and so forth) plus those that emerged during iterative development (mainly content codes, dramatic language and narrative progression codes) were combined into a hierarchy, which was integrated into *nVIVO* (a similar package for analyzing qualitative data could also be used such as *Transana*) (Figure 12.2). Within the hierarchy, sub-divisions cover the information sought attached to each specific utterance, such as who is speaking, what is the content of the utterance, to whom are they speaking, where in the game storyline is this taking place, is the speaker IC or OOC and so forth. During the

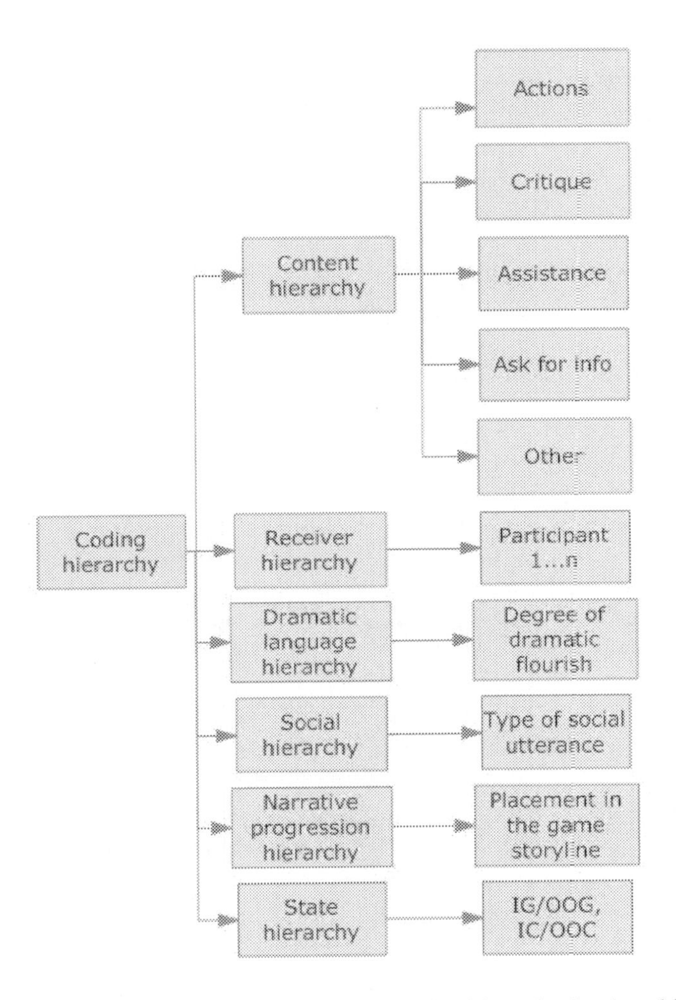

Figure 12.2 The top levels of the communication coding hierarchy developed for the analysis of player communication in multi-player RPGs.

coding process, each utterance is coded in terms of each sub-hierarchy. Note that utterances can have several components requiring several codes within the same hierarchy; for instance, a player describing both a character action and providing an environmental description at the same time.

Within protocol analysis, hierarchy codes are usually given as abbreviations. In this case, it proved more manageable in practice to utilize short descriptions of the content code categories than three- or four-letter abbreviations. Most commercial coding packages support this approach by auto-assigning higher level codes at the same time that lower-level codes are added. This also means hierarchies can be complex without prolonging the coding procedure.

Content hierarchy

Content codes form the bulk of the framework, defining the actual content of what the players are communicating, such as whether it is describing a character action, asking for help, critiquing the game or similar. The content codes are divided into five overall groups, depending on their overall purpose: assistance, critique, ask for info, actions and other (see Figure 12.1 and Table 12.2). Communication that can be content-coded relates to the game process (story or game play). The content code hierarchy can potentially be further sub-divided almost indefinitely to greater levels of detail.

Receiver hierarchy

Receiver codes define the players to whom the utterance is directed. This information is important for social network analysis. Utterances can either be directed at one other player or up to the entire group. Apart from shedding light about whether players communicate mostly in pairs or with the entire group, tracking the communication to and from the individual players provides the means for evaluating if variations and patterns are consistent across the involved groups of players, and not the result of one or a few participants acting in an unusual manner.

Table 12.2 The coding hierarchy: content code categories

Sub-hierarchy	*Code*	*Description/Example*
Assistance	Ask for advice	Asking for advice of any kind
	Ask for help	Asking for help of any kind
	Give info spontaneously or non-spontaneously	Giving information spontaneously or non-spontaneously (upon request)
	Give advice spontaneously or non-spontaneously	Giving advice spontaneously or non-spontaneously (upon request)
	Give help spontaneously or non-spontaneously	Giving help spontaneously or non-spontaneously (upon request)
	Confirm info/advice/help	Confirming the receipt of information/advice/help
Critique	Critique of self	Critique of oneself, either player or character Typically critique of an action
	Critique of one other person	Critique of another player or character Typically critique of an action
	Critique of group	Critique of a group, clan or faction of people, either players or characters. Typically critique of an action
	Critique of NPC/NPC group	Critique of an NPC or group of NPCs
	Game critique	Critique of a feature or aspect of the game in question
Actions	CAD	Describing the action of one or more player characters
	Order action (followed)	Ordering one or more other player characters to perform specific actions

world, relaying of information about objects found and lost, descriptions of the status of the player characters and so on, for example, in the following transcription snippet (player numbers refer to the numbers they were assigned in the group the transcript snippets are from, transcripts from both CRPGs):

'*Player 1:* *There's another corpse!*
Player 2: *Is there?*
Player 3: *I found one speared.*
Player 4: *Oh yes. I'll examine corpse. Are you examining it?*
Player 1: *Oh you can, sorry.*
Player 4: *That's all right. No, we can both examine it, it's fine.*
Player 2: *There's multiple corpses, we could all examine the corpse.*'

The constant flow of information between the players emphasize and illustrate the cooperative atmosphere of the *Champions of Norrath* game sessions, although it is important to note that some instances of *Give info* in fact may be interpreted as requests for help, such as:

'*Player 1:* *Uh, I almost burned up*'

In these cases, players are both providing factual information of importance to the group; however, the player may also be trying to elicit help from his/her teammates, for example, in the form of healing or better tactical cover in the future.

Apart from the underlying necessity of providing tactical information to each other, the prevalence of the *Give info* code is also explained by the many possible uses of providing information. A similar explanation can be attached to the high frequency of the *Ask for info* code (19.8 percent average, varies from 9 to 28 percent). This content code covers a variety of player communication, from general-level tactical questions such as:

'*Player 3:* *Okay, shall we progress or what?*
Player 1: *What is this next to this rock?*
Player 2: *No. What's on this side of the map?*'

The *Ask for info* content code also encompasses specific inquiries about game mechanics, for example, about the capabilities of items or other player characters, such as:

'*Player 2:* *Can't you cast healing magic?*
Player 3: *Can I walk through water?*'

The *Ask for info* code forms a secondary but significant part of the functional part of the in-game communication in multi-player CRPGs, reflecting the general level of collaboration evident in the participating player groups. Generally, the players were mutually helpful and willing to answer each other's questions. This includes situations where experienced and novice players operated in the same groups. Since *Champions of Norrath* and *Neverwinter Nights* pose considerable challenges of coordination (for instance, the players must continuously move in the same direction), players often proposed that others (or the whole group) act in a certain way, for instance,

'Player 1: Should we regroup? We should probably regroup.
Player 3: Let's regroup outside the pub.'

The two less common codes, *Give advice* and *Game critique*, cover expressed beliefs and advice rather than factual information, and comparisons with other games/critique of the game being played. The latter was much more common in both the CRPG contexts than in the PnP games sessions: *Game critique*-based utterances take up a few percent of the total verbal conversation (1–8 percent), whereas none of the critique-based utterances reach above 1 percent frequency in the PnP game sessions.

It is here speculated that this difference relates to the increased demands for real-time tactical coordination between the players in *Champions of Norrath* and *Neverwinter Nights*, as compared to the PnP situation, where the participants can effectively 'freeze' the events in the game world to discuss their tactics and approaches.

In the PnP games, the frequency distribution of the content codes in the coding hierarchy is very different from the CRPG games. The most commonly used content codes in *What a Lovely War!* were *Environment-world description* (18.4 percent average across the PnP sessions), *CAD* (17.5 percent), *Ask for info* (16 percent), *Social communication* (13.1 percent) and *Give info* (8.5 percent). Lower end has *Suggest-request action* (5.2 percent), *Order action* (4.6 percent), *Die roll result* (4 percent) and *Acknowledgement* (3.3 percent).

The initial observation is that there appears to be a broader distribution in the content codes being utilized by the PnP players; however, for both the CRPG and PnP situations, 12 content codes reached more than 1 percent frequency.

The pattern in the code frequencies are not immediately surprising – given that PnPs are based on shared, imagined game-world realities, frequent descriptions of the state of the game world are necessary in order to keep the participants up-to-date. Similarly, given that players need to state the actions of their character in order to perform actions in the game world, it is expected that this content code is frequent. The GM in PnPs spend almost 50 percent of his/her verbal communication on environment-world descriptions, and given that GMs make up roughly a third of the communication in these games, this fits well with their role as 'story managers'. For example:

GM: None of the other ships are setting down soldiers in the same area. So you're basically on your own. As you start walking towards the underground area you can see blasts coming down from the sky, it's a bit like fireworks, just the other way around. And whenever one of these blasts hit it goes (makes impact sound) (. . .).

GM: All right, you look a bit around and about five hundred metres away from where they were supposed to be, there is some sort of a metal trap door that you can open. It looks like an entrance into a sort of bunker system.

GM: *(. . .) Since everything that you can see out into the horizon is an artillery barrage, and the ground is still shaking under you, anything that has been alive is not alive anymore.*

CADs were commonly utilized by the players to describe the behaviour of their characters, for example:

'Player 2: I sneak to the edge to get the peak down ..
Player 2: Yeah, I'm following second so I can get a good camera shot.'

CADs can be used either directly ('I do this..', 'my character does this…') or indirectly ('let's all go down there . . .').

Ask for info and *Give info* are, similarly to the CRPG context, fairly frequent codes and reflect the need for exchanging functional information (i.e. game relevant) between the players. Players have a constant need for information and for providing information about the different elements of the game, their characters and the state of the game-world. Perhaps surprisingly, the *Give info* content code is much less frequent in the PnP context as compared to the CRPG context – even if the *CAD* codes is added (a form of information provision) – 41 percent average in CRPG compared to 26 percent for the PnP situation. However, if the *Environment-world description* code is added, which is arguably mostly used by the GM, the total for PnPs is 44.4 percent, a number comparable to the CRPG situation. Take some examples of information sharing in PnPs:

'Player 5: What sorts of weapons do they have?
GM: Most of them have little pistols in their giant hands (demonstrates little pistols), or they have rifles (demonstrates holding a rifle). And so you know, a long rifle, sort of like (demonstrates holding a large rifle).'

Finally, the players in the PnP sessions utilized substantial verbal resources on communication that had nothing to do with the actual game –13.1 percent of the communication was tagged as 'social' and without direct, functional game value. In comparison, non-functional communication in the CRPG game sessions was virtually non-existent.

On the surface, this difference does not appear to be related directly to the game format; however, it is important to note that in the PnP context, players have free visible access to each other – the game is based on direct player interaction. In both the examined CRPG contexts, players were somewhat mediated by a screen. In the *Champions of Norrath* situation, players needed to turn away from the screen to face their co-players, in *Neverwinter Nights*, players interacted fully via the monitors. This does not fully explain the difference in the need of players in the PnP context to chat socially; however, it should be noted that doing so is probably easier when you are sitting physically in front of the people you are talking to, with no monitor to distract your attention. PnP players will not only converse to a higher degree in-character but also carries the inherent encouragement to have informal chats as well. Possibly, this is a method for players to immerse themselves in their characters and gather information about the other characters (Tychsen *et al.* 2006). The PnP format also apparently promotes dialogue between individual players, as 40–50 percent of the total communication in the PnP context was aimed at the entire group, with more than 60 percent of the communication in the CRPG context being directed at the entire group.

While there is some variation between the five PnP sessions, the frequency of occurrence of the five most common content codes is fairly homogenous, varying a maximum of five percent between game sessions. In comparison, the variance in the CRPG context can be more than 20 percent (such as, for the *Ask for info* content code) – this indicates that there are direct differences in the variance of player communication in non-digital as compared with digital multi-player RPGs.

The relatively homogenous pattern of the most common content codes across the analyzed game sessions is not reflected in the more infrequent codes (i.e. *Suggest-request action, Order action, Die roll result* and *Acknowledgement*). For example, communication about rules (*Die roll* result and *Rule-interface-technical* comment) varied from virtually zero to 10 percent across the game sessions. This difference possibly relates to the relative greater freedom of PnP groups to define their playstyles relative to CRPG players, such as, in how they manage rules and how much the game rules should control the game. It was, for example, observed that some groups use rules to support the game-story and social interaction, while others used the game rules more frequently. The degree of variation indicates that the playing styles of the PnP groups varied substantially, such as, in terms of acknowledging the help or advice received by others, or the organization of character actions.

Communication structure

Using the Active Perl script, a series of descriptive statistical indicators were calculated for the three experimental conditions in order to evaluate the communication intensity (Table 12.3 and Figure 12.3).

These statistics show markedly lower communication intensity in the *Neverwinter Nights* game sessions: At 37.25 characters per utterance and just 546.12 characters per minute, the *Neverwinter Nights* players used the shortest overall utterance lengths and had the lowest average communication density (Figure 12.2). The maximal

Table 12.3 General statistics for the game sessions

Game	*Words/utterance*	*Characters/ utterance*	*Words/ minute*	*Characters/ minute*	*Max utterance*
What a Lovely War!	10.2	50.5	176	894.4	174 words 940 characters
Champions of Norrath	9.8	82.8	164.8	1314.4	54 words 499 characters
Neverwinter Nights	7.2	37.25	105.1	546.1	112 words 877 characters

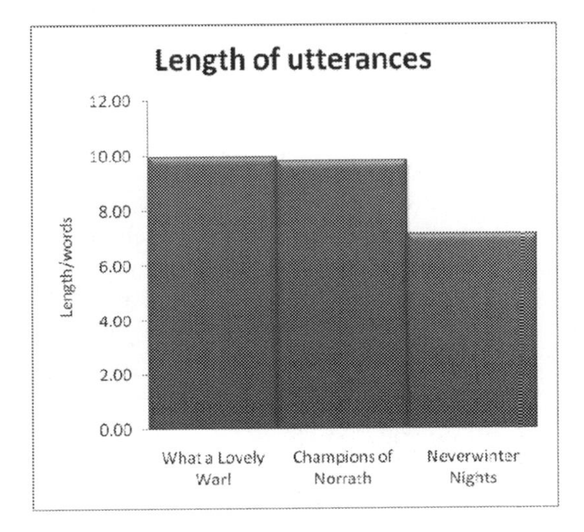

Figure 12.3 Average length in words of utterances in the three game contexts.

transcribed sentence lengths were highest for the tabletop RPG game sessions, in terms of numbers of character and words. Standard deviations of the utterance lengths are also highest for this format.

When it comes to the three scenes that were transcribed individually, the variation in the average words-per-minute (WPM) as a function of the specific scene varied minimally in the PnP game sessions – players communicated with almost equal intensity (varies from 151.1 to 225.1 WPM across all groups) and there was no statistically valid difference in the communication intensity as a function of the narrative content of the game. This lack of a difference could be explained by the fact that all events in a PnP take place through verbal communication, and therefore there is no immediate reason for why specific narrative content should stand out – there is the same need to communicate CADs, world state changes, request information and so forth. Given the lack of a visual interface, players have fairly constant needs in terms of the important content codes. Despite the fact that players were visually agitated during these intense scenes, and in their body language showed this readily, they did not increase their rate of communication.

In comparison, there was a substantial amount of variation in the average WPM for both CRPG situations (Figure 12.4). There appears to be a clear tendency for the players in the CRPGs to communicate the most during the initial shopping/NPC encounter scene (Scene 1), and least during the lethal combat scenes at the end of the gaming sessions (Scene 3).

A potential explanation for this variation is that the players of CRPGs are unburdened cognitively, and under no specific time pressure, during the initial shopping/social sequences, where in the lethal combat scene (Scene 3) they are required to concentrate on the task at hand to stay alive. This conclusion appears

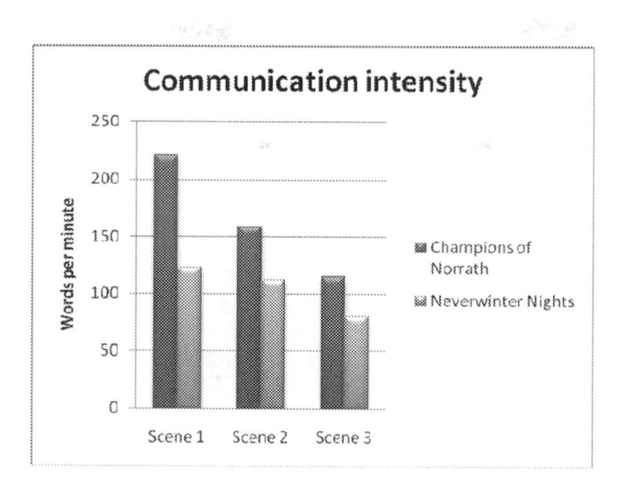

Figure 12.4 Communication intensity as a function of the narrative content in the two CRPGs.

to be backed up by the middling values of Scene 2, where the stress applied on the players is not as substantial as in the third scene.

On a final note, it was tested during the analysis to determine if there was a significant correlation between the experience of the participant, gender or Danish/Australian background (Pearson's correlation coefficient), but there was found no statistically significant correlation between these demographic features and the results. Similarly, there was no apparent pattern of difference caused by the variation in group size – whether there are four or five players in a multi-player RPG does not appear to impact on communication intensities (Drachen and Smith 2008).

Discussion

In this chapter, the various theoretical reasons for player communication in multiplayer (and MMO) games have been described, and a method is presented for how to evaluate player communication across digital and non-digital, off-line and online formats. Furthermore, selected results from a series of experiments applying the method (Tychsen *et al.* 2006; Drachen and Smith 2008; Tychsen and Smith 2008) have been presented to show how the coding matrix works in practice and the types of results that can be achieved.

Player communication remains an under-explored area of game studies, lagging behind the study of communication in, for example, Internet-based social environments. The case study presented here emphasizes the importance of game format on player communication – in essence, whether a game is digital or non-digital has a potential significant impact on verbal communication, and it is therefore dangerous to directly apply knowledge from traditional communication studies directly to the digital games domain. The results also highlight the variety of reasons for why players communicate, essentially focusing on strategic, social and functionalist perspectives at the same time.

A number of conclusions can be drawn from the experimental data:

First, the communication intensity in the three games analyzed is lower than the rate for 'normal' human conversation, which generally ranges from 190 to 230 WPM (30). This may be the result of the cognitive pressure being applied to players in a gaming situation – in essence, they need to divide their attention between the game and communicating with the other players.

Second, the game format – digital, non-digital, console-based or online/LAN setup – has a direct impact on the structure of the communication between the players. The intensity (speed) of communication as well as the length of utterances varied between the three game situations, although most profoundly between the *Neverwinter Nights* game and the two other RPGs (WPM rate 75 percent lower on average as compared to the PnP sessions, and 50 percent lower as compared to *Champions of Norrath*). In other words, the division in communication intensity does not happen across the change from non-digital to digital formats, but between the off-line and online/LAN format: In *Neverwinter Nights*, each player has his/her own computer, monitor, headset, and so forth – this prompts interaction through the game to a degree that is higher than in the other two gaming contexts, with the result of lowering the total amount of communication.

Other variables may impact on this relationship such as human factors; however, it should be noted that the experiments were run with a diverse set of participants. These results indicate that game designers should be careful when designing online games that require a high degree of verbal communication and/or coordination between the players.

Third, players generally communicate to the entire group of players, irrespective of the game format, although there is a tendency for relatively more communication between a subset of the group of players in the tabletop context. This may be the result of the multi-player games in question promoting collaborate interaction where players work together to achieve their goals.

Fourth, the amount of verbal communication apparently varies as a function of narrative content in computer-based RPGs, irrespective of whether these are console-based or online/LAN-based. The communication intensity was the most intense during periods of low stress in the game play, gradually dropping as the games became more intensive and stressful. This indicates that low-speech intensities in online games do not necessarily signify boredom, but rather that an increase in the cognitive load imposed by the game. It is possible that communication patterns among players are related to the experience of game content, which would mean that player communication analysis could be used to track user experience in games. The relationship between communication and user experience is not simple (a range of game- and player-related variables could have an impact), and further research is needed to fully explore this relationship.

Finally, it is worth noting that the coding framework described in this chapter was designed to be applicable to communication by participants in all forms of RPGs, and has also been tested on MMORPG chat logs. While the hierarchy is robust to RPGs, and it translates well to other types of multi-player games as well, additional utterance codes are usually necessary in these cases – for example, for specific action in a shooter-type game. As a method for analyzing player communication, protocol analysis is a detailed – if slightly cumbersome – approach.

Acknowledgements

The research presented in this chapter is part of the result of an international research collaboration running since 2005 between the author and Dr Michael Hitchens, Dr Jonas Heide Smith, Dr Susana Tosca, Dr Sandy Louchart, Dr Ruth Aylett and others. The author gratefully acknowledges the contributions of these esteemed colleagues in the continued work.

References

Aha, D. W., D. McSherry and Q. Yang (2005) Advances in conversational case-based reasoning. *Knowledge Engineering Review,* 20(3): 247–54.

Akin, Ö. (1986) *Psychology of Architectural Design*, London, Pion.

Austin, J. L. (1975) *How to Do Things with Words*, London, Clarendon Press.

Aylett, R., S. Louchart, J. Dias, A. Paiva, M. Vala, S. Woods and L. Hall (2006) Unscripted narrative for affectively driven characters. *IEEE Computer Graphics and Applications,* 26(3): 42–52.

Bartle, R. (2003) *Designing Virtual Worlds*, London, Pearson.

Björk, S., J. Falk, R. Hansson and P. Ljungstr (2001) Pirates! Using the Physical World as a Game Board. *Proceedings of INTERACT 2001*. Tokyo, Japan.

Bøckman, P. and R. Hutchinson (2005) Dissecting Larp. *Collected Papers for Knutepunkt 2005 – the 9th Annual Nordic Conference on Larp*, Tampere, Finland.

Castronova, E. (2005) *Synthetic Worlds: The Business and Culture of Online Games*, Chicago, University of Chicago Press.

Chen, L., M. Harper, A. Franklin, T. Rose, I. Kimbara, Z. Huang and F. Quek (2006) A multimodal analysis of floor control in meetings. In S. Renals, S. Bengio and J. Fiscus. (eds) *Machine Learning for Multimodal Interaction*, Berlin/Heidelberg, Springer: 36–49.

Cross, N., H. Christiaans and K. Dorst (1996) *Analysing Design Activity*, New York, Wiley.

Drachen, A. and J. H. Smith (2008) Player talk: The functions of communication in multplayer role-playing games. *Computers in Entertainment,* 6(4): 1–36.

Edwards, R. (2001) Gns and Other Matters of Role-Playing Theory. Retrieved 24 September 2009, from http://www.indie-rpgs.com/articles/1.

Eladhari, M. and C. Lindley (2004) Story Construction and Expressive Agents in Virtual Game Worlds. *Proceedings of Other Players Conference*, Copenhagen, Denmark.

Ericsson, K. A. and H. A. Simon (1993) *Protocol Analysis: Verbal Reports as Data*, Cambridge, MIT Press.

Fine, G. A. (2002) *Share Fantasy: Role-Playing Games as Social Worlds*, Chicago, University of Chicago Press.

Gabriel, G. C. and M. L. Maher (2002) Coding and modelling communication in architectural collaborative design. *Automation in Construction,* 11: 199–211. Retrieved from http://www.ingentaconnect.com/content/els/09265805/2002/00000011/00000002/art00098.
http://dx.doi.org/10.1016/S0926-5805(00)00098-4.

Hallford, N. and J. Hallford (2001) *Swords and Circuitry: A Designer's Guide to Computer Role Playing Games*, Roseville, Prima Tech.

Henry, L. (2003) Group Narration: Power, Information and Play in Role-Playing Games. Retrieved 29 September 2009, from http://www.darkshire.net/jhkim/rpg/theory/liz-paper-2003/.

Holmes, R. M. and A. D. Pellegrini (2005) Children's social behavior during video game Play. In J. Raessens and J. Goldstein. (eds) *Handbook of Computer Game Studies*. Cambridge, MIT Press: 133–144.

Hutchinson, A. (2003) Analyzing the Performance of Interactive Narrative. *Proceedings of the Fifth International Digital Arts and Culture Conference*, Melbourne, Australia.

Kim, J. (2003) Story and Narrative Paradigms in Role-Playing Games. Retrieved 29 September 2009, from http://www.darkshire.net /jhkim/rpg/theory/narrative/paradigms.html.

King, B. and J. Borland (2003) *Dungeons and Dreamers: The Rise of Computer Game Culture from Geek to Chic*, London, McGraw-Hill.

Lazzaro, N. (2004) Why We Play Games: Four Keys to More Emotion in Player Experiences. *Game Developers Conference 2004*, San Jose.

Lindley, C. (2004) Trans-Reality Gaming. *Proceedings of Second Annual International Workshop in Computer Game Design and Technology*. Liverpool, UK.

Lindley, C. and M. Eladhari (2005) Narrative Structure in Tran-Reality Role-Playing Games: Integrating Story Construction Form Live Action, Table Top and Computer-Based Role-Playing Games. *Proceedings of DIGRA 2005*, Vancouver Canada.

Mackay, D. (2001) *The Fantasy Role-Playing Game: A New Performing Art*, Jefferson, McFarland and Co.

Maher, M. L. and S. Simoff (2000) Collaborativly Desigining with the Design. *Proceedings of the Co-Desiging Conference*, Coventry, UK.

Manninen, T. (2004) *Rich Interaction Model for Game and Virtual Environment Design'*, Oulu, Department of Informatin Processing Science, Oulu University.

Mateas, M. (2003) Expressive Ai: games and artificial intelligence. *Proceedings of International DiGRA Conference*.

Murray, J. H. (1997) *Hamlet on the Holodeck: The Future of Narrative in Cyberspace*, Cambridge, MIT Press.

Padol, L. (1996) Playing Stories, Telling Games: Collaborative Storytelling in Role-Playing Games. Retrieved 29 September 2009 from http://www.recappub.com/games.html.

Peinado, F. and P. Gervás (2004) Transferring Game Mastering Laws to Interactive Digital Storytelling. *Proceedings of the International Conferece on Technologies for Interactive Digital Storytelling and Entertainment*, Darmstadt, Germany.

Salen, K. and E. Zimmerman (2004) *Rules of Play: Game Design Fundamentals*, London, MIT Press.

Searle, J. R. (1969) *Speech Acts: An Essay in the Philosophy of Languge*, Cambridge, Cambridge University Press.

Smith, J. H. (2006) *Plans and Purposes: How Videogame Goals Shape Player Behaviour*, Copenhagen, IT University of Copenhagen (Unpublished PhD Thesis).

Smith, M. J. (2009) *Online Communication: Linking Technology, Identity and Culture*, Mahwah, LEA Publishers.

Sudweeks, F. and S. J. Simoff (2005) Leading Conversations: Communication Behaviours of Emergent Leaders in Virtual Teams. *Proceedings of the 38th Annual Hawaii International Conference on System Sciences (HICSS'05)*. Big Island, Hawaii.

Tosca, S. P. (2003) The Quest Problem in Computer Games. *Proceedings of the International Conferece on Technologies for Interactive Digital Storytelling and Entertainment*, Darmstadt, Germany.

Trappl, R. (1997) *Creating Personalities for Synthetic Actors: Current Issues and Some Perspectives*, Springer.

Tychsen, A. (2006) Role Playing Games: Comparative Analysis Across two Media Platforms. *Proceedings of the 3rd Australasian conference on Interactive entertainment*, Perth, Australia.

Tychsen, A. and Heide Smith, J. (2008) Game format effects on communication in multiplayer games.*Proceedings of FUTURE PLAY 2008*, Toronto, Canada.

Tychsen, A., M. Hitchens, R. Aylett and S. Louchart (2009) Modelling Game Master based Story Facilitation in Multi-Player Role-Playing Games. *Proceedings of the 2009 AAAI Symposium on Intelligent Narrative Technologies II*, Stanford, USA.

Tychsen, A., M. Hitchens, T. Brolund, D. McIlwain and M. Kavakli (2007a) Group play: determining factors on the gaming experience in multiplayer role-playing games. *Computers in Entertainment*, 5(4): 1–29.

Tychsen, A., K. Newman, T. Brolund and M. Hitchens (2007b) Cross-Format Analysis of the Gaming Experience in Multi-Player Role Playing Games. *Proceedings of DiGRA*, Tokyo, Japan.

Tychsen, A., J. H. Smith, S. P. Tosca and M. Hitchens (2006) Communication in Role Playing Games. *Proceedings of the International Conferece on Technologies for Interactive Digital Storytelling and Entertainment*, Darmstadt, Germany.

Vera, A., T. Kvan, R. West and S. Lai (1998) Expertise and Collaborative Design. *CHI '98*, Los Angeles, USA.

Wright, T., E. Boria and P. Breidenbach (2002) Creative player actions in FPs on-line video games: playing counter-strike. *Game Studies*, 2(2). Retrieved from http://www.gamestudies.org/0202/wright/

Young, M. J. (2005a) Theory 101: System and the Shared Imagined Space. Retrieved 29 September 2009, from http://ptgptb.org/0026/theory101-01.html.

Young, M. J. (2005b) Theory 101: The Impossible Thing Before Breakfast. Retrieved 29 September 2009, from http://ptgptb.org/0027/theory101-02.html.

13 Recallin' Fagin

Linguistic accents, intertextuality and othering in narrative offline and online video games

Astrid Ensslin

This chapter examines how linguistic accents of English are functionalized in narrative offline and online computer games to portray a variety of character types and interrelationships, in terms of epistemic and political/military power, as well as ethnic, social and moral binaries. I examine spoken and multimodal discourse of strategically significant non-player characters insofar as it reveals underlying, seemingly unreflected and undocumented language ideologies (see also Ensslin 2010).

In a recent study on linguistic accents in mainstream Anglophone offline computer games (Ensslin 2010), I argue that the popularity of many, especially narrative as opposed to abstract, games (cf. Wolf 2001) derives not only from the specific mechanics and graphical representations that they exhibit, but also – in a more subliminal way – 'from the strategic reiteration of stereotypes relating to gender and race, and also language attitudes' (Ensslin 2010, 205). More specifically, stereotypes are often framed in terms of conventional and unconventional oppositions (Jones 2002; Davies 2007; Jeffries 2010) – standard (such as, good vs. bad and black vs. white) and non-standard (such as, Queen's English vs. Scouse) polarizations that take advantage of the 'economy of Manichean allegory' (JanMohamed 1995: 20),[1] thus perpetuating binary thought and stereotyping particularly with respect to morality, erudition, social class, gender and race.[2] One intended effect on the player is the speedy formation of straightforward othering processes, which help define players' moral, social and cultural world pictures by creating clear-cut boundaries between friend and foe, thus optimizing player identification and goal achievement.

A particularly powerful way of perpetuating such binary stereotypes is their combination with linguistic features, which are put in (unconventional) opposition to one another. For instance, the attribution of a strong East-London accent to a somewhat suspicious or even roguish character will cause a powerful recognition effect, which is based on underlying intertextual and intermedial associations with characters in previously encountered artefacts, such as popular films and TV series. Juxtaposing such a negatively connoted character with, for instance, a learned, benevolent master wizard, who speaks with a 'Queen's English' (known as 'Received Pronunciation' (RP)) or even a Shakespearean stage accent, will further strengthen the otherness of the roguish character and subconsciously direct the player in their in-game endeavours.

Linguistic anthropologists Irvine and Gal (2000) refer to such associative semiotic processes in terms of 'iconicity', 'recursivity' and 'erasure'. By 'iconicity' they mean 'the interpretation of linguistic form not just as a dependable index of a social group but as a transparent depiction of the distinctive qualities of the group' (2000: 19). In this respect, the abbreviated style of mobile phone texting or in-game chat might be used to iconically characterize and thus collectivise young people. 'Recursivity' is a process arising from such iconic readings in that 'an opposition salient at one level is projected onto other levels of a linguistic and social relationship' (p.19). This links back to the example given in the previous paragraph, where social and moral differences are associated with linguistic differences, polarized into binary opposites and thus subjected to processes of othering. 'Erasure', finally, is 'the rendering invisible of some sociolinguistic activities or actors in a way that bolsters the iconic reading of language differences' (p. 19). In other words, the simplicity caused by iconicity is further reinforced by ignoring any irregularities or complexities in peoples' or characters' social backgrounds and/or linguistic behaviour that would detract from the intended, ungrounded representation.

A specific type of language ideology is attitudes towards standard and non-standard varieties of English. These come to the fore in particular when studying the linguistic accents of non-player characters (NPCs) in introductory cut scenes and first-level gaming sequences of offline and online video games. Introductory game levels are crucial in that they set the scene of the game world, introduce the conceptual background and objectives, explain the main functionalities and interface elements, and situate the player-character in the game. In online games, this latter point is usually done by means of customizing the avatar, such as defining a name, a profession, a psychological and physical framework and his/her appearance. In offline games, avatar customization is usually (though not always) omitted as players do not encounter other players in-world. What both offline and online games share, however, is an element of training, which is often performed as part of the game narrative. More precisely, the player is often taken under the wings of one or more 'masters', who take them through a step-by-step training process, during which they learn how to navigate the game world and to use the mechanics of the game. The spoken language used – in combination with subtitles – in such training sequences is indicative of the way in which players are programmed into subscribing to a particular moral in-game world picture.

In what follows, I shall first outline relevant implications of recent sociolinguistic, accent-attitudinal research. In a second step, I shall apply those insights to a number of level-one sequences taken from the offline games *Return to Castle Wolfenstein*, *Fable: The Lost Chapters* and *Black and White 2*, as well as the massively multiplayer online role-playing game *Wizard101*. All four games epitomize the afore-mentioned Manichean allegory (JanMohamed 1995: 20) by creating clear-cut moral binaries and commissioning the player to improve, optimize or save the respective in-game worlds by controlling or eradicating evil powers.

In particular, I shall look at how accents of English are functionalized to portray a variety of character types and interrelationships, in terms of epistemic and political/military power, as well as ethnic, social and moral oppositionality. I examine spoken and multimodal discourse of salient game characters insofar as it reveals underlying, seemingly unreflected and undocumented language ideologies (see also Ensslin 2010). I aim to show how Standard British and North American as well as non-standard regional and foreign accents are used in combination with other semiotic modes and paralinguistic features (such as pitch and intonation) to construct and maintain dominant language attitudes towards varieties of global English (such as, Kachru 1985, 1986). Further to that, accent, voice, pitch and intonation are employed to create parodistic effects by referring intermedially and intertextually to canonical audio-visual narrative, such as popular film, literature and TV series. It will become evident that game designers' deliberate fictionalization of emotionally and politically charged accents confirms recent accent-attitudinal research (such as, Lippi-Green 1997; Coupland and Bishop 2007; Bayard *et al.* 2001), particularly with respect to creating heroic versus roguish, strong versus weak, and provident versus flippant characters. A further insight that can be gained is that, even in online games, which are socially and communicatively oriented and tend to gain their attraction from human-computer-human rather than just human-computer interaction (cf. Taylor 2006), the intertextually and intermedially informed implementation of 'recognizable' accents and voices can contribute significantly to players' identification with the game.

Language ideologies and accent-attitudinal research

To map out the linguistic framework underlying the analyses in the following two sections, it is useful to outline some significant observations made by recent research into language attitudes towards 'Global Englishes'. Perhaps unsurprisingly, language ideologies as revealed in computer game discourse seem to reconfirm the insights gained by accent-attitudinal research into varieties of English, a trend that may either be intuitive or based on targeted research undertaken by game designers and corporate researchers. Early research into the attitudinal evaluation of variants of English in the 1970s revealed a significant degree of conservatism, with subjects favouring standard RP whilst denigrating ethnic and urban varieties in terms of both prestige and attractiveness (Giles 1970). Shuy and Williams (1973) found in a comparison between British RP and Standard North American (SNAm) that RP was considered by subjects to be more complex and active, yet less potent and valuable than SNAm, a trend that seems to have been perpetuated and expanded given the influence of an increasingly globalized, Americanized media especially on young people (Bayard *et al.* 2001). In a survey involving US undergraduates, Stewart *et al.* (1985) established that RP was attributed a higher social status than subjects' own accents yet was regarded as less intelligible and as arousing more discomfort. Of particular relevance to the present chapter is Lippi-Green's (1997) comprehensive survey of accents used in Hollywood and Disney characters, in which she finds that, while SNAm and RP accents are particularly prevalent in lead roles, non-standard, foreign accents are used mostly as negative stereotyping short-cuts for indexing marginal and roguish characters. As the following section will show, the same observation can be made for mainstream online and offline video games.

Following on from Shuy and Williams' (1973), research carried out by Donn Bayard in the 1990s has found that RP is rated highly for status and power variables yet is closely followed or even being gradually replaced and supplemented by SNAm. Of further importance is the phenomenon of 'dialect/accent loyalty' (Cargile *et al.* 1994; Giles and Powesland 1975), which refers to subjects' positive attitudes towards their own local and regional accents, mostly for emotive features such as warmth, friendliness and sense of humour. This is confirmed by Bayard *et al.* (2001), who claim that the so-called Pax Americana – the hegemony in the sense of normalized acceptance rather than imperialism of SNAm – is, at least amongst their subjects (students from New Zealand, Australia and the United States), gradually replacing accent loyalty, with far-reaching global implications. Finally, and perhaps of most interest for further research into language ideologies in video games, is Coupland and Bishop's (2007) observation that, among British subjects, positive ratings of RP rise proportionally to age, and that both younger and female respondents tend to be more positively inclined towards diversity. Generally low in prestige are (and always have been), according to Coupland and Bishop (2007), urban vernaculars, especially from Birmingham, Glasgow and Liverpool.

As the following analyses will show, the default use of SNAm in Anglophone games reconfirms the 'attitude hegemony' described by Bayard *et al.* (2001) in terms of 'Pax Americana', a concept referring to increasing degree of social attractiveness attributed to SNAm by native speakers of English at a global level (Ensslin 2010). The somewhat less popular British RP or, in some cases, a theatre stage version of it, is often used to create a symbolic distance to the implied player, by attributing it to NPCs whose age and wisdom are far more advanced than the players. Non-standard varieties, on the other hand, are used rarely in video games, and if they are, they are used to foreground specific character traits in NPCs. That said, even SNAm and RP are functionalized by some game producers to portray negatively connoted NPCs, as we shall see in the analyses of *Wizard101* and *Return to Castle Wolfenstein*, respectively.

Functionalized accents of english in offline video game discourse

Turning from sociolinguistic research to video game discourse, the following trends are noticeable. First, quasi-standard varieties – mostly SNAm, but sporadically also RP – are used for both voice-over narrative and instructions given to the player. While it could be argued that this makes for maximum levels of clarity and efficiency in information transfer, it also raises the question of authenticity and naturalness. From a stylistic point of view, this phonological prevalence causes any deviation to be understood as foregrounding and hence to deserve specific analytical attention.

As indicated previously, even in British video game products, SNAm is often used to index – in a process that may be described in terms of Irvine and Gal's (2000) 'iconization' – military and physical prowess in NPCs. These qualities are by default positively connoted as the objectives of games and are mostly related to success in physical battle and strategic warfare. The use of SNAm as a scientifically proven, most socially attractive variety of English, is therefore plausible. That said, the analysis of *Wizard101* will show that the default accent-attitudinal connotation of SNAm may be reversed and put in opposition to positively connoted RP, which is used to evoke associations of learnedness and moral integrity.

Of particular interest to a discourse analyst is, of course, the phenomenon of mixing standards in processes of 'recursivity' (Irvine and Gal 2000), which appears, for instance, in the introductory cut-scene of *Return to Castle Wolfenstein*. Here *political* and *military* power on the side of the 'goodies' is represented by characters who speak either standard or 'posh' Queen's English: The central, decision-making character's major characteristics are positively connoted, in terms of rationality, decision-making ability and sharpness of political reasoning, and he speaks an unmarked RP variety to further signal his moral integrity. Some more marginal yet still politically powerful characters with less decision-making power, on the other hand, are characterized, in their shallowness of argument and assumed lack of reasoning power, by a potentially unattractive clipped British accent, which is further enhanced through paralinguistic features, such as, pitch and body language, thus evoking an image of snobbery and superficiality. The central opposition in this scene, however, is formed by the juxtaposition between RP and SNAm. As opposed to political power attributed to Britishness, cognitive, intellectual, heuristic and hence strategic power, on the other hand, is represented by a young researcher speaking with a SNAm accent, whose general attire, relative youth and self-effacing yet confident body language further attract the viewer's sympathy.

One of the so far few games featuring non-standard varieties of British English as means of characterization is Guildford/UK-based Lionhead Studios *Fable: The Lost Chapters*. It deserves close attention not merely because it foregrounds non-standard accents but because it does so in thought-provoking ways. In an example from the first level, a Cockney/North-East London (urban, hence socially unattractive) accent, reminiscent of the rogue character Fagin from Charles Dickens' *Oliver Twist*, played by Ron Moody in Carol Reed's *Oliver!* and by Ben Kingsley in Roman Polanski's more recent film adaptation, is used to characterize an uncanny 'trader' set against the idyllic, pastoral village of Oakvale, where the game is set. The intertextual and intermedial allusion to Fagin is made deliberately as a cognitive and emotive device implemented to produce a recognition effect in the player. Furthermore, Fagin's ambiguous personality, which combines fatherly elements with those of a morally decrepit outcast commissioning under-age children to commit theft, arouses suspicion in players of *Fable*, whose apprentice heroes have to take a number of decisions between morally good and evil, which will impact their individual character development. In the novel, Fagin is portrayed as the benevolent criminal, thus forming a contrast to the unfailingly malicious, murderous antagonist, Bill Sykes. Fagin is hence rendered more trustworthy and humane than his evil counterpart, although the overall discourse of Dickens' novel clearly conveys Victorian moral values, which dictate Fagin's eventual conviction and execution. Interestingly, the anti-Semitic connotations evoked by Dickens' choice of describing Fagin as 'a very old shrivelled Jew' with a 'villainous-looking and repulsive face' and 'dressed in a greasy flannel gown, with his throat bare' (Dickens 2008) are semiotically deleted, or 'erased' (cf. Irvine and Gal 2000) in the game, thereby adapting it to a twenty-first century associative framework, which is informed by notions of neoliberalism, egalitarianism and political correctness.

The ambiguity of character evoked intertextually renders the trader in *Fable* morally dubious and potentially dangerous as he offers his goods to the still inexperienced young hero. At the same time, the trader – a nameless, formulaic 'type' – is juxtaposed phonologically to the hero's father and sister, the two 'commissioners' and undoubtedly most trustworthy roles in Level One, both of whom speak with an RP accent. That said, whereas the trader's accent is in the first instance used to demonize him and his 'trade' in general – in the Christian sense of 'temptation' – similar East London accents occur in a range of morally and functionally unmarked NPCs, who are browsing the village seemingly aimlessly, thus demystifying and naturalizing the trader's indexed untrustworthiness.

Despite those foregrounding tendencies of non-standard accents that are either positively or negatively connoted (a 'cuddly' Northumbrian accent is used in Level Two to signal the need for sympathy and help in a female villager), the game's main discourse remains firmly anchored in what Nielsen (1988) refers to as 'white [standard RP] discourse as a set of self-confirming propositions'. The all-important Guildmaster, a Dumbledorian, Gandalfian character of infinite wisdom, martial experience, untainted moral integrity and a white Western phenotype, speaks with a standard RP stage accent featuring the characteristic alveolar trill – or 'rolled [r]'. The voice is thus reminiscent of such popular Shakespearean actors as John Gielgud and Laurence Olivier and the authoritarian, often majestic parts they are commonly associated with. Furthermore, the hero himself – also a white male – speaks with an RP accent, thus adding to the standard 'awe-inspiring', 'prestigious' (cf. Coupland and Bishop 2007) RP matrix underlying the macrostructure of the game.

The deliberate choice of British English over SNAm matches the decisively British (albeit fantastical) setting of the game, which reconstructs a quasi-medieval 'Albion'. The game world conveys a quasi-touristic nostalgia for all things Anglo-Saxon, and it may be argued that the contextually coherent, stylized use of unpopular yet memory-evoking British urban vernaculars contributed to the fact that the game received positive reviews throughout and the original *Fable* (Lionhead Studios 2004) was one of the top-selling games of 2004.

Generally speaking, game designers tend to use conventional and unconventional semantic oppositions (see, e.g., Jones 2002; Davies 2007; Jeffries 2010) to construct simple Manichean binaries of morally good or bad behaviour, and to merge them with artificial linguistic (such as, accent) oppositions. In this respect, perhaps the most obvious example of simple oppositional gaming discourse is Lionhead Studio's (2005b) *Black and White 2*, a strategy or 'god' game in which the player assumes the role of an either good (generous, lenient and protective) or evil (malevolent, aggressive and tyrannical) deity. Most interesting from a language-attitudinal point of view is the embedding of an unconventional accent opposition in a conventional visual, mythopoetic, religious opposition: in the introductory cut-scene of the game, the player is initiated into the game world by two 'masters' rather than one: a white, elderly, angelic character speaking with an RP, over-articulated accent mixed with a formal register is set against a dark-skinned, thick-lipped, devilish character speaking with a SNAm accent, mixed with an informal register. The two visually juxtaposed characters symbolize the Faustian conflict of 'two souls [. . .] dwelling in [his] breast', which 'will wrestle for the mastery there' (Goethe 1960: 145), and the player has to decide which of the two inner voices to obey in defining their divine trajectory.

It is not surprising that, when I asked my first year undergraduate students, after watching the same introductory cut scene, what accents they had heard, at first nobody seemed to recall anything. Upon further reasoning, one student said that he thought the dark character was speaking with a 'Black New York accent'. This response seems to suggest that reverse recursivity (Irvine and Gal 2000) with regard to accent attribution takes place, where accent attribution is triggered by visual stimulants. Furthermore, what happens here semiotically is precisely what Nielsen (1988: 6) points out: that 'only a suggestion of blackness need appear for the entire structure to be articulated'. In other words, the metonymic transfer from *pars* to *toto* happens without critical reflection, whereby possible linguistic alternatives are precluded and stereotypes thus perpetuated.

Nevertheless, the importance of both 'Black' and 'White' as complementary concepts – a concept reminiscent of Chinese rather than Western philosophy – lend the game its holistic agenda: in order to encourage players not to avoid but rather to confront or even opt for a malevolent moral framework, the allegory of evil is placed at the same level of desirability as that of good. A closer multimodal look at the two personified inner voices suggests that evil is aestheticized to the degree of cuteness. Both guides are small and of non-athletic physical constitution: pyknic in the case of the 'dark' guide; asthenic in the case of the 'white' one. The 'dark' guide further reveals the following de-mystifying processes: his informal register is aligned with the ideal[3] gamer's colloquial register. Furthermore, the relaxed, dynamic paralinguistics conveyed by the dark character is designed to appeal particularly to young players: he exhibits an adolescent, 'cool' body language, a dynamic relationship to the space framed by the computer screen, and an unconventional relationship to culturally prescribed proxemics (he changes his distance to the player viewer from long shot to extreme close-up), thus seeking their attention in a non-adult way. This performed affinity with the ideal player is further accentuated by the visible age difference between him and the much older 'white' guide. His SNAm accent follows, inadvertently or not, the 'Pax Americana' highlighted by Bayard *et al.* (2001), which can be expected to attract the sympathy of most players.

In sum, the dark character's functional role as the inner voice of evil prominently in a game whose title symbolically juxtaposes good and bad yet simultaneously puts them on the same level of importance and prestige. The quick-paced, harmlessly trickling musical soundtrack accompanying the sequence further reinforces the overall image of harmony, thus legitimizing both moral extremes for the sake of enjoyable and successful game-play.

Functionalized accents in online gaming: the case of *Wizard101*[4]

Importantly, video games allow their players to act creatively within the limits of the game world, for example, by hack-cheating their way into otherwise inaccessible locations, or to mod the source code to create new costumes, accessories or weapons. That said, in offline games, which are dominated by human-computer rather than human-computer-human interaction (cf. Barnes 2002), natural language cannot normally be used in a productive-creative way. Players are subjected to the use of language coded into the gaming interface, which results in the inadvertent processing of either simple or – albeit more rarely – complex pragmatic choices.[5] *Online* games, on the other hand, are socially and communicatively oriented and tend to gain their attraction from human-computer-human rather than just human-computer interaction (cf. Taylor 2006). Players communicate with each other as well as interacting with the game's internal mechanics, which opens up an extensive playing field for linguistic, communicative and artistic creativity. Early research into intercultural and interlinguistic communication in the Massively Multiplayer Online Role Playing Game *Final Fantasy XI*, for instance, has found that players of diverse linguistic backgrounds develop and acquire new types of pidgins to facilitate in-game interaction (Nolen 2009).

The importance of player-to-player interaction and communication notwithstanding, this paper is mostly interested in the ways in which linguistic accents are encoded in game narratives and interfaces to achieve certain emotive and cognitive effects in players. Similarly, despite the importance of in-game computer-mediated communication between players, I aim to demonstrate that the intertextually and intermedially enriched implementation of 'recognizable' accents and voices in online games can contribute significantly to players' identification with the game.

A preliminary survey of short-term (such as, *Bejeweled* and *Slingo Millennium*) and long-term online games (such as, *Silkroad Online* and *Aion*) suggests that the need to spice up game narratives with specific characterization devices such as non-standard accents is either non-existent – as in short-term games – or indeed of such little importance that NPC accents are kept at an unmarked SNAm. Intertextuality and intermediality in many if not most graphically and narrationally complex online game worlds operate less at the level of personalized voices in NPCs than in their allusions to other narrative worlds, such as, for instance, those of epic and fantasy narratives (such as, *Age of Conan* and *Aion*) and popular American movie genres (such as, *Borderlands*). In *Aion: The Power of Eternity*, for instance, in-game narrative and directive accents follow the Pax Americana by adhering to an unmarked SNAm throughout. Player-NPC communication takes place by means of communication windows, where the speech of NPCs is transcribed, and the player's responses in the sense of pragmatic choices (such as, 'accept' vs. 'decline') are displayed as options in the same window. The acceptance of a quest is usually followed by a short-spoken affirmation or greeting on the part of the NPC, such as 'take care'. These utterances do not convey any accent variation either, thus contributing to the game's phonologically unmarked linguistic framework.

To demonstrate a rare example of accent variability in online games, I now turn to an analysis of strategically employed phonological aspects embedded in the first two levels of *Wizard101*, an ESRB-rated massively multiplayer online adventure game in which players assume the roles of apprentice wizards enrolled in the Ravenwood School of Magical Arts and develop a number of skills and experience in battling fantasy creatures using turn-based card games and interacting with other players. The ultimate aim is to become a Master Wizard and expel all evil powers from the game world. In the introductory sequence, which is a cut scene interspersed with player interaction (pressing the 'Next' button), the player learns from that he/she is a 'student with amazing potential . . . enough, perhaps to save Wizard City'. The master figure of the game, Headmaster Merle Ambrose, initiates – with the assistance of Gamma, the Owl – the previously customized wizard-avatar to the basic moves, duelling techniques, communication aids, and recreation tools. The same scene introduces the main antagonist and embodiment of the Manichean 'other', the Death Professor and in-game villain, Malistaire Drake.

The game's success is largely based on its rich intertextual and intermedial frames of reference. Even more compellingly than in the case of *Fable*, telling names are used to evoke unmistakable associations with popular fiction and visual narratives. The intertextual references to Hogwarts School of Witchcraft and Wizardry, the House of Ravenclaw, master wizards such as Albus Dumbledore and dubious characters like Draco Malfoy (such as, Rowling 1997) are unmistakable and produce strong anagnoritic effects in the target player. IP experts may in fact argue that some elements and characters in *Wizard101* verge on plagiarism, not least because the intertextual links and other sources are not referenced or otherwise made explicit.

Indeed, the game abounds with allusions to artefacts of popular culture – most evidently through telling names. One of the quest-giving NPCs, for instance, is called Private Ryan, who wishes to be saved from looting Skeletal Pirates. Ice Professor Lydia Greyrose is audio-visually modelled on the Fairy Godmother from Disney's *Cinderella II: Dreams Come True*, and Storm Professor Halston Balestorm – a frog in disguise – bears curious audiovisual resemblance with Kermit the Frog from *The Muppet Show*. The effect of alluding to characters from popular culture in a game that conveys a child-friendly environment for learning and achievement – as *Wizard101* does – is that of parody, thus causing comic effects in players with suitably developed intertextual frames of reference.

When looking at the spread of accents and voices, the Manichean opposition between the two most powerful characters is represented iconically by the use of an RP stage accent (see above) for Headmaster Ambrose and contrasted with a SNAm accent for Malistaire Drake. Hence, the Pax Americana (Bayard *et al.* 2001) is reversed for the thematic opposition between moral integrity and the malice of the fallen angel, or here the faculty member fallen from grace. Malistaire's voice is both lower and more languid than Ambrose's, and paired with his long drawn-out facial features, the character inspires associations with the antagonist Scar from Disney's *Lion King* as well as numerous other villains from popular audio-visual narratives. Interestingly, however, the simple Manichean opposition implemented at the opening of the game is lifted when the player enters Wizard City, and it turns out that all other NPCs speak with varieties of SNAm, thus positing the Manichean self in harmony with Anglo-Americanism.

Conclusion

Further to the observations made in Ensslin (2010), the following conclusions can be drawn from this study. Firstly, the default use of SNAm in the human–computer communication of offline *and* online computer games yet again reconfirms the 'attitude hegemony' described by Bayard *et al.* (2001) as 'Pax Americana'. This strategic, profit-driven choice on the part of game designers appears to support the increasing degree of social attractiveness attributed to SNAm by native speakers of English at a global level. Non-standard varieties, on the other hand, are used rarely – even less in online than in offline games, as this study seems to suggest – and if they are, they tend to be functionalized for character portrayal along the lines of Manichean binaries. As my analysis of *Fable: The Lost Chapters* has shown, non-standard accents of English are embedded in an overarching 'white' linguistic and multimodal matrix to functionalize, emotionalize and demonize individual characters and their moral outfits, thus again reconfirming recent research into stereotypical language attitudes (such as, Coupland and Bishop 2007).

The fact that video game discourse tends towards the construction of unconventional opposites and generally seems to take advantage of the ideological benefits arising from their embedding in conventional Manichean binaries – as the archetypal examples of *Black and White 2* and *Wizard101* have revealed – further confirms Lippi-Green's (1997) findings with regard to Disney characters. The 'economy of the Manichean allegory' (JanMohamed 1995: 20) is used extensively by game designers, not least because the convenience of pigeon-holing and straightforward othering helps to channel players' concentration onto motoric and cybernetic interaction with the game rather than triggering critical reflection and debate. This results in heightened degrees of susceptibility to underlying ideological content, which tends to go unnoticed yet continues to be 're-implanted' multimodally in players' subconscious minds.

Finally, the fact that massively multiplayer online games are by definition socially and communicatively oriented and tend to gain their attraction from human–computer–human rather than just human–computer interaction (cf. Taylor 2006) seems to render complex varieties in NPCs' voices and accents unnecessary. Players communicate with each other as well as interacting with the game's internal mechanics, which opens up an extensive playing ground for linguistic, communicative and artistic creativity. As my analysis of *Wizard101* has revealed, binary oppositions are drawn, if at all, on the basis of emotionally 'reliable', unmarked linguistic accents, and the juxtaposition of stage RP and SNAm to iconize and anthropomorphize good versus evil in the game world safely prevents any critical engagement with the semiotic and ideological implications of such unnatural linguistic varieties. Far more important for attractive game play is the strategic implementation of recognizable voices, telling names and other intertextual and intermedial features that will cause recognition effects in players at whom the game is pitched.[6] Mixed with distinct notions of humour and parody, which are again evoked by intertextual means, these anagnoritic devices can contribute significantly to players' identification and prolonged engagement with the game.

Notes

1 The term 'Manichean allegory' denotes discourses of moral oppositions attributed to stereotypical representations of self and other in popular culture, in particular the 'colonialist remythologizing aspects' (Everett 2005: 132) conveyed by a large number of mainstream video games. Postcolonial theorist JanMohamed (1995: 21) explains that 'the imperialist is not fixated on specific images or stereotypes of the Other but rather on the affective benefits proffered by the manichean allegory, which generates the various stereotypes. [. . .] The fetishizing strategy and the allegorical mechanism not only permit a rapid exchange of denigrating images which can be used to maintain a sense of moral difference; they also allow the writer to transform social and historical dissimilarities into universal, metaphysical differences. ' [sic]

2 For related studies on the use of non-native and non-standard accents in popular visual narrative, that is, Disney and Hollywood movies and the use of explicit and implicit metalanguage and metacommunication in American TV series, see Lippi-Green (1997) and Richardson (2006), respectively.

3 The term 'ideal gamer' is used here to refer to any member of the game's main target audience.

4 I would like to thank Susana Sambade Casais for her helpful comments on this section.

5 In *Desperate Housewives: The Game*, for instance, players can choose between three or more pragmatic intentions per conversational move, which are represented by text choices in speech bubbles and accompanying emoticons.

6 See also, for instance, the success story of *Shrek*, which is mostly due to its metafictional status as 'fairy tale about fairy tales' (French 2001) and its plethora of parodistic, intertextual references to popular culture.

References

Barnes, S. B. (2002) *Computer-Mediated Communication: Human-to-Human Communication across the Internet*, Boston, Allyn and Bacon.

Bayard, D., A. Weatherall, C. Gallois and J. Pittam (2001) Pax Americana? Accent attitudinal evaluations in New Zealand, Australia and America. *Journal of Sociolinguistics*, 5(1): 22–49.

Cargile, N., H. Giles, E. B. Ryan and J. J. Bradac (1994) Language as a social process: A conceptual model and new directions. *Language and Communication*, 14(3): 211–26.

Coupland, N. and H. Bishop (2007) Ideologised values for British accents. *Journal of Sociolinguistics*, 11(1): 74–93.

Davies, M. (2007) The Attraction of Opposites: The ideological function of conventional and created oppositions in the construction of in-groups and out-groups in news texts. In L. Jeffries. (ed.) *Stylistics and Social Cognition*. Amsterdam, Rodopi: 71–100.

Dickens, C. (2008) Oliver Twist. Project Gutenberg e-book (November 12) Retrieved from http: //www.gutenberg.org/files/730/730-h/730-h.htm.

Ensslin, A. (2010) "Black and White": Language ideologies in computer game discourse. In

S. Johson and T. Milani. (eds) *Language Ideologies and Media Discourse: Texts, Practices, Policies.* London, Contiuum: 205–222.

Everett, A. (2005) Serious play: playing with race in contemporary gaming culture. In J. Raessens and J. Goldstein. (eds) *Handbook of Computer Game Studies.* Cambridge, MIT Press: 312–25.

French, P. (2001) Shrek. *The Guardian* (November 12) Retrieved from http: //www.guardian.co.uk/film/2001/jul/01/philipfrench.

Giles, H. (1970) Evaluative reactions to accents. *Educational Review*, 22(3): 211–27.

Giles, H. and P. F. Powesland (1975) *Speech Style and Social Evaluation*, London, Academic.

Goethe, J. W. v. (1960) *Faust*, Part One (Trans. P. Wayne), Harmondsworth, Penguin.

Irvine, J. and S. Gal (2000) Language ideology and linguistic differentiation. In P. V. Kroskrity. (ed.) *Regimes of Language: Ideologies, Polities and Identities.* Oxford, James Currey: 35–83.

JanMohamed, A. R. (1995) The economy of the Manichean allegory. In B. Ashcroft, G. Griffiths and H. Tiffin. (eds) *The Post-Colonial Studies Reader.* New York, Routledge: 18–23.

Jeffries, L. (2010) *Opposition in Discourse: The Construction of Oppositional Meaning. Advances in Stylistics*, London, Continuum.

Jones, S. (2002) *Antonymy: A Corpus-Based Perspective*, London, Routledge.

Kachru, B. (1985) Standard, codification and sociolinguistic realism. In R. Quirk. (ed.) *English in the World.* Cambridge, Cambridge University Press: 11–30.

Kachru, B. (1986) *The Alchemy of English: The Spread, Functions and Models of Non-Native Englishes*, Chicago, University of Illinois Press.

Lippi-Green, R. (1997) *English with an Accent: Language Ideology and Discrimination in the United States*, London, Routledge.

Nielsen, A. L. (1988) *Reading Race: White American Poets and the Racial Discourse in the Twentieth Century*, Athens, University of Georgia Press.

Nolen, C. (2009) *Overcoming Language Barriers in New Media: Observations and Questions Regarding Innovative Communication Strategies in Multilingual Contexts. Proceedings of the Language in (New) Media: Technologies and Ideologies Conference*, Seattle, USA.

Richardson, K. (2006) The dark arts of good people: How popular culture negotiates "spin" in Nbc's the west wing. *Journal of Sociolinguistics*, 10(1): 52–69.

Rowling, J. K. (1997) *Harry Potter and the Philosopher's Stone*, London, Bloomsbury.

Shuy, Roger W. and Frederick Williams. 1973. Stereotyped attitudes of selected English dialect communities. In Roger W. Shuy and Ralph W. Fasold, (eds), *Language attitudes: Current trends and prospects*, 85–96. Washington DC: Georgetown University Press.

Stewart, M.A., Ryan, E.B. and Giles H. (1985) 'Accent and social class effects on status and solidarity evaluations', *Personality and Social Psychology Bulletin*, 11: 98–105.

Taylor, T.L. (2006) *Play Between Worlds: Exploring Online Game Culture*, Cambridge, MA: MIT Press.

Wolf, M.P. (ed.) (2001) *The Medium of the Video Game*, Austin, TX: University of Texas Press.

14 *Second Life* as a digitally mediated third place

Social capital in virtual world communities

Fern M. Delamere

As virtual worlds and digital games expand, so too does our cultural understanding of them. This is particularly true given the academic interest they have garnered and the calls for further research. While not meeting the definitional criteria of a game, the ludic virtual world of *Second Life* (*SL*) is akin to massively multiplayer online role playing games (MMORPGs) both in the sociality and the playful environments of each. The virtual world of *SL* and the communities of interest found in it are culturally significant places. Disability and health groups in *SL* are specifically presented in this chapter as crucial places where meaningful social engagement, support, and advocacy occur and human capital is developed.

Oldenberg's (1999) concept third place is defined as a neutral social space, different from home or work, and an open space where public gathering and social discourse occur. Drawing upon this concept this writing bolsters support for the exploration of online virtual worlds as digitally mediated third places. The concept of third place is being applied to digitally mediated online places, suggesting that these online places, like other physical third places, act as a social gathering place different from home and work where civic discourse take place. Third place is used as a framework, in conjunction with what Putnam (2000) terms social capital, to better understand the role virtual worlds play in fostering relationships and developing human potential. As used here, social capital entails structures of networks, norms and relational trust that collectively work together for some mutual benefit. This chapter presents collaborative ethnographic research and participant observation conducted with disability and health groups in *SL*. Literature and theoretical conceptualizations are discussed to frame the analytical principles used in this research. The work presented here is primarily conceptual rather than empirical in nature. Preliminary observations and descriptions of various disability and health groups are provided and linked to these concepts.

The decline of social capital in modern Western society (Putnam 2000) and the lack of public third place locations where these relationships can be forged (Oldenberg 1999) are two areas of social concern being voiced in scholarship. In addition, Oldenberg (1999) and Putnam (2000) similarly suggest that media participation fosters solitary and socially isolating leisure pursuits in private rather than public locations. These authors suggest that the media is a contributory factor in the decline of third places (Oldenberg 1999) and to the decreased potential for the development of social capital (Putnam 2000).

Social capital: foundations, critiques and online applications

The origins of the concept of social capital are found in political science and the early nineteenth century work of Alexis de Tocqueville (Mansfield and Winthrop 2002). De Tocqueville connected social capital to associational life and the building of democracy. An expanded reading of social capital by Bourdieu (1986) highlights important theoretical linkages of social capital to issues of social class. Recent popularization of social capital within academe is most often attributed to the work of Putnam (2000). Putnam explained social capital as social networks and norms of reciprocity and trustworthiness that arise from the formation of social bonds. Social capital is also viewed as investments in social relations by individuals that, intended or not, facilitate collective actions with returns that exceed those that an individual might achieve acting independently of others (Warren 2008).

Social capital consists of two conceptual components, bonding social capital and bridging social capital (Putnam 2000). According to Putnam, bonding social capital is the social glue that binds together homogeneous groups of people in strong bonds, while bridging social capital is viewed as a social lubricant that helps diverse groups of people to form loose ties with one another. While the positive outcomes of social capital are most often cited, it is important to recognize possible antagonistic outcomes of close groups whose purposes lie outside typical social norms; such as street gangs.

Clearly Putnam's main thrust, however, is centred on the positive outcomes of social capital. His book is a statistical enumeration of the various ways that social capital has been on the decline in Western society since the mid-1950s. To bring this point home, Putnam (2000) asserts that the increasing privatization of leisure time has changed the way that we interact – to the detriment and decrease of strong social relationships and social connectedness. Creatively using this image as the title of his book, Putnam points to the progressive decrease in the number of bowling leagues as a primary example of the decline in social capital, and the resulting social vacuum. Putnam's work is well documented and supported by empirical data, although it is not without criticism.

Some of the criticism is directed toward social capital's general lack of acknowledgment of the competing interests of divergent groups, the inequity of social access to power, and the depoliticized approach resulting from this (Muntaner *et al.* 2002; Siisihainen 2000). This has clear implications for understanding how this concept applies to those with limited access to power based on race, class, sexuality, gender, and disability. I suggest that infusing a Bourdieuian perspective of social capital, one that highlights the recognition of social class as an important factor, counterbalances and addresses many of these criticisms.

Putnam's (2000) claim that isolated leisure participation with digital media contributed to the void in social capital is also a point of criticism. As a counter-point, many suggest that the human need for sociality and connectedness is in actuality being fulfilled through technologically mediated networks (Blanchard and Horan 1998; Jones 1997; Rheingold 1994; Wellman and Haythornthwaite 2002). In fact, recent scholarship in media communications and game studies is now exploring games and virtual worlds as computer-generated third places that nurture human connectedness (Bergstrom 2009; Ducheneaut *et al.* 2007; Soukup 2006; Steinkuelher and William 2006; Wadley *et al.* 2003).

A (third) place to hang: a matter of community

Oldenberg (1999) states, 'third place is a generic designation for a great variety of public places that host regular, voluntary, informal, and happily anticipated gatherings of individuals beyond the realms of home and work' (p. 16). *SL* is being conceptualized and proposed in this study to be a computer-generated third place. From this perspective it is being viewed as a publically accessible social place and context for examining online communities of practice and communities of interest (Wenger 1991).

Third places are described as socially equal, playful, homey and congenial environments that are easily accessible with available hours for people to meet and partake in informal conversations (Oldenberg 1999). In a third place, the status of guests is levelled; it is a neutral ground, not 'my place,' not 'your place.' A third place provides an interactive setting where grassroots issues are discussed and community is built (Oldenberg 1999). English pubs, German beer gardens, corner coffee houses, French cafés, and Victorian gardens are Oldenberg's nostalgic examples of traditional third places. In this regard, third places are thought to break down barriers that exist in everyday life by offering a space in which to 'hang' and interact, a place we can all call our own, a place of community. The social concept of third place is well supported in architectural research (Alexander 1977).

Virtually . . . communities of interest

Until the advent of digital technology, definitions of community have focused on close-knit groups in a single, often local, geographic location (Hand and Moore 2006). Wenger's (1991) approach allows us to move beyond community as a fixed geographical construct and towards conceiving of community as a set of social relations. To rethink traditional notions of a community, we understand it as 'a set of relations among persons, activity, and world' (Wenger 1991: 98). We must explore the places where these interactions happen, including online spaces.

According to Wenger (1991), communities of practice are formed by people who engage in a process of collective learning in a shared domain of human endeavour. Wenger views online communities as important interactive spaces where the formation of social practice takes place, which I believe is inclusive of virtual communities such as those discussed here.

> New technologies such as the internet have extended the reach of our interactions beyond the geographical limitations of traditional communities, but the increase in flow of information does not obviate the need for community, In fact, it expands the possibilities for community and calls for new kinds of communities based on shared practice.
>
> (Wenger 2008, online)

Transference of traditional notions of community into online places and focusing on communities of interest in the virtual worlds are productive avenues for exploring the social meaning and impact of virtual communities in both online and offline realms. Virtual community has become an accepted concept in communication, media, and game studies. It is recognized for the important role it plays as a common forum where a large portion of current social interactions and human connectedness takes place (Jones 1997; Katz and Rice 2002; Preece and Maloney-Krichmar 2005; Wellman and Haythornthwaite 2002). This aligns well with Oldenberg's conceptualization and description of third place.

Rheingold (1994) suggests that, rather than contributing to the loss of informal public places, involvement with digital media through online communities has burgeoned in response to the increasing unavailability of geographic places for people to engage in convivial conversation. A recent reprint of an original 1987 essay by Rheingold provides this working definition of virtual community:

> A virtual community is a group of people who may or may not meet one another face-to-face, and who exchange words and ideas through the mediation of computer bulletin boards and networks. Like any other community, it is also a collection of people who adhere to a certain (loose) social contract, and who share certain (eclectic) interests.
>
> (Rheingold 1994: 20)

Recent works highlight how early Internet platforms – PC email, mobile phone email, real-time chat (instant messaging and IRC), mailing lists, bulletin boards – help to build social capital within and beyond these virtual communities (Katz and Rice 2002; Kobayashi *et al.* 2006; Wellman *et al.* 2003; Wellman and Haythornthwaite 2002). Virtual community research is founded on these early Internet applications, now being applied to other Internet applications such as digital games and virtual worlds. Given the connection between earlier virtual community research and the current exploration of communities formed within virtual worlds, we are reminded that what appears as new communication research, may not be new (Bell and Consalvo 2009). In comparison, however, virtual worlds as a new application add a complexity of experience due to additional aspects that early online text-based communities did not have including, visual richness, game character (avatar) embodiment, and a ludic (playful) environment.

Virtual worlds of online play

'Networked social games are a wholly new form of community, social interaction, and social phenomenon that is becoming normative faster than we have been able to analyze it, theorize it, or collect data on it' (William 2006: 1). Current efforts are now being directed towards the exploration of communities founded in digital games and the virtual worlds of MMORPGs, also referred to as massively multi-player online games (MMOGs or MMOs). *SL*, while not defined as a game, has similar playful opportunities, games within it, and ludic experiences analogous to MMOGs. For this chapter, the term massively multiplayer online virtual world (MMOVW) will be used as an inclusive term to collectively refer to all networked, online, avatar-based game and non-game virtual worlds.

Bell (2008) presents a comprehensive, albeit recognizably evolving, definition of virtual worlds. Virtual worlds are 'a synchronous, persistent network of people, represented as avatars, facilitated by networked computers' (Bell 2008). Involvement in a MMOVW is no longer considered a fringe pastime. Economic indicators act as evidence of growth in the gaming industry, but act as a substitute for the more interesting cultural feature that increasing numbers of people are using technologically mediated sociality in their everyday lives. The social side of MMOVWs – that is, what happens with and between players, friends, family, and communities – is of greatest interest here. Taylor (2006) states that virtual worlds and networks found in the MMOVW *EverQuest* are grounded in practices of technology that engage participants in their everyday lives, and in turn, their everyday social networks and communities. Online experiences are neither vacuous nor separate from the rest of life: rather, they are an interwoven and integral part of life (Rosenberg 2009).

Digitally mediated third places and social capital

Third place has recently proven to be a useful framework from which to examine digital play and online social behaviour (Boellstorff 2008; Ducheneaut *et al.* 2007; Soukup 2006; Steinkuelher and William 2006; Wadley *et al.* 2003). William (2006) ponders if games will become a modern third place and, if so, how this might affect our social understanding of human interaction and behaviour.

Recent research supports the utility of studying MMOVWs as a conceptual third place. A study of online networked games has reported the importance of sociality and the sense of community they foster in players' enjoyment of game-play (Wadley *et al.* 2003). Steinkuelher and William (2006) conclude that Oldenberg's concept of third place is also useful in understanding MMOGs' role in developing a sense of place and community. Ducheneaut *et al.* (2007) study of the MMOG *Star Wars Galaxies* as a third place discussed the importance of game design and creating places within games where the sociality needs of players can be met.

Framing social capital as an analytical tool for virtual world research has also proven useful (Fielder 2008; Kobayashi *et al.* 2006; Malaby 2006). Survey research in *SL* indicates that socializing is a major motivation for participation, and highlights the general correlation between high levels of social capital in real-life and high levels of social capital in that world (Holmberg and Huvila 2009). The movement of human actions between the boundaries of the real and the virtual must be accounted for.

> . . . boundaries that only appeared to separate the real and the virtual are fading fast, from both sides, and it is the social actors on the ground who are making use, in every new moment, with every new challenge, of the increased scope that these new domains afford.
>
> (Malaby 2006: 160)

Malaby (2006) cautions against the common misreading of social capital in virtual world research in terms of market only, as this often creates misunderstanding of the net value of human exchanges and outcomes related to social connectedness, reciprocity and learning. Malaby (2006) defines human capital as the first resource coming from human efforts; it is through this effort and over time that human capital is thereby transformed into other capitals including material, social and cultural capital. Malaby proposes that virtual world research must examine all forms of human capital as an important step to better understand the online and offline implications of these virtual worlds.

Connections in and beyond virtual worlds: disability and social capital online

Computer-generated communications in virtual worlds do decrease barriers of time and space. However, we are reminded not to think of these technologies in purely utopian ways, doing so, we neglect important socio-political analysis inherent in disability studies (Seymour and Lupton 2004). The disability studies literature also recognizes a digital divide and constraints related to some people's with disability capacity for involvement. This divide is based on numerous factors, such as the type of disability, the associated usability of the technology, and economic access (Bush 2006; Chaudry 2005; Dobransky and Hargittai 2006). It is therefore acknowledged here, that virtual worlds have been shown to present both opportunity and challenge for people with disabilities (Carr 2009; Dobransky and Hargittai 2006; Trewin *et al.* 2008). While it is important to acknowledge these problematic issues and the difficulty of access, the focal point here is on those who do interact with technology and participate in virtual world communities.

This said, it is clear that people with disabilities do want to participate in virtual worlds, and in fact may be over-represented in them compared to their population share (20 percent of casual gamers have a disability vs. 15 percent of the general population) (Information Solutions Group 2008). Further to this, many features and universal design principles can be built in to enable access (Carr 2009; Krueger *et al.* 2010; Mancuso and Cole 2009; Smith 2009).

The social relevance of virtual worlds for people with disabilities and others involved in disability groups in *SL* is important ground for exploration (Forman *et al.* 2009; Smith 2009; Trewin *et al.* 2008). As virtual worlds become a new paradigm in which to operate, many are just beginning to understand the implications and possibilities for our social understanding, including the social constructions of disability (Smith 2009). People with disabilities who are active participants in virtual worlds accrue beneficial individual and socio-cultural outcomes. Virtual participation in techno-sociality presents a personal avenue for fulfilment, as well as a platform for political action (Seymour and Lupton 2004). 'Virtual worlds have the potential to transform the way society operates' and views disability (Trewin *et al.* 2008: 177), perhaps more centred on capabilities and personhood rather than incapacities. In addition, Huang's (2005) study examining social capital and online disability communities reports the need for more research in this area to better understand their socio-cultural and political implications.

Empirical research suggests that relationships built online positively relate to common indicators of social capital (Best and Krueger 2006; Blanchard and Horan 1998; Holmberg and Huvila 2009; Huang 2005). These indicators include general trust, reciprocal support, social participation, and friendships.

People with disabilities meet and form communities in virtual worlds such as *SL* in virtual third spaces. Over 6 months of ethnographic observations, of the day-to-day culture and work of several of the most active health and disability groups in *SL*, support the theoretical arguments made in this chapter. It is argued that these communities form social capital that has value to their participants and is parlayed into their online and offline lives in meaningful ways, impacting on them and society. One clear example of this was described by the leader of the Virtual Ability community group in *SL*. She explained how one member of this group overcame some significant problems with social anxiety through her online involvement, which even led to her acquiring a job. Through the efforts of her friends in a virtual peer support community, the woman with severe social anxiety learned new coping strategies and IT-related job skills in *SL*. She used a resume highlighting these skills, and a letter of recommendation by her *SL* mentor, to obtain gainful employment offline. Related to her newly developed sociality, she later said, 'People need a reason to get together and talk. Besides, it's fun!' This example supports the theoretical foundations of how computer-generated third spaces assist to build social capital and create important networks of support for people with disabilities, beyond easier access to social interaction opportunities and leisure entertainment value.

People with disabilities have the freedom in virtual worlds to 'escape their bodies, if they so choose, or to celebrate the contradistinction of their unique gifts in the presence of peers' (Smith 2009). Such choices also act as a 'levelling ground' for them, whereby they are addressed according to the merit of their character (Bowker and Tuffin 2002), rather than disability serving as their first, only, and often stigmatized component of identity (Bedini 2000; Cahill and Eggleston 1995; Joachim and Acorn 2000; Scambler 2009).

As discussed in the previous section, Malaby (2006) emphasizes the importance of parlaying human capital – collectively social, market and cultural capital – and understanding the conversion from one to the other, allowing better understanding of how online social actions shape reality and human experience.

> By thinking in terms of the forms of capital within and beyond synthetic worlds, researchers will be able to chart how human actors move within and among different domains of all kinds, converting different forms of capital into one another.
>
> (Malaby 2006: 160)

Data from the research thus far applies to Malaby's notion of human capital and the conversion from one domain of social capital to another. First, I have observed how some individuals with disabilities who have learned functional game skills through the support of the Virtual Ability community (social capital) then leveraged those skills to learn building and scripting, creating commodities (such as, scripted vehicles or elevators, clothing and so forth) that they sell or give away for free (market capital) in *SL*. Another example form this research is Helen Keller Day, held 27 June 2009, in *SL*. Organized by the Virtual Helping Hands group, the community event's overall purpose included information acquisition, education, exploration of employment opportunities (market capital), social engagement, enjoyment of arts and entertainment (cultural capital) and the unveiling of Max the virtual guide dog, a free product developed in *SL* for use in-world by blind and visually impaired users (Linden Lab 2009).

Information presented to me by a participant of this study also represents how social capital bridges and bonds people together with unexpected outcomes. As described by the participant, a man who was using *SL* as respite from care giving for his severely disabled adult child found additional personal support within the virtual world, and eventually brought the child into *SL* with him. This was not successful, as the child, who has autism, interacted randomly and did not respond to any communication with other players. Ready to give up on having his child successfully interact and enjoy the virtual world, the first man met another *SL* participant who had learned scripting (using computer programming language to make objects become animated). This person used his scripting skills to create a tool that helps guide his blind wife's avatar to move independently in the virtual environment, thereby allowing her fluid advancement through the metaverse and the ability to attend *SL* music concerts with him (social capital). Recognizing another application of that tool, the two men are now collaborating to design an open source parental control device that would allow a guardian to determine movements of a dependent such that they could, with the assistance of their parent, successfully interact in the virtual world (freely available market capital).

A further example that can be drawn from the research findings is how the communities of Deaf people in *SL* created bonds of friendship and a social network which they put to use when they rallied together expressing opposition to the exclusionary nature of the introduction of voice communication in *SL* . They thus parlayed social capital into the collective domain of cultural capital through social activism actions. Viewing social capital from a social action perspective and socio-political lens is important and fully takes into account issues related to access to social power (Huang 2005; Muntaner *et al.* 2002; Seymour and Lupton 2004). Using social capital in the analysis of these observations supports how making interconnections between various forms of human capital helps to understand the blurring of boundaries between worlds.

This is also supported by previous literature which has shown that individual and collective empowerment (Fernback 1997; Hopkins *et al.* 2004; Kobayashi *et al.* 2006; Meekosha 2002; Ospina *et al.* 2009; Smith 2009) and what has been termed e-empowerment by Zielke *et al.* (2009) are benefits for those who engage in social activities of online networks and communities. A conversation I had with Simon Stevens (known in *SL* as Simon Walsh), a founder of *Wheelies* social club in *SL* that is open to all but geared towards people with disabilities, supports this point. Stevens stated, 'When I started Wheelies I had no idea the impact it would have here and in my real life. You know I was given an award in the UK for founding Wheelies, right?' In 2008, Stevens was given a UK Catalyst Award for social action and technology, sponsored by the Department for Business, Enterprise, Regulatory Reform and National Endowments for Science, Technology and the Arts.

The literature also shows learning as another substantive benefit of online disability communities in virtual worlds (Zielke *et al.* 2009). Research by Zielke *et al.* (2009) used adult learning models to show the benefits of teaching functionality and mastery of virtual-world skills to participants with disabilities in *SL*. This researcher has observed other outputs of learning in *SL* for people with or without disabilities. These include learning about disability-related topics, access and utilization of in-world and offline resources, self-efficacy and empowerment, and advocacy/self-advocacy related skills, to mention but a few. Benefits of learning have also been found in another MMOVW, *World of Warcraft* (Oliver and Carr 2009), linking learning to Wenger's (1991) ideas about communities of interest. On a collective level, 'researchers are only now beginning to appreciate the impact that virtual worlds are having in helping patients adapt to their disability and discover a sense of community' (Smith 2009).

Digital communities focused on disability support and advocacy have the potential to demystify societal conceptualizations and fragment many prevalent misconceptions of people with disabilities. In fact, dominant discourse surrounding disability can be resisted through online communities (Ospina *et al.* 2009) and may be transformed through the deconstruction of typically stigmatizing discourses (Kang 2009). Transformation through resistance is highlighted as a key element in the disabled people's movement and in the imagining of a 'politics of hope' (Peters *et al.* 2009). As Goggin and Newell (2007) note, it is important to reframe disability as a central category of power and identity, and to explore the ramifications of new information communication technologies (ICT) so as not to replicate and reproduce common and oppressive disability discourses. A central question to be asked here is; what does social capital in the computer-generated third place of *SL* contribute to the framing of disability in the virtual community and offline?

Conclusion

As spatial, temporal and social locations, virtual worlds have within them the potential to develop virtual communities filled with socio-cultural meaning for individual members. As new technologies such as MMOVWs shrink distance and erase the limitations of geography, the creation of computer-generated third places where empowered publics emerge and communities of interest create social capital, results in intended or unintended social change. As seen here, framing *SL* as a third place opens the door for better understanding of the social functions that online communities such as disability and health groups in *SL* have. As a community the disability and health groups have organized collective networks and built social trust that has facilitated co-operation and coordinated mutually beneficial activities. Examples include shared and similar mandates, such as, information sharing, mentoring new members, offering social and recreation events, building job skills opportunities, scheduled public education seminars, care-giver support and socio-political-oriented disability advocacy.

Fernback (1997) noted that social bonding and the communal spirit developed online may be purely instrumental and remain there, but it may also extend outward in a manner whereby communities manifest themselves in actions having very real effects on socio-political issues. It is suggested that civic engagement and the accumulation of social capital in online communities will only be strengthened as that social capital becomes linked to other offline applications and spaces (Blanchard and Horan 1998). Preliminary observations indicate that *SL* disability and health groups are instrumental in creating strong social relations, and the action of these groups does create social capital both within and outside of virtual worlds.

Chairman Emeritus Nicholas Negroponte of Massachusetts Institute of Technology's Media Lab forecast that the interactive, entertainment and information worlds would eventually merge (Negroponte 1995). In the case of virtual worlds, these domains have now merged, and we are only beginning to understand the limitless social implications and the blurring of the boundaries between online and offline effects.

References

Alexander, C. (1977) *A Pattern Language*, New York, Oxford University Press.

Bedini, L. A. (2000) 'Just sit down so we can talk': Perceived stigma and community recreation pursuits of people with disabilities'. *Therapeutic Recreation Journal*, 34(3): 55–68.

Bell, M. (2008) Toward a definition of 'virtual worlds'. *Journal of Virtual Worlds Research*, 1(1). Retrieved from http: //journals.tdl.org/jvwr/article/view/283/237.

Bell, M. and M. Consalvo (2009) Culture and virtual worlds: The not-quite-new experiences we study. *Journal of Virtual Worlds Research*, 1(3). Retrieved from http: //www.jvwresearch.org/index.php?_cms=default,2,2.

Bergstrom, K. (2009) *Adverturing Together Exploring Romantic Couples of MMOS as Part of Their Shared Leisure Time*, Calgary, University of Calgary.

Best, S. J. and B. S. Krueger (2006) Online interactions and social capital: distinguishing between new and existing ties. *Social Science Computer Review*, 24(4): 395–410.

Blanchard, A. and T. Horan (1998) Virtual communities and social capital. *Social Science Computer Review*, 16(3): 293–307.

Boellstorff, T. (2008) *Coming of Age in Second Life*, New Jersey, Princeton University Press.

Bourdieu, P. (1986) The forms of capital. In J. Richardson. (ed.) *Handbook of Theory and Research for the Sociology of Education*. New York, Greenwood Press: 241–58.

Bowker, N. and K. Tuffin (2002) Disability discourses for online identities. *Disability and Society*, 17: 327–44.

Bush, M. L. (2006) *Beyond Web-Accessibility: Barriers to ICT for Disabled People*, London, Aidis Trust.

Cahill, S. E. and R. Eggleston (1995) Reconsidering the stigma of physical disability. *Sociological Quarterly*, 36(4): 681–98.

Carr, D. (2009) Virtually accessible (second life and disability). *Access: The Inclusive Design Journal*: 23–5.

Chaudry, V. (2005) Rethinking the digital divide in relation to visual disability in India and the United States: towards a paradigm of "information inequity". *Disability Studies Quarterly*, 25(2). Retrieved from http: //www.dsq-sds.org/article/view/553/730.

Dobransky, K. and E. Hargittai (2006) The disability divide in internet access and use. *Information, Communication and Society*, 9(3): 313–34.

Ducheneaut, N., R. J. Moore and E. Nickell (2007) Virtual 'third places': a case study of sociability in massively multiplayer games. *Computer Supported Cooperative Work*, 16(1–2): 129–66.

Fernback, J. (1997) The individual within the collective: virtual ideology and the realization of collective principles. In S. Jones. (ed.) *Virtual Culture: Identity and Communication in Cybersociety*. London, Sage: 36–54.

Fielder, J. D. (2008) Synthetic Democracy: Associations of Social Capital in Virtual Worlds. *Proceedings of International Society of Political Psychology Annual Meeting*, Paris, France.

Forman, A. E., P. Baker, J. Pater and K. Smith (2009) Beautiful to Be Me: Identity, Disability and Gender in Virtual Environments. *International Communication Association (ICA) Conference on Women in a Digital World: Conceptual Models of Inclusion*, Chicago, USA.

Goggin, G. and C. Newell (2007) The business of digital disability. *The Information Society*, 23(3): 159–68.

Hand, M. and K. Moore (2006) Community, identity and digital games. In J. Rutter and J. Bryce. (eds) *Understanding Digital Games*. London, Sage: 241–66.

Holmberg, K. and I. Huvila (2009) Social Capital in Second Life. *Proceedings of AOIR 9.0*, Copenhagen, Denmark.

Hopkins, L., J. Thomas, D. Meredyth and S. Ewing (2004) Social capital and community building through an electronic network. *The Australian Journal of Social Issues*, 39(4): 369–79.

Huang, J. (2005) Building social capital: a study of the online disability community. *Disability Studies Quarterly*, 25(2). Retrieved from http: //www.dsq-sds.org/article/view/554/731.

Information Solutions Group. (2008) Disabled Gamers Comprise 20% of Casual Video Game Audience. *PR Newswire* (September 25, 2009) Retrieved from http: //www.infosolutionsgroup.com/press_release_E.htm.

Joachim, G. and S. Acorn (2000) Stigma of visible and invisible chronic conditions. *Journal of Advanced Nursing*, 32(1): 243–48.

Jones, S. (ed.) (1997) *Virtual Culture: Identity and Communication in Cybersociety*. London, Sage.

Kang, J. G. (2009) A teacher's deconstruction of disability: a discourse analysis. *Disability Studies Quarterly*, 29(1). Retrieved from http: //www.dsq-sds.org/article/view/173/173.

Katz, J. E. and R. E. Rice (2002) *Social Consequences of Internet Use: Access, Involvement and Interaction*, Cambridge, MIT Press.

Kobayashi, T., K. I. Ikeda and K. Miyata (2006) Social capital online: collective use of the internet and reciprocity as lubricants of democracy. *Information, Communication and Society,* 9(5): 582–611.

Krueger, A., A. Ludwig and D. Ludwig (2010) Universal design for virtual worlds. *Journal of Virtual Worlds Research,* 2(3). Retrieved from http: //journals.tdl.org/jvwr/article/view/674.

Linden Lab. (2009) 'Helen Keller Day.' (June 27, 2009) Retrieved from http: //wiki.secondlife.com/wiki/Helen_Keller_Day.

Malaby, T. (2006) Parlaying value. *Games and Culture,* 1(2): 141–62.

Mancuso, K. and J. Cole (2009) Gimpgirl Community's Best Practices for Facilitating an Accessible Community in a Virtual World. *Proceedings of IEEE Accessing the Future Conference,* Boston, USA.

Mansfield, H. C. and D. Winthrop. (eds) (2002) *Democracy in America (Trans. H.C. Mansfield and D. Winthrop).* Chicago, University of Chicago Press.

Meekosha, H. (2002) Virtual activists? women and the making of identities of disability. *Hypatia,* 17(3): 67–88.

Muntaner, C., J. W. Lynch, M. Hillemeier, J. H. Lee, R. David, J. Benach and C. Borrell (2002) Economic inequality, working-class power, social capital, and cause-specific mortality in wealthy countries. *International Journal of Health Services,* 32(4): 629–56.

Negroponte, N. (1995) *Being Digital,* Toronto, Random House.

Oldenberg, R. (1999) *The Great Good Places: Coffee Shops, Community Centers, Beauty Parlors, General Stores, Bars, Hangouts, and How They Get You through Your Day,* New York, Marlowe and Company.

Oliver, M. and D. Carr (2009) Learning in virtual worlds: using communities of practice to explain how people learn from play. *British Journal of Educational Technology,* 40(3): 444–57.

Ospina, A., J. Cole and J. Nolan (2009) 'Gimpgirl Grows Up: Women with Disabilities Rethinking, Redefining, and Reclaiming Community'. *Proceedings of AOIR 9.0,* Copenhagen, Denmark.

Peters, S., S. Gabel and S. Symeonidou (2009) Resistance, transformation and the politics of hope: imagining a way forward for the disabled people's movement. *Disability and Society,* 24(5): 543–56.

Preece, J. and D. Maloney-Krichmar (2005) Online communities: design, theory, and practice. *Journal of Computer-Mediated Communication,* 10(4). Retrieved from http: //dx.doi.org/10.1111/j.1083-6101.2005.tb00264.x.

Putnam, R. D. (2000) *Bowling Alone,* New York, Simon and Schuster Paperbacks.

Rheingold, H. (1994) *The Virtual Community,* London Secker and Warburg.

Rosenberg, A. (2009) A Cultural Sociological Approach to Second Life. *Proceedings of AOIR 9.0,* Copenhagen, Denmark.

Scambler, G. (2009) Health-related stigma. *Sociology of Health and Illness,* 31(3): 441–55.

Seymour, W. and D. Lupton (2004) Holding the line online: Exploring wired relationships for people with disabilities. *Disability and Society,* 19(4): 291–305.

Siisihainen, M. (2000) Two Concepts of Social Capital: Bourdieu vs. Putnam. *Proceedings of ISTR 4th International Conference,* Dublin, Ireland.

Smith, K. (2009) The Use of Virtual Worlds among People with Disabilities. *Proceedings of the 24th Annual International Technology and Persons with Disabilities Conference,* Los Angeles, USA.

Soukup, C. (2006) Computer-mediated communication as a virtual third place: Building Oldenburg's great good places on the world wide web. *New Media and Society,* 8(3): 421–40.

Steinkuehler, C. and D. Williams (2006) Where everybody knows your name: online games as 'third places'. *Journal of Computer Mediated Communication,* 11(4): 885–909.

Taylor, T. L. (2006) *Play between Worlds: Exploring Online Game Culture*, Cambridge, MIT Press.

Trewin, S., V. L. Hanson, M. R. Laff and A. Cavender (2008) Powerup: An Accessible Virtual World. *Proceedings of the 10th International ACM SIGACCESS Conference on Computers and Accessibility*, Halifax, Nova Scotia, Canada.

Trewin, S. M., M. R. Laff, A. Cavender and V. L. Hanson (2008) Accessibility in Virtual Worlds. *CHI '08 Extended Abstracts on Human Factors in Computing Systems*, Florence, Italy.

Wadley, G., M. Gibbs, K. Hew and C. Graham (2003) Computer Supported Cooperative Play, Third Places and Online Videogames. *Proceedings of the 13th Australian Conference on Computer Human Interaction*, Brisbane, Australia.

Warren, M. E. (2008) 'The nature and logic of bad social capital', in D. Castigilione, J. W. V. Deth and G. Wolleb (eds) *The Handbook of Social Capital*, New York: Oxford University Press.

Wellman, B. and C. Haythornthwaite (2002) *The Internet and Everyday Life*, Malden, Blackwell Publishers.

Wellman, B., A. Quan-Haase, J. Boase, W. Chen, K. Hampton, I. Díaz and K. Miyata (2003) The social affordances of the internet for networked individualism. *Journal of Computer-Mediated Communication,* 8(3). Retrieved from http: //dx.doi.org/10.1111/j.1083–6101.2003.tb00216.x.

Wenger, E. (1991) *Communities of Practice: Learning, Meaning and Identity*, Cambridge, Cambridge University Press.

Wenger, E. (2008) 'Communities of practice.' (July 11, 2008) Retrieved from http: //www.ewenger.com/theory/.

William, D. (2006) Why game studies now? Gamers don't bowl alone. *Games and Culture,* 1(1): 13–6.

Zielke, A., C. Roome and A. B. Krueger (2009) A composite adult learning model for virtual world residents with disabilities: a case study of the virtual ability Second Life® Island'. *Journal of Virtual Worlds Research,* 2(1). Retrieved from https: //journals.tdl.org/jvwr/article/view/417/461.

15 Representations of race and gender within the gamespace of the MMO *EverQuest*

Keith Massie

This chapter explores the ways in which graphic depictions of race and sex are constructed within a popular, online video game and examines the values and relationships produced by the racial and gendered dimensions of avatars – both playable and non-playable – within the online game *EverQuest* (a subscriber-based, fantasy-themed, online video game). I will argue that the ways in which avatars are depicted within the game further reinforce and reify social disparities found in 'real life' (offline subject positions). I will examine both explicit differences of representation (such as, the demographics of various sites within the game) as well as depictions of race and sex that are subtly imbedded within the game's format (such as, the suggestion within the game that bestiality is more preferable than miscegenation). By researching not only the explicit *rate* of representation of racial and gendered images but also *how* such individuals are subtly depicted regarding issues of power, I will demonstrate that women and Black and Minority Ethnic characters are relegated to weaker roles than their 'White' male counterparts. Thus, this work will show how various social biases surrounding identity are perpetuated within this popular game.

'Avatars are', states Nakamura (2002, p. 31), 'the embodiment, in text and/or graphic images, of a user's online presence in social spaces'. *EverQuest* has both player character (PC), as addressed by Nakamura, and non-player character (NPC) avatars (representations of characters within the game that are used to interact with, aid, and/or inhibit action of PCs). Both types of avatars (PC and NPC) in *EverQuest* were analysed to determine how various character types were represented. To do this, the study focused on three primary spaces within the game: the character creation, the tutorial, and the central 'city' within the game known as the Planes of Knowledge (POK).

The research that informs this chapter was conducted over a 6-month period when the game expansion known as 'The Prophecy of Ro' was the primary frame of analysis; however, near the end of the study in September 2006, another expansion, 'The Serpent's Spine', was released. As such, this chapter makes only peripheral comments on the character type (the Drakkin) opened by that release and, instead, focuses the bulk of analysis on those already in play during the earlier period. Appropriating the terms of Swiss linguist Ferdinand Saussure, this study is synchronic rather than diachronic in nature. This means that while this study addresses avatars within a fixed time comparing avatars in the space to each other, future research may implement a diachronic analysis to determine if the representations found herein have changed or remained the same.

A mixed methodological approach informs this chapter. The first section reviews the literature addressing the size and scope of *EverQuest*, Each subsequent section has its own methodology. The second section addresses the character creation phase of the game through a rhetorical lens – including both textual and visual rhetoric (an analysis of both the words present and the images within the gamespace as well as the interconnected nature of the two). The third section, which examines women characters in the game, uses a similar approach but also includes Strauss' phenomenological understanding of posture as it relates to being human. The fourth section uses a simple quantitative methodology by using percentages, percent differences, degrees of likelihood, and a split half design. Combined, these methods – which will be further explicated in the following sections – should expose some hidden elements within what the rhetorician would label the text, the social scientist would call the data, and we will simply call '*EverQuest* as game'.

Massively Multi-Player Online Role-Playing Games, like *EverQuest*, have been abbreviated in numerous ways. They are commonly abbreviated as MMORPGs. Krotoski (2004) calls them MMOGs; however, both Massive Magazine – which is solely dedicated to discussing such games – and VirginWorlds.com (an online source, including a weekly podcast, dedicated to such games) calls them MMOs. Throughout this chapter, I will use the abbreviation MMO to denote such games. Within the first section addressing MMOs, I will also highlight previous scholarship on the intersections of: women and video games; females and *EverQuest*, and video game play and the racial 'other' – a decentred non-white group. I will, then, attempt to draw out how race and sex are constructed or (re)presented within *EverQuest*. In order to do this, I will examine, in the second section of this chapter, how sex and race are rhetorically constructed during the character creation phase of game play. The third section will focus on how – within the actual game environment – women are sexualized as well as displayed as 'weak'. In the fourth and final section (which will be my central focus), I will discuss the cyber-demographics of the game space to show how women as well as the racial 'other' are devalued through a lack of representation in the preconstructed identities of NPCs. This chapter argues that MMOs are not so much fantasy worlds (where 'true' equality can exist) but rather are merely, spaces in which racial and gendered stereotypes are constructed, re-constructed, and maintained.

Everquest and mmos: their size, impact and significance

Discussing MMOs in general, Spool (2006, online), citing Ed Castronova, points out the economic impact of MMOs:

> Games now generate the same revenue as Hollywood blockbusters [yet] Hollywood box office sales have been flat for 3 years [,but] gaming software grew exponentially over the same period.

In addition, Castronova (2005) claims that more than 10 million people play MMOs. With so many people playing these games, as well as the amount of revenue generated, it seems apparent that MMOs are a significant cultural artefact and worthy of analysis.

Why study *EverQuest*? To answer, we need only look at the economics tied to it as well as the number of participants and their time spent within its world. Castronova (2005, online) found that 'the Gross Domestic Product [GDP] of the world of *EverQuest* comes in at about $2000 per capita . . . which ranks above China at about $500 per capita'. Becker (2002), citing previous work by Castronova, states that *EverQuest* would produce a Gross National Product (GNP), if it were a nation unto itself, of '$2,266 [per capita] – comparable to the 77th richest country on Earth and ranking it between Russia and Bulgaria' (2005, online). Though these economic elements suggest the cultural and social importance of the game, it seems necessary to address the 'per capita' aspect of these figures by looking the game's population.

Sony, the makers of *EverQuest*, provided subscription numbers for the game until January 2004. In September of 2003, Sony claimed that there were more than 450,000 subscriptions; however, Sony's last report in January of 2004 points out that there were more than 430,000 players. Such numbers become more concretized if we take a moment to consider *EverQuest* as a city and see where it ranks among cities globally. In 2006, the fourth largest French city, Toulouse, had a population of approximately 437,715, and the 16th largest German city, Bochum, was around 376,586 people. In both Italy (Bologna; population 374,561) and the United Kingdom (Bristol; population 416,400), *EverQuest* as a city would be about the seventh largest city in the country. Moving away from European comparison, we still see significance. As a city, *EverQuest's* population would rival Yokosuka, Japan (population 422,737) to be the 33rd largest city in Japan, and it would compete with Kansas City, Missouri (population 444,965) to be the 40th largest city in the United States. *EverQuest's* population is not only large, but players spend an exorbitant amount of time within the games' realm. Table 15.1 shows a few of the top players according to the eq.com website.[1]

It seems astounding that some of these individuals (as well as countless others not listed) have spent more than a year of their lives absorbed by the *EverQuest* realm.

Table 15.1 Information as of 5 November 2006

Name	*Level/race/class*	*Hours logged*
Rremix	75 Vah Shir Bard	55 days 12 hours 44 minutes
Elrusion Startail	75 Half Elf Paladin	289 days 4 hours 1 minute
Kumulas Aggro	75 Erudite Wizard	327 days 17 hours 27 minutes
Ryuichl Dragonslayer	70 Dark Elf Shadowknight	548 days 14 hours 58 minutes
Kyrina	75 Half Elf Bard	462 days 6 hours 29 minutes

Given that it is such a significant cultural phenomenon/artefact, surprisingly, few scholars have addressed this issue. Regarding women, Postigo (2003) has pointed out that '43 percent of . . . videogame players [are] women'; although, Krotoski (2004) places the figure slightly lower at 39 percent. Even so, Krotoski (2004) claims that women are 'flocking' to MMOs like *EverQuest* because of their community and cooperation aspects; however, such research highlights the sex of the player and not the representation(s) of sex found within the game's virtual world. Notably, researchers Chee and Smith (2003) as well as Chappell *et al.* (2006) looked at the possible addictive element of *EverQuest*. Oddly, when discussing their participant observation (through having constructed a character), Chee and Smith (2003) *never* tell the reader what *sex* was selected – though they do point out the race and class of the character. Additionally, all three of their interviewees were male, which seemingly biases their data (though they do claim that the information is not really 'generalizable'). Focusing on female players, Krotoski (2004: 7) claims that some MMOs 'support enthusiastic female majorities of up to 60 percent.', yet this rate appears to be for such games as *Sims Online* and a more accurate range of 12 to 20 percent matches fantasy-based games like *EverQuest*. In addition, Bauman *et al.* (2006) graphically illustrate through MMO graphics differences between male and female players on such factors as age distribution, willingness to gender-bend, and whether they play the game with a significant other. In contrast, the BBC (2003) reported on findings by Edward Castronova that *EverQuest* characters that were auctioned off online garnered a higher price when such characters were male than when they were female. According to the BBC (2003), this – along with the fact that only 'attractive' female character types are played – shows 'that *EverQuest*, for all its fantasy elements, is not free from sexism'. While this addresses the sex of the character rather than that of the player, it fails to discuss in detail the representation or lack of representation of female character types.

The aforementioned research addressed to some extent women as players/characters, but there is little consideration of the racial 'other' as a player/character. Remarking on cyberspace generally, Kolko *et al.* (2000: 5) argue, 'the bulk of the growing body of literature in cyberspace studies has focused on only a handful of issues and arguments, in ways that have effectively directed the conversation on cyberculture away from questions of race'. Unfortunately, this seems true of the specific context of online video games. Kilman (2005) describes how Larry Fitcheard (an African American male), who plays MMOs, is considered an anomaly due to his race. Fitcheard, in the article, not only tells us that he is consistently told by others online that they are surprised that he is Black but he also points out 'in some of the earlier games, you couldn't pick a Black character' (2005, online). While this highlights the fact that representation may be absent for the racial 'other', it fails to address an analysis of representations when they are possible or present. For example, Hayot and Wesp (2004) address the representation of the Erudite race (a 'Black' race) in *EverQuest*. Their reading of the race as 'inverting a number of pernicious racist stereotypes of African or African-American and attaching them to *EverQuest's* only "Black" race [giving the race high intelligence but low strength]'

(p. 415) may be a fair *initial* analysis but, later in the present chapter, this will be shown to be somewhat faulty.

There has been almost no research into the representations in *EverQuest* of women and racial 'others' but such research is necessary. If Kotoski (2005) is correct that women are 'flocking' to MMOs, then it seems significant to examine how women are depicted to inform future, female players about the format they are 'buying into' (literally, given the initial cost of the game as well as the monthly subscription). Kilman (2005, online) points out that not only did a 'survey by the Kaiser Family Foundation [find] that black and Latino youth played video games an average of 23 minutes longer per session than their white counterparts' but also a 'recent study from Nielsen Entertainment identifies black and Latino players as an emerging market for the video game industry' (2005, online). Like their female counterparts, Black and Minority Ethnic players should be made aware of the format they are 'buying into'. With that, let us begin to illuminate the representations of both groups as depicted in *EverQuest*.

The first phase: creating (a) character

Before engaging the *EverQuest* cyberscape, a player must log into an account and create a character. Rhetorically, 'character' functions as a unique, polysemous term. On one hand, it refers to what could be called a 'detached other' that is removed from the self and can be viewed as an object to some extent (such as, a 'character' in a play or book). On the other hand, 'character' can be seen as a sum total of an individual's personality (such as, she has strong character). Thus, what could be labelled the 'object/attribute distinction' of the term is easily determined based on whether the term is functioning as a noun (object) or as an adjective or object (attribute). Though these two meanings seem clear in most contexts, a new *EverQuest* player may find them blurring when he/she begins to create a character. This blurring occurs because the top of the screen states 'Character Creation' and the button to click when a player has finished selecting such things as sex, race, eye/hair colour, and class reads 'Create character'. The phrase 'Create character' subtly depicts an abject state since it can mean *both* an external image (the visual avatar on the screen) as well as an internal attribute (the specific player's integrity, honesty, and other traits that make up 'character'). While this rhetoric – the graying of the 'object/attribute distinction' – may encourage the player to 'embody' the selected avatar for *EverQuest*, such phenomenological dimensions of the event must be left to future study. For this chapter, we must instead focus on the available selections during 'Character Creation' and what such selections say about women and racial 'others'.

During the 'Character Creation' stage, a player must select the sex, race, and class of his/her character or avatar. After sex and race are selected, the player may also choose to select various attributes such as eye colour, hair length, facial hair, hair colour, face shape, etc., but this is not a necessary step to playing the game. Even though it is a fantasy game, there are only two sexes – female and male. There are, however, 16 'races' (though this signifier will be challenged later in this chapter) as well as 16 'classes'. Table 15.2 provides quick overview of the various 'races'.

Table 15.2 'Races' playable in *EverQuest*

Race	*Brief description*	*Determining male from female*
Barbarian[2]	Appears to be a large, muscular human and always has 'white' skin	Easy to determine because female has large breasts, shows cleavage, and is somewhat smaller
Dark Elf	An adolescent-sized human with pointed ears and 'dark blue' or 'black' skin	Same as Barbarian
Dwarf	Extremely stocky, short human[3]	Same as Barbarian
Erudite	'Black' human with *slightly* elongated forehead	Same as Barbarian
Gnome	Extremely short human, pointed ears, and 'white' skin	Same as Barbarian (though due to size, female has smaller breasts)
Half Elf	A young adult human with semi-pointed ears and 'white' skin	Same as Barbarian
Halfling	Short human with big feet and 'white' skin	Same as Barbarian
High Elf	An adolescent-sized human with pointed ears and 'white' skin – more pale than Wood elf	Same as Barbarian
Human	Human with 'white' skin	Same as Barbarian
Iksar	Lizard-like biped with tail	Female has smaller frame and different facial structure (no breasts)
Ogre	Extremely large, ugly, human-like appearance	Same as Barbarian
Troll	Tall, ugly, rotting-looking skin	Same as Barbarian (though female does not have large breasts)
Vah Shir	Cat-like biped with tail	Female smaller, colours of sexes distinct, and clothing styles different (no breasts)
Wood Elf	An adolescent-sized human with pointed ears and 'white' skin – less pale than High elf	Same as Barbarian
Froglok	Frog-like biped that hops about to move	Only distinction between sexes is coloration (quite difficult to distinguish)
Drakkin[4]	Humans with tattooing and tiny horns on face and skull	Same as Barbarian

Perusing Table 15.2, you should see that overall female characters in the game tend to possess large breasts and show cleavage; stereotypes that reinforce harmful notions that a woman's value is connected to her cup size. Such sexualized representations are not new. Previous researchers have shown how the iconic heroine Lara Croft of *Tomb Raider* was given sexualized characteristics and designed for the heterosexual male gaze (Krzywinska and King 2006; Richard and Zaremba 2005). Showing cleavage for female characters is not isolated to this initial phase of *EverQuest*. While I will address this in more depth later in the chapter, I believe a quick example may be useful here. Within the game, I had two characters that own the same magical robe. The first character is an Erudite, male wizard named Foucault, and the second character is a Human, female magician name Debeauvior. Although the robes are *identical* (I received them for doing the same task), Debeauvior's robe has a triangle piece of fabric missing in order to show her cleavage, but Foucault's completely covers him (see Figures 15.1 and 15.2). This

Figure 15.1 Debeauvior.

Figure 15.2 Foucault.

conforms to Taylor's (2006: 13) assertion: '[female players report] an assortment of fairly stereotypical sexualized bodies. Female avatars in *EQ*, especially those derived from a basic human form, wear very little clothing and often have large chests and significant cleavage'.

Maybe more subtle is the construction and representation of 'race'. Any observant player will quickly notice that to select 'human' is to be 'white', but more is occurring than this simple neglecting of the racial 'other'. If I wish to play a 'black' character, I must choose a different 'race' – Erudite. This may appear obvious (that is, a 'black' character would be a different 'race'); however, when one investigates the concept of 'race' as constructed in the game and listed on Table 15.2, you see something else entirely. Taylor (2006: 12) argues, 'In the game[,] players choose 'races' that probably are best thought of as species types'. Given such choices as lizard-like, cat-like, or frog-like characters, I would concur that the list is not one of 'race' but rather one of 'species'. As such, the game rhetorically suggests that to be 'Black' (Erudite) is to be a different *species* than 'Human'.

Dark elves are to High elves and Wood elves as Erudite are to Humans. In short, Dark elves are the 'Black' version of elf. Such a claim runs counter to previous researchers. Taylor (2006:114) argues that dark elves are 'blue' and sides with Hayot and Wesp (2004) that Erudites are the only 'Black' race. The blueness/blackness of the race can be seen in Figure 15.3.

Though monitor settings may play a role in how dark elf avatars are coloured, it is important to realize that even when seen as 'blue' they are visually and rhetorically constructed to contrast the 'White' elves (High and Wood). Given this, I think Taylor (2006) and Hayot and Wesp (2004) are incorrect when they profess that the Erudite is the *only* 'Black' race. There are two 'Black' races in *EverQuest*: Erudite and Dark elf. With this similarity established, I would suggest three racial categories for PC types within *EverQuest*: black, white, and animal. Table 15.3 illustrates how such a categorization would work.

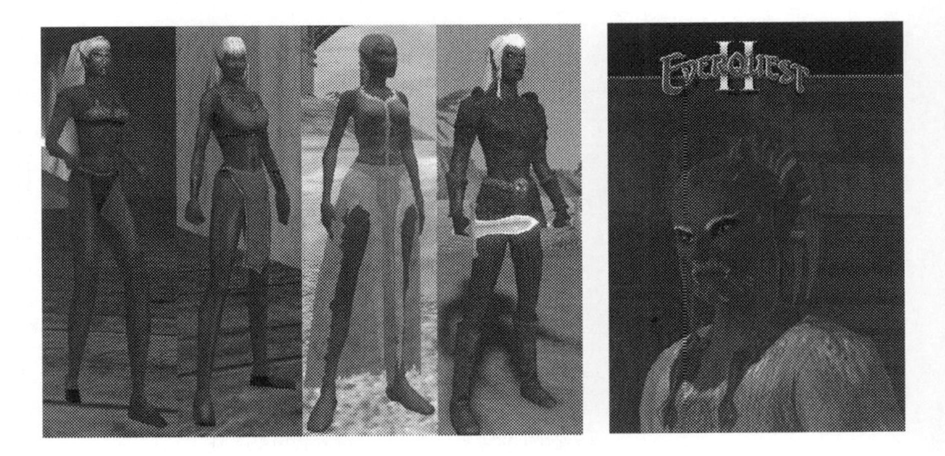

Figure 15.3 The colour of Dark Elves.[5]

Table 15.3 Racial categorization

Category	*Races included*
Black	Erudite and Dark elf
White	Barbarian, Dwarf, Gnome, Half elf, Halfling, High elf, Human, and Wood elf
Animal (or Other)	Iksar, Ogre, Troll, Vah Shir, and Froglok

The 'Character Creation' stage not only has a way to reinforce that 'Black' characters (Erudite and Dark elf) are a different 'species' but also that it does so through *privileging bestiality over miscegenation*. One need only compare two character types to make such a profound assertion: Half elf and Drakkin.

A Half elf, by its description within the game, is a character created by blending the attributes of the Human race with those of an elf. The game even professes that 'many tales of romance between these two races' are not uncommon. This, however, is confusing. What two 'races' are they addressing? There is only one Human 'race', but elves have three 'races': High, Wood, and Dark. The game's assertion would seem to suggest that Humans can mix with any Elvin race, but this does not seem to be the case because we find only 'white' Half elves (no apparent mixing of Human with Dark elf). Therefore, the romance is between 'white' Humans and 'white' elves (either High or Wood). Such a construction suggests that the game promotes an anti-miscegenation ideology. This anti-miscegenation further pushes 'Black' characters (both Erudite and Dark elf) to the margin as a different 'species' because (1) Humans can 'procreate' with 'white' elves, but Erudite cannot, (2) Humans cannot 'reproduce' with Dark elves, (3) Dark elves cannot even blend with *other elves*, and (4) Erudite and Dark elf – while both 'Black'– do not gain the advantage of being able to 'reproduce' with each other.

To make matters worse, the game seemingly suggests that bestiality is preferred to miscegenation. According to the game, the Drakkin, as a 'race', are 'humans touched by dragon's blood'. As such, they are a blending, a combining, or a 'procreation' of a Human with a dragon. In addition, a Pan-like creature (a half man and half goat entity similar to that found in Greek mythology) can be found within the game in the library of the Plane of Knowledge (POK). Such representations presumably whisper that animal bodies are more aesthetic, valued, and/or desired than 'Black' bodies.

Still in the 'Character Creation' screen, a player who has selected his/her race and sex must still chose a class before being able to click the 'Create Character' button and move into the cyberscape of *EverQuest*. 'Class', Taylor (2006: 13) points out, this 'might be best thought of as ability or vocation'. As an attribute, 'class' is whether a character is a wizard, warrior, monk, etc. During my study, there were a total of 16 'classes' from which to choose; however, a player's selected 'race' will dictate which of the given 16 choices are still accessible. For example, the game does not allow for Human Beastlords, Dwarf Monks or Erudite Rangers just to name a few. Certain classes are more rare (fewer races are allowed to select them).

For example, the Monk class only allows for one of three 'races' (Human, Iksar or Drakkin)[7]. The average number of choices of class available to any given race is seven; Humans have the most options with 13 class choices and numerous races are tied for the least with five. We need not only look at the number of choices to see that 'White' Humans are privileged, but we also need to address what I will call 'class rarity'.

Class rarity can be seen as how often a player is able to select a given class. As stated previously, a player must select one of three, specific races or he/she will not be able to play a Monk; in contrast, only two races *cannot* be a Warrior (they are Erudite and High elf). Taking the total number of playable combinations ($n = 112$) and dividing it by the number of playable races ($n = 16$), you arrive at the average number of class choices for any given race (i.e., seven). Using a pseudo-split halves method, I broke classes into two categories: rare and common. If a class was available to seven or more races, then that class was deemed 'common'. On the contrary, if a class had less then seven races that could select it, I called it 'rare'. Furthering this framework, I determined – by dividing the number of 'rare' classes available to a race by the total number of classes available to that race – a variable I call '% rare'. The '% rare' variable illuminates to what extent a given race has access to classes that other races do not (how privileged is the race). For example, the Erudite race can select from seven different classes. Of these seven classes, two are 'rare' and five are 'common'. Dividing the number of rare ($n = 2$) by the total number ($n = 7$), we find that Erudite has a '% rare' of approximately 29 percent. The '% rare' variable has little meaning in itself, but if we determine the average '% rare', we can find how a race compares to its peers.

Combining the data for all races, I found that the average '% rare' is 40 percent. Using this as a baseline, some interesting findings emerge from the data. First, 'white' Humans not only have the most possible selections of classes but they also have an above average '% rare' (at 46 percent) – having the largest number of 'rare' classes (6) available to them of any race. In contrast, the 'Black' races (Erudite and Dark elf) have a below average '% rare'. Erudite at 29 percent and Dark elf at 25 percent are both significantly below the average '% rare'. In addition, the two races combined only have four 'rare' classes available to them. It appears that 'Black' races, to some extent, lack privilege or access.

Having struggled to select the sex, race, and class of his/her character, a player clicks the 'Create Character' button and anxiously awaits his/her entrance into the cyberscape of *EverQuest*.

Entering the tutorial

Seemingly regaining consciousness, you open your eyes to find yourself in a jail cell. In the cell with you, you see a Barbarian male named Arias. Arias, an NPC, not only shows you how to escape the jail but also is the leader of the revolt in the Mines of Gloomingdeep, which serves as your initial training grounds (notably, only those players who select the Vah Shir race will *not* use this tutorial). It seems almost comically stereotypical that your first encounter would be with a brawny, 'white' male who, as the leader, is the central figure of this realm. Within this realm, there are two elements that should be addressed before moving on to the 'main' realm: demographics and women resting.

The demographics of the Mines of Gloomingdeep quickly display common biases. While a reader may believe demographics to be the total population of the space, it is important to realize that players come and go frequently; thus, demographics, for the purposes of this chapter, are composed solely of the NPCs found within the given cyberspace. Such demographics are enlightening because they demonstrate possible (un)conscious biases constructed by the game designers.

The Mines of Gloomingdeep are basically composed of two areas: the revolt camp and that which is outside of the camp, which I will refer to as 'the mines'. Table 15.4 illustrates the demographics of the camp as well as those of the total area.

Table 15.4 clearly shows that female and 'Black' NPCs are relatively rare – such as, females make up a mere 17 percent of characters within this entire area, and Black characters fair even worse at only 13 percent. Although lack of representation creates one gauge by which to critique the space, it is also interesting *how* females or 'Black' characters are represented.

While females are smaller than their male counterparts of the same race, they, for the most part, match the strength (a character attribute) of males of the same race, level, and class. On the surface, this appears to be parity; however, a quick look at how bodies – both male and female – perform within the space may highlight how female PCs are ever so subtly depicted as weak (or at least weaker than their male counterparts). After a battle, a character rests to regain health and mana (magical energy). The fastest way to regenerate these essentials is to sit. Oddly, the performance of sitting is different for male and female characters. Male PCs sit either cross-legged with arms resting on their legs or kneeling with knees together and arms resting on thighs. In contrast, all female characters appear to sit with their legs to side of the centreline of the body and right arm outstretched for support.

Strauss (1966: 155-56) states, 'One should not forget that the human arm neither supports the trunk nor has to hold up the body, a function still assigned to the

Table 15.4 Demographics of the mines of Gloomingdeep

Location	*Category*	*Number*	*% of total in given space*
Revolt Camp[6]			
	Male	15	79%
	Female	4	21%
	Black	3	16%
	White	10	53%
Camp plus Mines			
	Male	19	83%
	Female	4	17%
	Black	3	13%
	White	12	55%

forelimb of tree-inhabiting primates'. In addition, Strauss argues that it is the 'upright posture' (the title of his chapter) that make humans human. Using Strauss's logic, I would posit that the male sitting position affirms his strength by having an upright posture. In contrast, the female sitting position performs weakness requiring an arm to 'support the trunk [and] hold up the body'. It should be noted that while the two cyber-representations may have the same strength (they could have the same number for their strength attribute), such an attribute of their 'being' is personal (within the game, another PC cannot know this number); though, the visual representation of sitting is a social performance that may non-verbally and subtly display weakness.

After your initial training and an array of small tasks, you are ready to leave The Mines of Gloomingdeep. You leave out a dark tunnel and head to the POK.

Entering the plane of knowledge

The POK is a gathering point for adventurers in *EverQuest.* It is a large city with various buildings and numerous portals that teleport a player to other spaces within the cyberscape of *EverQuest.* As a central hub where players consistently pass through and/or interact, POK is one of the most significant 'cybergraphic' locations in the game. As such, this final section will focus on the demographics of POK, the visual representations of female NPCs, how female and 'Black' NPCs are neglected in some important roles, and how cyber-segregation is a sad fact of POK.

POK can be conceived of having a central (octagon-shaped) section – where the library, bazaar, main bank, and guildhalls are – and three subsections that extend off of this central area. Each subsection is distinct with one to the North, to the East, and to the South.

Throughout POK, I found a total of 301 NPCs; however, eight of those NPCs were not a playable character race, so I removed them from the data because I could not clearly determine their sex. Using the remaining 293 NPCs, Table 15.5 illuminates the NPC demographics of this important game space.

Since women are approximately 50 percent of the world population and almost 51 percent of those in the United States, it becomes quite obvious that as a group they are comparatively underrepresented in the POK demographics. Such an assertion may be perceived as accepting McQuail's (2004: 171) claim that 'media

Table 15.5 Demographics of NPCs in POK

Category	*Number*	*% of total population*
Male	222	76%
Female	71	24%
Black	45	15%
White	195	67%
Animal/Other	53	18%

should *reflect* . . . societies (and communities) . . . in a more or less, proportional way'; however, I only seek to be descriptive and leave the evaluative element (regarding this) to my reader. Since they are inversely related to females, males are obviously overrepresented. In contrast to these previous two groups, 'Black' people in the United States are approximately 15 percent of the populations, and, oddly enough, make up 15 percent of NPCs in POK. We will find, however, that it is not *that* 'Black' people are represented but *how* they are excluded and positioned that matters. Before bringing those elements to light, which will also illuminate similar elements for female NPCs, I will address briefly the visual representations of female NPCs.

Previously in this chapter, I pointed out how the majority of women characters in the game are depicted as busty and have cleavage displayed regardless of what clothing or armor is selected. Such attributes of female cyberbodies in *EverQuest* are also found in the female NPCs' depictions. Although busty, cleavage showing female NPCs are common throughout POK, I would like to especially note one named Illusionist Lobaen. As an NPC, Illusionist Lobaen can be found standing in the 'Basement' of the library. Her role as an NPC is to sell Enchanter spells; however, her representation appears more like a stripper ready to sell lap dances. For she appears to be wearing what resembles a black 'teddy' covered by a white, transparent robe.

This example (as well as the first two images of Figure 15.3) shows how female representations are constructed to be sexually provocative, but I could find no similar example of male representation (there are not male NPCs wearing a pair of Speedos and bow tie like a Chippendale dancer). Having illustrated, the visual depictions of female NPCs, I turn now to highlight how important NPC roles lack any significant female or 'Black' representation.

Two of the most central roles for NPCs in POK are the Banker and the Guildmaster. Players *must* interact with Bankers constantly because they not only hold the player's currency (which, if a large volume, can encumber and slow down the player who does not place it in the bank) and exchange currency (such as, converting silver into gold to diminish how much weight is being carried) but they also serve as a safety deposit box for placing additional equipment. In POK, we find three Bankers (public bankers that can be used by any player compared to guild bankers that only members of the given guild have access to); unfortunately, we also find that all three are 'white' males. Given that female and 'Black' characters combine to make up more that a third of POK population, probabilities would suggest that one of the Bankers would be either 'Black' or female. Does this subtly suggest that 'Black' people and women cannot be trusted with money and other's belongings?

Women and 'Black' people are not Bankers, but do they get fair representation in the other central NPC role of the Guildmaster? Before answering this, let me explain what a Guildmaster is. A Guildmaster is, for the most part, a highly skilled teacher or mentor. We could compare them to a college professor or any skilled labourer who would take on apprentices. After each gained level, players seek out their appropriate Guildmaster (based on class) and use earned 'experience points' to purchase lessons and, thereby, improve a given skill for his/her character. For example, a character who always uses a two-handed sword (a 2H Slashing weapon) may, when he/she has 'levelled up', find his/her Guildmaster and 'purchase' a lesson in 2H Slashing, which will improve his/her skill in that area and increase the likelihood of hitting an enemy when using a 2H Slashing weapon. As can be seen, it is imperative to a character's improvement and success that he/she interacts with the NPC Guildmaster.

In POK, there are a total of 37 Guildmasters. Of the 37 Guildmasters, four are female and six are 'Black'. Given this, females make up only 11 percent of all NPC Guildmasters, and 'Black' characters are 16 percent. It seems somewhat disturbing that, while women NPCs make up 24 percent of the POK population, only 11 percent of Guildmasters are female. Maybe, since women NPCs are a small group, females may have a higher rate of representation as Guildmasters (the number of female Guildmasters divided by the total number of women in POK will give us what percent of women are Guildmasters or the rate of representation, and we can, then, compare this to men). Finding that 15 percent of all male NPCs and only 6 percent of all female NPCs are Guildmasters, such an investigation only furthers the idea that women are underrepresented. Women, it appears, lack significant representation in two of the most central NPC roles: Bankers and Guildmasters. In contrast, the 16 percent of Guildmasters that are 'Black' seems to accurately parallel the fact that 15 percent of NPCs in POK are 'Black'. Before professing that Black and Minority Ethnic people may have obtained some parity, it is pressing that we conclude by looking at the cyber-segregation that occurs in POK.

As noted previously, POK is divided into four parts: the octagon-shaped centre, the North area, the East area, and the South area (refer to Image 16.5). With such an arrangement, I contend that you can analyze it *as if* it were a city (the centre) with three suburbs. To hone in on 'race' and gender, I examined the three suburbs and bracketed out all NPCs (within the specific regions) that would fit within the 'animal' race. By placing the 'animal' races to the side, I condensed the data, which serves to make it richer. Thus, the demographics of the various suburbs suggest that cyber-segregation occurs in POK (see Table 15.6).

While Table 15.6 highlights cyber-segregation since the North is only 7 percent 'Black' and the South is 75 percent 'Black', I think that a somewhat unseen variable buried in the data may aid in solidifying the claim of cyber-segregation. Note that with 17 percent of the six female NPCs found in the South being 'White', you find that five of the females in the South must be 'Black'. This alone has little

Table 15.6 Cyber-segregation in POK

Suburb	*White*	*Black*	*% Black*	*Male*	*Female*	*% Female*	*%White Female*
North	27	2	7%	20	9	31%	100%
East	24	3	11%	19	8	30%	75%
South	5	15	75%	14	6	30%	17%

value, but if you add it to the fact that only nine 'Black' female NPCs exist in all of POK, you begin to further solidify the claim that cyber-segregation is occurring in POK.

It appears that 'Black' NPCs are relegated to the South suburb. This aspect of POK alone is somewhat morally questionable. I need only highlight one fact that distinguishes the South from all other parts of POK. Unlike the Central, North, and East regions of POK that all have lively flora and vibrantly green grass, the South has almost no grass and a few dead trees. To some extent, it is as if the game designers intentionally constructed a cyber-ghetto and, then, placed the highest percentage of 'Black' NPCs into that space. This cyber-ghetto element may explain why in Table 16.6 the percentage of 'White' Female NPCs rapidly decreases for the South region. Is it possibly due to the expectation that a 'White' Female would not be safe in the 'Black' ghetto? Or maybe it is to keep 'White' women away from 'Black' men (remember the previous rhetoric in the game that seemingly argued against miscegenation)?

Conclusion

This chapter has demonstrated that, unlike previous research that centres on the player's sex or race, virtual representations of sex and race in *EverQuest* reveal how biases and ideologies are merely perpetuated, reinforced, and (re)inscribed in, often, subtle ways within a new space or arena., I would argue that taken together, the previous findings create a challenging critique of the virtual representations of both women and Black and Minority Ethnic people in the game *EverQuest*. Future research should consider the findings herein as a baseline for comparison over time to examine whether such depictions continued in later expansions of the game. Additionally, researchers may appropriate the methods of analysis within this chapter (such as, the '%rare' variable or the rate at which character types have access to a game's classes or job roles) to investigate other online games (such as, *World of Warcraft*) to determine if the findings within this chapter are consistent with other games or simply a trait of *EverQuest*. Since game companies have the power to create and construct any conceivable world, researchers may consider studying how such companies determine the construction of their avatars and how such stereotypical representations, as noted in this chapter, find their way into the gamespace.

Notes

1 It should be noted that Table 15.1 consists of just a *small*, random sample of individuals listed on the 'Leaderboards' found at http://eqplayers.station.sony.com/leaderboards.vm

2 According to the famous bioethicist Peter Singer (2000), ancient Greeks and Romans did not guarantee the right-to-life to slaves, deformed children, or barbarians. The right-to-life is the most basic right of personhood and is the precondition for any meaningful existence. As such, one should note that while historically 'barbarians' could essentially be any race, *EverQuest* makes only 'White' ones. The etymology of the term 'barbarian' in Proto-Indo European, 'barbar' suggests a 'non-Aryan'. However, the *EverQuest* avatars display typical Aryan features.

3 Interestingly, the game professes that 'some females [dwarves] have well-trimmed beards'; however, during all my game play, over 150 hours, I did not see a player character (PC) or non-player character (NPC) who was a female dwarf with a beard.
4 The Drakkin were added as a race in the September 2006 expansion called *The Serpent's Spine*. They are supposed to be a blend of Dragon blood with human blood. I will only discuss them briefly in this piece because the expansion was released as I was ending this study.
5 These images were obtained through publicly displayed images on Yahoo images.
6 From this point on, I will no longer acknowledge the Drakkin as a race in the game because the expansion that introduced them was released as I was concluding my research.
7 The total number of NPCs in the Revolt camp is 21; however, two of the NPCs are not playable character types (one is a small dragon, the other a kobold). Given this, they have been excluded from the data. Both characters appear somewhat androgynous, yet other NPCs use the pronoun 'he' when referring to them.

References

Bauman, S., A. Conklin, T. Martin, C. Yans and N. Yee (2006) What's next? *Massive Magazine,* 1: 20–32.

Becker, D. (2002) EverQuest Spins Its Own Economy. *CNET News*. Retrieved 21 November, 2006, from http: //news.cnet.com/2100-1040-823260.html.

BBC (2003) *EverQuest exposes cost of sexism.* Online at: http://news.bbc.co.uk/1/hi/technology/3016434.stm

Castronova, E. (2005) Gold from Thin Air: The Economies of Virtual Worlds. *Podtech Presentation.* Retrieved 3 November, 2008, from http: //cdn.itconversations.com/ITC.PopTech2005-EdwardCastronova-2005.10.21.mp3.

Chappell, D., V. Eatough, M. Davies and M. Griffiths (2006) *EverQuest* – It's just a computer game right? an interpretative phenomenological analysis of online gaming addiction. *International Journal of Mental Health and Addiction,* 4(3): 205–16.

Chee, F. and Smith (2003) Is Electronic Community an Addictive Substance? *Proceedings of DiGRA and Level Up 2003*, Utrecht, The Netherlands.

Hayot, E. and E. Wesp (2004) Style: Strategy adn mimesis in ergodic literature. *Comparative Literature Studies,* 41(3): 404–22.

Kilman, C. (2005) Video Games: Playing against Racism. Retrieved 5 November, 2006, from http: //www.tolerance.org/news/article_tol.jsp?id=1228.

Kolko, B. E., L. Nakamura and G. B. Rodman (2000) Erasing @Race: Going white in the inter(face). In B. E. Kolko, L. Nakamura and G. B. Rodman. (eds) *Race in Cyberspace.* New York, Routledge: 1–13.

Krotoski, A. (2004) Chicks and Joysticks: An Exploration of Women and Gaming. *Entertainment and Leisure Software Publishers Association.* Retrieved 21 November, 2006, from http: //www.elspa.com/assets/files/c/chicksandjoysticksanexplorationofwomenandgaming_176.pdf.

Krzywinska, T. and G. King (2006) *Toob Raiders and Space Invaders – Videogame Forms and Contexts*, London, IB Taurus.

McQuail (2004) *Mcquail's Mass Communication Theory.* 4th edn, Thousand Oaks, Sage Publications Inc.

Nakamura, L. (2002) *Cybertypes: Race Ethnicity and Identity on the Internet*, London, Routledge.

Postigo, H. (2003) From pong to planet quake: post-industrial transitions from leisure to work. *Information, Communication and Society,* 6(4): 593–607.

Richard, B. and J. Zaremba (2005) Gaming with Grrls: Looking for sheroes in computer games. In J. Raessens and J. Goldstein. (eds) *Handbook of Computer Game Studies.* Cambridge, MIT Press: 283–300.

Singer, P. (2000) *Writings on an Ethical Life*, New York, Harper Collins Pubishing Inc.

Spool, J. (2006) Good Listen – Edward Castronova: Gold from Thin Air – the Economy of Virtual Worlds. Retrieved 2 November, 2006, from http: //www.uie.com/brainsparks/author/jared/page/10.

Strauss, E. (1966) *Phenomenological Psychology*, New York, Basic Books.

Taylor, T. L. (2006) *Play between Worlds: Exploring Online Game Culture*, Cambridge, MIT Press.

16 *Wordslinger*

Visualizing physical abuse in a virtual environment

Kate E. Taylor

For many, the world of the video game is far removed from the world of the domestically abused man or woman. Indeed there would seem to be no greater contrast than to place the virtual movements enacted in the world of computer graphics against the very real and physical hardship suffered by an alarming number of people. Weekends and holidays for example, which is for many the time to engage intensively in their gaming, is also the period of time where domestic violence is at its most widespread (Lawless 2001). However, in the online video game *Wordslinger*, the world of domestic violence enters into the world of the video game as an important part of a self-help site geared towards aiding abused women. During the game, the female player[1] will pick an abuser from the list of well-documented profiles and seek to move inside the world of the game targeting typical abuse statements to score points and develop her online character. The dominant construction of games as 'masculine technoplay' (Kennedy 2007) arguably makes a computer game structure an unlikely forum to voice a debate on male on female domestic violence. However, as this chapter will demonstrate, the process of individual action and achievable goals, which are a central element of most video games (Carr *et al.* 2006), can also be seen as a method through which those who have suffered abuse can begin to actively articulate and develop from their experiences. What this chapter is seeking to explore is how the *Wordslinger* video game interacts with the real life embodied state of the female players to create an intensely affective experience that aims to aid a woman in her own recovery.

Before I discuss the game in detail it should to be noted that in any engagement with questions of abuse there is a need to examine help websites for academic, philosophical and educational reasons. This needs to be balanced with the requirement to maintain the safe environment where many women can find solace and comfort. For this reason I will be concentrating on a specific area of the 'you are not crazy' domestic violence support website, rather than engaging in debates about the individual narratives which are posted. As an ethical point I am only discussing the open access areas of the respective site (where *Wordslinger* is located) that do not need a user name and/or individual registration or login to be viewed.

This chapter develops from the disciplines of 'gender' and 'body studies' rather than the rather newer 'video game studies'. My own personal interaction with video games is actually extremely limited. Indeed, other than a brief period as a child in the late 1980s where I desired nothing more from my parents than a Nintendo Game Boy (which never materialized), my whole attitude towards video games, either in an arcade setting or on the computer screen, has been fairly dismissive. Compared to the world of literature, games seemed to provide little scope for enjoyment unless you were a 12-year-old boy with delusions of being James Bond. The desire to shoot aliens or drive a car on an imaginary circuit (the sum total of all video games as far as I could tell) was far removed from my realm of interest. It was then with great surprise that I found myself actively engaged in playing a video game in, of all places, a domestic violence self-help website. It should be noted that *Wordslinger* is, when compared to the aesthetically impressive and highly intricate video games available in the modern market, a very basic and simplistic game in terms of the tasks you are asked to undertake and the graphics used. However, what this chapter will discuss is not the merits of *Wordslinger* as an example of a video game *per se*, but rather how the video game format interacts with the notion of embodied narrative[2] and thus, acts as an affective tool for an individual players development.

In their introduction to *The Video Games Theory Reader 2* (2009), Perron and Wolf list several areas on which games studies should focus in order to develop as a discipline. Pertinent here is their call for 'Interaction of interdisciplinary approaches' (14). As with all disciplines, games studies will benefit and grow from interconnectedness with a multitude of approaches from both the social sciences and the humanities. As cultural artefacts, as much as technological ones, video games are part of the increasing convergence that is taking place between different medias such as film, television and the Internet. To this end, methodologies and critiques from other fields will offer contributions when used in engagement with the relatively new game studies specific approaches. In line with this, the methodology that this chapter employs will be flexible and interconnect with a variety of thinkers and approaches.

Wordslinger is part of a larger website (www.youarenotcrazy.com) geared towards helping and aiding women who have suffered abuse in all its forms including physical, verbal and mental. A useful summary of domestic violence is provided by Lawless who suggests it occurs when 'a man is beating, kicking, violating and raping a woman in the privacy of their shared home' (2001: 45). This premise of domestic violence as a violence that takes place in a home environment means that the Internet has an important potential role to play to helping women. The Internet opens up the possibility of communication, connection to the outside world and a (relatively) safe forum for debate and the telling of individual narratives. Optimistically the Internet represents a potentially powerful platform for women's subversion of gender stereotypes and for the development of a virtual public sphere that reflects women's needs and aims (Harcourt 1999). In short, new technologies are used for individual and collective empowerment. *Wordslinger* is unique (as far as I am aware) in that it combines the protocols and format of a typical older style video game with the desire to empower women (in this case those who have suffered abuse) through the use of technology. You enter into the site via a portal that then provides information about abuse and some traits that personify an abusive relationship. *Wordslinger* is signposted as part of this site and clicking on it navigates you away from the main site into the world of the video game.

I wish to focus in this chapter on the interaction between the video game and the notion of an 'embodied narrative' and how *Wordslinger* actively invites the female player to undergo an affective experience, which she engages with, in a multitude of ways.

Embodied narrative and the world of the game

'We think of ourselves as separate from the world – our skin as the limit of ourselves . . . yet the body is pierced with a myriad of openings. We project our bodies into the world – we speak, we breathe, we write, we leave a trail of cells and absorb the trail of others. The body enfolds the world and the world enfolds the body' (Dunning and Woodrow 2001: online).

As the above quote states, the body is central to our experience of the world in all its forms, including those 'virtual' experiences seen in media forms such as video games. Any engagement with the relationship between the 'virtual' and the 'real' needs a philosophical construction of the body that engages with the notion of fluidity and flux rather than one that is constrained by the restrictive models built on the legacy of Cartesian dualism. (This in basic terms saw a clear division between mind and body. The body was seen as highly inferior to the mind and nothing more that the physical vessel though which the mind could work.) The intricacies and complications that haunt this 'post-modern' body can be seen (although in sometimes conflicting ways) in the work of writers such as Butler (1990, 1993), Derrida (1974, 2002), Deleuze and Guattari (1987) and Harraway (1991) from Butler's Performative body to Harraway's Cyborg, the focus on the body in a variety of visual and textual strategies (White 2006) has dominated contemporary theory regarding the body and its positioning in cultural and social narratives. This, however, needs to be balanced with a methodology that does not neglect corporeal materiality or reduce the body to an 'undifferentiated surface of inscription' (Price and Shildrick 1999). The notion of the 'lived body' rather than cold abstractions (Williams and Bendelow 1998) needs to be kept at the centre of the debate. Many of the debates seen in the questioning of the interplay between the body and video games are related to the wider debate about the virtual body and how the human element narrates and acts inside the virtual space. For Hayles (1999: 38) narrative interplay is something that brings the machine and the human together. 'Cyberspace is created by transforming a data matrix into a landscape where narratives can happen . . . Data are thus humanized, and subjectivity is computerized, allowing them to join in symbiotic union whose result is narrative'.

In this way, narrative becomes embodied. The subject is simultaneously constructing, and being constructed, in their own personal (in-body) narratives. The question of what constitutes 'embodied narrative' is one that is open to a vast amount of debate. The act of embodiment is a multi layered one that asks us to consider, often simultaneously, not only our literal in-body processes (I hurt) but also our own sense of self-aware agency (I hurt because I cut myself), or lack of (I hurt because he can hurt me). In any interaction, the sense of being literally 'in-body' is open to a myriad of interconnected narrative and subjective considerations. As Butler states, 'there is no reference to a pure body which is not at the same time a further formation of that body' (1993). Weiss speaks of 'self image' as central to embodiment where an overlapping and simultaneous 'multiplicity of body image' is 'constructed through a series of corporeal exchanges that take place both within and outside of specific bodies' (1999: 2). These 'exchanges' for Liora Brezler, are illuminated by the notion of 'I-Thou', where connection becomes central to the notion of the self as an embodied subject (2006: 21). Merleau-Ponty's (1962: 146) assertion that 'the body is our general medium for having a world' is relevant here. He goes on to state that:

> sometimes it is restricted to the actions necessary for the conservation of life, and accordingly it posits around us a biological world; at other times, elaborating upon these primary actions and moving from their literal to a figurative meaning, it manifests through them a core of new significance: this is true of motor habits such as dancing. Sometimes, finally, the meaning aimed at cannot be achieved by the body's natural means; it must then build itself an instrument, and it projects thereby around itself a cultural world
>
> (Merleau-Ponty 1962: 146)

For him, the embodied subject opens up to the world in a variety of ways and emotions, actions and interaction are all part of the embodied processes. In his reference to an instrument, the development and popularity of interactive media such as video games have resulted in an added dimension to the human embodied experience. With the convergence of the physical and the virtual, the interplay between the two has become central to modern theorizations. Although for some the desire to reject the 'real' in favour of a 'virtual' existence has been documented, there are some clear issues with this hope. Lupton (2000) in her examination of the embodied computer/user focuses on the tension between the desire for a disembodied and virtual cyber body with the reality that 'meat', a frequent motif for the body in cyber literature and culture, is impossible to disregard. Lupton (2000: 480) concludes that 'while an individual may successfully pretend to be a different gender or age on the internet, she or he will always have to return to the embodied reality of the empty stomach, stiff neck, aching hands, sore back and gritty eyes caused by many hours in front of the computer terminal'. As Gramola (2000) using Barbara Brooke's work on the disabled body (1999) has noted, physical pain cannot be disassociated from the lived body regardless of the desire to engage in cyber virtual narratives. Stone notes that 'cyberspace developers foresee a time when they will be able to forget about the body. But is it important to remember that virtual community originates in, and must return to, the physical . . . Even in the age of the technosocial subject, life is lived through the bodies' (2000: 525).

In order to be able to theorize a matrix which allows for the interplay between real and virtual the status of the embodied subject needs to be central. For

Katherine Hayles, 'bodies disappear into information with scarcely a murmur of protest; embodiment cannot, for it is linked to the circumstances of the occasion and the person. As soon as embodiment is acknowledged, the abstractions of the Panopticon disintegrate into the particularities of specific people embedded in specific contexts' (1999: 198). Although the worlds of specific video games allow for an individual to loose themselves in a myriad of virtual gender, class and even 'species' (such as 'furries' in *Second Life*) subject positioning, they cannot escape the reality of their own embodiment and it will continue to inform their relationship to technology. What the exact relationship is between the online character interacting with the online world and the off line player is one that is continually open to debate. Lahti (2003) states that the video game has become 'a paradigmatic site for producing, imagining, and testing different kinds of relations between the body and technology in contemporary culture' (158).

The notion of agency here is key. The video game player is more than just responding to stimuli provided by the game. Often games involve highly complex processes that involve manipulation and exploration of the virtual environment. Gee (2003) sees the online gaming character becoming the 'surrogate' for the offline player: there is a melding of the characters goals and game world values with the player who must attempt to inhabit the world of the game to succeed. 'We act in word or deed in terms of that identity' (Gee 2003: 85). Gregersen and Grodal state that the experience of the video game player is one of 'an embodied awareness in the moment of action, a kind of body image in action – where one experiences both agency and ownership of virtual identities' (2009: 67).

In relation to female players, studies such as Kennedy's examination of female *Quake* players (2007) and Pelletier's (quoted in Carr *et al.* 2006: 171) series of interviews with teenage female players, have illustrated the complexity and negotiation that takes place in the game world with regards to female players and their agency both inside and outside the world of the video game. The debates regarding women and video games have often focused on the comparative marginalization of women from the gaming world. Cassell and Jenkins state in their introduction in the main book devoted to women and gaming *From Barbie to Moral Combat* that 'Too often, the study of computer games has meant the study or boys playing computer games (1999: 5). In recent years the development of the gaming industry has resulted in more and more attention being aimed at the female player and women designers and 'hardcore' players are becoming slightly more commonplace (Dovey and Kennedy 2006). However, positive female characters remain limited which often results in the alienation of female players (Bryce and Rutter 2002) and women face continual problems in trying to articulate an identity for themselves in the often hyper-macho online gaming world (Taylor 2003). Lourdes Arzipes' question of 'what happens to gender when it goes through hardware' (1999: xii) is pertinent here as many game narratives focus on aggression and violence and clearly this is a problem for a game aimed at victims of abuse. The bodily experiences that are undergone in a gaming process are mental (thinking inside the world of the game), physical (literal actions taken in the controlling the game), and very importantly they are affective. Games produce feelings and emotions and it is the engagement with the affective experience produced in *Wordslinger* that is central to the use of the game as a 'self-help' tool for the player.

Wordslinger

> It's not that women don't understand video games, it's that video games don't understand women
>
> (Girard 2006, online)

The most important part of *Wordslinger* involves the introductory process. Entering into the game the opening statement tells us that we have met the perfect person: kind, caring, loving. However, gradually the sounds and images of blue sky, birds chirping and sun-filled days disintegrate into darkness, the sound of thunder and howling wolves ending with the statement that the man you love is an abuser. This is reflective of many abuse patterns; there is a 'honeymoon' period before the abuse starts (Lawless 2001). It is not immediate and the narrative 'patter' presented in *Wordslinger* is one that is highly recognizable to those who have been abused. As the player reads the screen and listens to the noises, she maps her own abuse narrative onto the games outline.

I turn here to Derrida's notion of auto-affection. Auto-affection is adroitly summarized by Ticinito Clough as 'giving-onesself-a-presense-or pleasure', 'hearing oneself speak' in the closed circuit of mouth and ear, voicing and hearing (17). Auto-affection occurs when the affect is conducted by the affect, in short when you or I, affect yourself, or myself. Derrida linked the notion of auto-affection with that of autobiography, the desire to articulate your own presence to yourself (as well as to others). When you 'speak' inside your own head, your inner voice if you will, you are seeking to validate your own existence. The auto-affective process means that you presume the unity of 'speech and the precommunicated though, giving the subject an inner presence, an inner voice, so that the subject when it speaks, is presumed to speak its own voice . . . and to express its inner being' (Clough 2000: 17). In the world of the computer game or any form of interaction with a machine, we are arguably seeking to engage in an act of auto-affection. For example, when I post an online update status I engage in an act of autobiography. I seek to grant myself 'a presence', to cause an affective moment.

Derrida (1974: 165) states that:

> Within the general structure of auto-affection . . . the operation of touching-touched receives the other within the narrow gulf that separates doing from suffering. And the outside, the exposed surface of the body, signifies and marks forever the division that shapes auto-affection. Auto-affection is a universal structure of experience. All living things are capable of auto-affection. And only a being capable of symbolizing, that is to say of auto-affecting, may let itself be affected by the other in general.

For Derrida, auto-affection is the process that allows an individual to suppose a unity between speech and the unspoken through the desire to voice a narrative and declare it ones own. The opening statement in *Wordslinger* acts to offer a moment of auto-affection. As we read inevitably inside our head rather than out loud, the statements we engage in giving ourselves a 'presence' in our own subjective thought. We hear ourselves speak in the 'closed circuit' and the story that we are being told links directly into the woman's own history, her embodied narrative as she experienced it. This is distinct from representation. Representation, as it has historically related to women is 'a act of violence, perpetrated by the self-present and knowing subject against, one can only assume, the others that the subject desires to know and control. Thus, representation is a form of colonization, an imperial move on the part of the subject' (Robinson 1991: 190). When those in power represent those who they are seeking to control they are engaging in a narrative that results in the further disenfranchisement of those that are representing. Representation in its 'pure' form is the sign of one thing standing in place of another (Mitchell 1995) and this process has often privileged the experiences of the dominant (white and male) subject. For example historically under many colonial empires such as France, Britain and Japan, the native inhabitants are often represented as savage and childlike, 'legitimizing' arguments for the colonial process (McLintock 1995). The usage of symbols often results in the disenfranchisement of the marginalized embodied individual experience. Rather then *representing* a narrative of abuse, *Wordslinger* is inviting the participant to *engage* in the narrative of abuse, not as victim but as an active and articulating presence. For those who are suffering from the event or the legacy of domestic abuse the construction of their narrative tale encounters several key issues. As Lindermann Nelson (2001: xii) notes 'the connection between identity and agency poses a serious problem when the members of an abusive power system are required to bear the morally degrading identities required by that system'. In the case of those that are likely to be engaged in *Wordslinger*, the harmful and restrictive narrative of victim is one which will likely loom large in the consciousness. Instead a counter-story should be developed to combat this negative subject positioning. The process of a counter narrative construction is what *Wordslinger* is attempting to promote. A counter story is one that seeks to resist an oppressed identity and replace it with one that promotes respect and the possibility of development (Lindermann Nelson 2001). Key in this process is the identification of the 'master narratives', the dominant stories that inform our social and cultural structures. For many women, the master narrative of abuse results in disempowerment and the inability to enact a personal agency, either due to physical or internalized mental constraints. The act of auto affection is one that is imbued with the legacy of the abuse. Derrida states 'auto-affection is the condition of an experience in general' (1974: 165). Experience is brought to bear in the narrative articulation of the woman player and in the construction of a counter story an affective change must take place in order to 'infiltrate her consciousness' and to 'change her self-understanding . . . to put greater trust in her own moral worth (Lindermann Nelson 2001: 7).

This affective change takes place as the narrative of *Wordslinger* develops. From the statement that your lover is an abuser the game changes to a screen where you are invited to pick your own abuser. You are offered such options as 'Lord of self-righteousness', 'explosive real man' and 'swaggering player'. They are all 'classic' abuser profiles taking many of the most recognizable manifest traits such as jealousy, aggression, manipulation and emotional blackmail. When you are asked to select your abuser, an act of embodied narrative consideration takes place. The woman is asked to focus on her own experiences and select a description which most aligns with her own narrative. This can be what Lindermann (2001) casts as an evolving story. One of the key and sustaining elements of domestic violence is secrecy and the inability for various reasons for the woman to tell her narrative. Fear of the abuser, the legal system and social shame all play a role in silencing and constraining women who have suffered domestic violence. The selection of an abuser in the game allows the woman to focus her narrative on her own experiences but is spared the process of articulating her own abuser in real or intensely personal terms. This classic video game format when you select your player is reconfigured to provide the player a chance to engage with her own narrative but in the 'safe' environment of the game. The personal selection process allows for 'the participant to take on the goals of the games as one's own' (Oatley 1992: 355). In the selecting of an 'enemy' the woman is asked to emotionally connect with the game via a literal physical action. One of the most basic and central elements that affect our interaction with the world of the video game is the literal mapping of our own actions 'by various technological means' (Gregersen and Grodal 2009: 69). Although there is a wide range of complex technologies that allow you to map onto the game world such as Wii-remotes, dance mats, joysticks and a variety of weaponry and musical instruments; *Wordslinger* just uses basic keyboard and mouse functions. This is not a game which is seeking to build technical skill levels such as seen in more complex games (such as those on the Wii gaming platform) but the physical interaction of the player and the game is important to give a sense of agency to the female player. It is part of what the late Teresa Brennan calls the 'transmission of affect', the process that is 'social in origin but biological and physical in effect' (2004: 3). As the player maps her presence into the world of the game, she becomes more and more involved in the processes taking place and interplays them with her own physical history – her embodied narrative.

I have chosen to take 'Manipulative Power Child', as the random example for this chapter. When you select this profile you are offered the following interaction:

> My Love, I planned on dinner with you, my mate. The fact that you insist on taking your bar exam instead of spending quality time with me just proves how selfish you are. You constantly ignore my feelings and put yourself first. THIS is why nobody wants you around for very long, including me. Would you rather have a macho jerk who treats you like a sex object?

In the enhancement of the emotional connection, the character speaks directly to the player. We are drawn into a moment of self-recognition linking into past events and actions. It is an act that allows the player to claim and appropriate her own narrative experience by transference into the world of the game. This emotional connection is further developed when the abuser character presents a series of typical statements common to abuse narratives:

> **Which verbal abuse do they typically sling at you?**
> **What do you wish you'd never hear again, edit below:**
> People say you are annoying and wonder why I am with you.
> AAARRGGH! I'M NOT YELLING!!!!
> Thank God I am here to clear up your messes.
> If you're not at my beck and call, what use are you?
> **Add your own statement.**

The edit below function allows you to enter your own statement, for example, entering 'You don't know how much I control my temper. You should be grateful that I try', results in that statement becoming part of the game. Returning to the work of Oately, he states 'emotions depend on evaluations of what has happened in relation to that persons goals and beliefs' (1992: 19). The act of narrative interplay allows for the woman to align herself even further with the world of the game as her own auto-affective response can enter into the game processes. The game gains added emotional resonance as the players embodied position becomes planted in the world of the game.

The phrases that are used become targets in the next section of the game. The female figure dressed as a typical fairytale princess, is supplanted into a barren space dominated by a volcano. As the female figure moves around thorns, above each thorn are the target statements. Using the basic up and down key strokes and space to fire, the figure sprays the thorns and they begin develop into flowers and the barren landscape changes into one of growth and transformation.

Presenting the woman figure in the guise of a princess opens up the debates to a wide range of female stereotypes. For many in the Western world, the narrative structures that surround the notion of 'princess' are ones based on a beautiful woman who will meet, fall in love, be rescued from some threat, and then live happily ever after with 'prince charming'. A good example of this is the *Desperate Housewives* game where the aim is to be the perfect wife and mother or the *Disney Princesses* games where the player will seek the 'happily ever after' ending that is central to the traditional damsel in distress narratives. Dependency on a male figure marks the Princess myth structures and it is passivity that so often informs the master narratives of gender with relation to women. The counter story being told in *Wordslinger* is of an active character that is systematically transforming the events that surround her. *Transforming* rather than *eradicating* is key. This is not a typical 'shoot em up' game; the aim is not destruction rather it is growth. The woman figure does not hold a gun (phallic symbol extraordinaire), rather she wields a watering can, a nonaggressive symbol of potential. The aim of the creation of a counter story is not to eradicate the master narrative since that is impossible. Rather, the focus is on retelling the story to grant the oppressed a great agency and power than previously established. The granting of agency and a sense of empowerment inside an embodied narrative allows the possibility to develop that 'if the counter story moves her to see herself as a competent moral agent, she may be less willing to accept others' oppressive valuations of her, and this too allows her to exercise her agency more freely' (Lindermann Nelson 2006: 7). Skelly notes that 'One of the primary reasons we play games is to gain a sense of being effective in the world, even if that world is on the other side of a window through which we cannot pass' (2009: xiii). The world of the game is clearly not real; it will not prevent further abuse or heal the scars left. What can occur is a sense of affective empowerment inside the embodied narrative of the player. As Gregersen and Grodal (2009: 66) state 'we may have an acute sense of body ownership and still have the distinct non-agentic feeling if we believe that we lack the ability to influence states around us'. The lack of ability to influence is something that is clearly dominant in those who have been abused, but at the end of *Wordslinger* the ground is covered with roses and the player is informed that they have won, 'they are a survivor'. The game format, as basic as it is, allows for a sense of achievement, the thorns change into roses. The site then navigates the woman towards a secure blog site where she can engage in conversation with other women.

This transferral to the blog site is an important part of the affective process. The act of successful agency demonstrated in the game is then hopefully transferred to a space where the woman can actively, in more personal detail, articulate her own counter story and thus start the road to the recovery of her own agency. In reference to the pleasure aspects of video games, Krzywinska states that 'Transformational elements do not simply operate in terms of identity play; becoming more skilled at playing the game, making for a greater sense of agency and acting as an apparent foil to the forces of determination, is also a form of pleasure-generating transformation. (2007: 117). For those playing *Wordslinger*, the transformational elements are the main 'pleasure' providing elements of the game. The woman literally has the chance to transform some of the main tropes of her abuse into a narrative in which she survives and develops. An affective and active articulation of the more personal side of her abuse narrative into the blog will hopefully lead to her eventual recovery and rejection of further abuse. Her embodied narrative will use her counter story rather than base itself on the previous master narrative that is for the main part in relation to those who have suffered abuse, restrictive and damaging.

Conclusion

During the writing of this chapter *The Wordslinger* game was removed for upgrading. At the date of submission it has not yet been re-released so I cannot comment on any changes that may have occurred. However, the usage of a video game as a method of affective development and self-help is one that I hope will continue. Video Games, be it the online variety or those related to the various gaming platforms, are a vastly growing industry and their impact on all levels of popular culture is steadily growing. In her work on Cyberfeminism, Fiona Wilding expresses the hope that 'while affirming new possibilities for women in cyberspace, cyberfeminists must critique utopic and mythic constructions of the Net' (n.d.). Video Game studies perhaps needs to hear this call and focus on the potentially transgressive and alternative strategies, which women are using in their interaction with the gaming world. *Wordslinger*, as basic as it is, offers an affective process for a woman to engage in, and as such presents the potential for video games to develop in alterative and socially important ways. Video games are already being used in the education of children and the affective and active processes that players undergo in the processes of playing games can, as demonstrated in *Wordslinger*, be used to aid individuals in a new and successful fashion.

Notes

1 It should be noted that men as well as women suffer from domestic violence. However male-on-female violence is far more prevalent than female-on-male and youarenotcrazy.com where *Wordslinger* is located is geared towards women only.
2 Embodied Narrative is literally a narrative that is felt and experienced 'in-body'. Semin and Smith (2008) state that the definition of embodiment that is built up on the core idea 'that the nervous systems evolved for the adaptive control of action – not for abstract cognition'. Embodiment can also be considered as an 'interaction of layers: there is the ability to sense a movement or moving from within, there is also the ability to 'think physically', and then a capacity to move with an awareness of the whole physical self' (Chappell 2006).

References

Arzipes (1999) Freedom to create: Women's agender for cyberspace. In Harcourt. (ed.) *Women @ Internet: Creating New Cultures in Cyberspace*. London, Zed Books: Preface.

Bell, D. (2001) *An Introduction to Cybercultures*, London, Routledge.

Bell, D. and B. M. Kennedy (2000) *They Cybercultures Reader*, London, Routledge.

Brennan, T. (2004) *The Transmission of Affect*, Ithaca, Cornell University Press.

Bresler, L. (2006) Embodied narrative inquiry: A methodology of connection. *Research Studies in Music Education*, 27(1): 21–43.

Brooke, B. (1999) *Feminist Perspectives on the Body*, London, Longman.

Bryce, J. and J. Rutter. (2002) Killing Like a Girl: Gendered Gaming and Girl Gamers Visibility. Retrieved 11 September, 2009, from http: //www.cric.ac.uk/cric/staff/Jason_Rutter/papers/cgdc.pdf.

Butler, J. (1990) *Gender Trouble*, London, Routledge.

Butler, J. (1993) *Bodies That Matter*, London, Routledge.

Carr, D., D. Buckingham, A. Burn and G. Schott. (eds) (2006) *Computer Games: Text, Narrative and Play*. Cambridge, Polity Press.

Cassell, J. and H. Jenkins (2000) *From Barbie to Mortal Combat: Gender and Computer Games*, Cambridge, MIT Press.

Chappell, K. (2006) Embodied Narratives. Retrieved 1 January, 2010, from http: //www.wellcome.ac.uk/stellent/groups/corporatesite/@msh_peda/documents/web_document/wtx050349.pdf.

Deleuze, G. and F. Guattari (1987) *A Thousand Plateaus: Capitalism and Schizophrenia*, London, Continuum.

Derrida, J. (1974) *Of Grammatology, Gayatri Chakravorty Spivak*, Baltimore, John Hopkins University Press.

Derrida, J. (2002) *Without Alibi*, Stanford, Stanford University Press.

Dovey, J. and H. W. Kennedy (2006) Games Cultures: Computer Games and New Media, New York, Open University Press.

Dunning, A. and P. Woodrow. (2001) The Einstein's Brain Project. Retrieved 11 September, 2008, from http: //www.ucalgary.ca/~einbrain/new/main.html.

Gee, J. P. (2003) *What Video Games Have to Teach Us About Learning and Literacy*, Basingstoke, Palgrave Macmillan.

Girard, N. (2006) Explaining Disconnect between Women, Video Games. *CNET* Retrieved 5 November, 2008, from http: //news.cnet.com/2100-1043_3-6082459.html.

Gramola, D. (2000) Pain and subjectivity in virtual reality. In D. Bell and B. M. Kennedy. (eds) *They Cybercultures Reader*. London, Routledge: 598–608.

Gregersen, A. and T. Grodal (2009) Embodiment and interface. In B. Perron and M. J. P. Wolf. (eds) *The Video Game Theory Reader 2*. New York, Routledge: 65–84.

Harcourt (1999) *Women @ Internet: Creating New Cultures in Cyberspace*, London, Zed Books.

Harraway, D. (1991) *Simians, Cyborgs and Women: The Reinvention of Nature*, New York, Routledge.

Hayles, Katherine (1999) *How We Became Posthuman: Virtual Bodies in Cybernetics, Literature and Informatics*, Chicago, IL: University of Chicago Press.

Kennedy, H. (2007) Female Quake players and the politics of identity.In B. Atkins and T. Krzywinska. (eds) *Videogame, Player, Text*. Manchester: Manchester University Press: 120–138.

Krzywinska, T. (2007) Being a determining agen in (the) world of warcraft: Text/play/identity. In B. Atkins and T. Krzywinska. (eds) *Videogame, Player, Text*. Manchester, Manchester University Press: 121–32.

Lahti, M. (2003) As we become machines: Corporealized pleasures in video games. In M. J. P. Wolf and B. Perron. (eds) *The Video Games Theory Reader*. London, Routledge.

Lawless, E. L. (2001) *Women Escaping Violence: Empowerment through Narrative*, Columbia, University of Missouri Press.

Lindemann Nelson, H. (2001) *Damaged Identities Narrative Repair*, Ithaca, Cornell University Press.

Lupton, D. (2000) The embodied computer/user. In M. Featherstone and R. Burrows. (eds) *Cyberspace, Cyberbodies and Cyberpunk*. London, Sage Publishing: 97–112.

McLintock, A. (1995) *Imperial Leather: Race, Gender and Sexuality in the Colonial Context*, London, Routledge.

Merleau-Ponty, M. (1962) *Phenomenology of Perception*, London, Routledge.

Mitchell, W. (1995). In F. Lentrichhi and T. McLaughlin. (eds) *Critical Terms for Literary Study*. Chicago, University of Chicago Press: 11–22.

Oatley, K. (1992) *Best Laid Schemes: The Psychology of Emotions*, Cambridge, Cambridge University Press.

Perron, B. and M. J. P. Wolf. (eds) (2009) *The Video Game Theory Reader 2*. London, Routledge.

Price, J. and M. Shildrick (1999) *Feminist Theory and the Body: A Reader*, London, Routledge.

Robinson, S. (1991) *Engendering the Subject: Gender and Self Representation in Contemporary Women's Fiction*, Albany, State University of New York Press.

Semin, G. and E. Smith (2008) *Embodied Grounding: Social Cognitive, Affective and Nuroscientific Approaches*, Cambridge, Cambridge University Press.

Skelly, T. (2009) Foreword. In B. Perron and M. J. P. Wolf. (eds) *The Video Game Theory Reader 2*. New York, Routledge: vii–xix.

Stone, R. A. (2000) Will the Real Body Please Stand Up? Boundary Stories About

Virtual Cultures. In D. Bell and B. M. Kennedy. (eds) *They Cybercultures Reader*. London, Routledge: 504–528.

Taylor, T. L. (2003) Multiple pleasures. *Convergence: The International Journal of Research into New Media Technologies*, 9(1): 21–46.

Ticineto Clough, P. (2000) *Auto Affection: Unconscious Thought in the Age of Teletechnology*, Minnesota, University of Minnesota Press.

Weiss, G. (1999) *Body Images: Embodiment as Intercorporeality*, London, Routledge.

White, M. (2006) *The Body and the Screen: Theories of Internet Spectatorship*, Cambridge, MIT Press.

Wilding, F. (no date) Where Is the Feminism in Cyberfeminism? Retrieved 24 September, 2010, from http: //www.lilithgallery.com/feminist/cyberfeminism.html.

Williams, R. and G. A. Bendelow (1998) *The Lived Body: Sociological Themes, Embodied Issues*, London, Routledge.

Part IV

Conclusion

17 It's not just a game

Contemporary challenges for games research and the internet

Garry Crawford, Victoria K. Gosling and Ben Light

In this final chapter, a number of key issues are raised that came to our attention whilst putting this book together. In doing this, we also reflect upon the seven challenges for video game theory that Perron and Wolf (2009) put forward in the introduction to the second video game theory reader given it is probably one of the most recent assessments in the area at the time of writing. These challenges are concerned with Terminology and Accuracy, History, Methodology, Technology, Interactivity, Play and the Integration of Interdisciplinary Approaches. These issues will be brought up throughout this chapter, but not necessarily in a mutually exclusive fashion.

To begin we consider issues of terminology and accuracy, specifically arguing that we should be careful and explicit about how we define online gaming. Following this we suggest that games of all kinds should be flexibly interpreted in different ways and in taking this approach, histories beyond histories of play are required. For online games, we would suggest that some kind of historical contextualization of such cultural artefacts, in terms of the Internet, and connected artefacts, which are contemporary and precursors to it, should be considered as possibilities. We then move on to consider the issue of methodology, critiquing the idea that games need different research approaches and methods than other artefacts, especially when one considers the calls for interdisciplinarity in game studies. We suggest that for online games at least work undertaken in Internet Studies may prove useful. Our section on technology and the non-human follows where we argue for a (re)consideration of the importance of that deemed technological in game studies. As part of this thesis, we suggest a need to consider not only what we mean by technology but also other non-human things that may play a part in studies of games. Our penultimate discussion concerns the trope of interactivity where we again consider the exclusivity of games and whether there is need for game-specific theories of interactivity. Here we point to other areas, such as, science and technology studies, audience studies and human-computer interaction studies as potentially fruitful bodies of work to draw upon to understand interactivity in games. Also in this section we consider the issue of convergence and how this muddies the water, where games are marked as different without good reason. This further serves to support our critique of games as other things.

Terminology and accuracy

Perron and Wolf (2009) argue first that a set of agreed-upon terms have been slow to develop throughout research in the area. They ask us to consider what we are talking about – videogames, video games, computer games, digital games and so on. Further they point to names for the field, noting that although game studies are broad enough to include boardgames and the like, often it does not. They then move on to argue for accuracy to improve clarity and move the field forward in terms of its accepted terminology. A specific set of examples they give concerns the utilization of the right capitalization and name construction with respect to the names of games under study.

Throughout the course of editing this book, issues of terminology and accuracy have arisen for us. As editors, we have had several conversations about which term to use – online games or online video games, and so forth. Video games obviously tends to be the term favoured in game studies, yet for us, in some ways this tends to privilege the screen – and we know that there is much more to gaming life than what happens there. We have thus chosen the term online gaming in recognition of this. Even then, we are forced to consider, what exactly are we talking about when we discuss online gaming? In order to understand this fully, we have to consider what we mean by online.

The term online has a variety of meanings. Most obviously in common parlance we are referring to engagement with the Internet, or more accurately, the World Wide Web, in some way. But more generally we might think of online as being connected to a computer or a state of something being ready for use. The main point here, is that we are talking about a state of play that is mediated, usually by the Internet, but may be mediated by other forms of connectivity such mobile phone networks or *ad hoc* wireless connections. Whilst the papers in this book generally rely on the Internet, the one thing we have pushed ourselves to do, is to not read online gaming as solely concerned with MMORPGs and even more specifically *World of Warcraft* (*WoW*). Whilst MMORPGs, and indeed *WoW*, are clearly a major part of the online gaming experience – and our contributors reflect this – other kinds of Internet-mediated gaming and play spaces are becoming popular. Importantly, as with other forms of gaming, we know that there is more to online gaming than what happens online. Moreover, as the authors in this book point to, it is important to recognize, and emphasize, the interconnectedness of the virtual and the physical. However, we do not propose to set up 'the' classification of what should count as an online game. As, whilst an ideal type classification system might have a consistent, mutually exclusive and complete set of categories, this is implausible. We know that classification systems are never perfect, they are contentious, and as Bowker and Star (1999) state, they can valorize one point of view whilst silencing others. The challenge therefore, is to be clear about our classification processes and the politics feeding into these.

There is very good reason to challenge the conflation of MMORPGs and online gaming. It is problematic for a number of reasons. First, it politicizes gaming in inappropriate ways. It is easy to say that certain games do not count as 'proper games' if they do not meet the criteria as set out by MMORPGs. Not only does this exclude games that do not fit the mould, it excludes the players of those games, such as *Second Life* and *The Sims* (as discussed in chapter one). Particularly, we would argue, a discourse of deep-seated masculinity can run through such classification processes – proper games are really the games that boys and men play. Other games are for 'girls and sissies'. Second, it leads to problematic analyses. We can miss what is specific to a game or genre of a game and what is a more general tendency across sets of gaming arrangements. However, this is not such an easy task to achieve. For example in this book, Esther McCallum-Stewart discusses social mangles where grinding and chat takes place in *WoW*. Frans Mäyrä also discusses *Fastr* whereby in the multiplayer version of the game chat exchanges among players are possible. We have here two gaming situations where there is, arguably, a shared tendency for the simultaneous enactment of game playing and more general socialization. However, we cannot ignore the specifics of the situations at hand in each game setting. For example, *WoW* involves the use of avatars whereas *Fastr* does not. Thus, whilst there may be general tendencies across games, these may not be instantly apparent as they are obscured by the specifics of a gaming situation. Moreover, the specifics and more general tendencies may not be mutually exclusive as shown in the examples here. Nevertheless, although difficult, it is important to work through such problems.

Games and their histories

Another challenge that Perron and Wolf (2009) set out relates to history. They argue that most academic research in the area of games played in the home and online has focussed upon games developed in the last 5 years; suggesting there is a need for further work that pays attention to older games and the ways in which franchises/series and so forth develop over time. This then proposes there is much to be gained from not just focussing on the latest version of games as if these existed with no predecessor, but rather considering their, and related (such as hardware), developments and histories. We completely agree. Several of the papers in this book consider evolution of game play. For example, Douglas Brown points to the development lineage between *WoW* and *EverQuest* and Neil Randall discusses lineage in a similar fashion, but considering the links between traditional boardgames and their transfer to online environments. Encouragingly, others are also engaging in undertaking these kinds of studies. For instance, Griffiths and Light's (2008) study of *Habbo Hotel* theorizes scamming ethics within the environment, in the context of prior technologies of play, such as *Pokemon* cards and *Beanie Babies*, being associated with the cause of harm as demonstrated in Cook (2001), and also more generally with children's leisure activities, as discussed by Avedon and Sutton-Smith (1971) and Fine (1987).

However, one might argue that for certain 'new' games, there is no precursor, certainly the rise of so-called social/casual gaming has resulted in titles heralding them as a revolution – consider Jesper Juul's *A Casual Revolution: Reinventing Video Games and Their Players*. In the introduction to this book he states 'In the short history of video games, casual games are something of a revolution – a cultural reinvention of what a video game can be, a reimagining of who a can be a video game player' (Juul 2010: 5–7). However, whilst Juul's book title implies revolution and discontinuity, he clearly talks about continuity too, about gamers who stopped gaming and started again for example. He suggests linkages, for example, based on the amount of effort required to learn to play games such as *Pac Man* and *Guitar Hero*. We do not necessarily agree with all of his arguments, but the effort is made here to make historical linkages. Other prior work that interrogates so-called social/casual games also makes historical linkages. Fletcher *et al.*'s ethnographic work around *SingStar* for example, has analysed the game's progression from its origins on the Sony PlayStation 2 platform, to the PlayStation 3 and beyond (Fletcher *et al.* 2008). It has also considered it as a mechanism for socialization in much the same way as boardgames like *Monopoly* (Fletcher and Light 2008). However, perhaps, most helpfully, they discuss *SingStar* and its relationship to music (Light and Fletcher 2009). They suggest that we need to recognize the place of gaming in relation to other media, suggesting that to date, music and gaming are only usually discussed in relation to the improvement of the 'in game' experience for the player. Instead, they conceptualize *SingStar* with direct reference to karaoke as a polystratic intertextual form and unpack it, not only as a game, but as glue, facilitating sociality; a mechanism for identity work; a vehicle for performance, celebrity and social grooming, and a gateway to, and influencer of, musical tastes. Such an analysis demands that we consider multiple perspectives on the histories of games. At present we would argue that histories of games have a 'from *PONG* to PS3' mentality. Whilst such analyses are incredibly valuable, and studies of games should continue to be rooted in histories of play, we argue this only takes us so far, given the comments we make here and also in the light of the emergence of serious games such as *Wordslinger* as discussed by Kate Taylor in this book. Specifically, we feel that online gaming would benefit from being contextualized and theorized in the context of the history of the Internet, and indeed, to complicate things further, the potential histories associated with the Internet as a cultural artefact.

Methodology and the terms of engagement

Perron and Wolf point to methodology as being a challenge for game studies. For instance, they report that methodologies will vary depending upon the purpose of the research, that different games will offer different gaming experiences, that gamers will have different experiences of games and that we need to ask how analysis might be affected if we do not consider a particular feature of a game? We do no disagree with any of these points, in as much as methodology is a challenge for any area of study. However, we believe that we need to undertake further work here if game studies is seemingly being positioned as needing to 'get its methods right'. It seems to us that Perron and Wolf's discussion is constructed in a way that privileges the game, and the act of game play, and we argue that this is not always the most helpful approach. For example, they state that 'before exercising analytical or interpretive skills, one has to draw on one's ability to play a game (or know someone with the ability)' (Perron and Wolf 2009: 11). Aarseth (2003: 3) also reinforces this view:

> For any kind of game, there are three main ways of acquiring knowledge about it. First, we can study the design, rules and mechanics of the game, insofar as these are available to us, e.g., by talking to the developers of the game. Second, we can observe others play, or read their reports and reviews, and hope that their knowledge is representative and their play competent. Third, we can play the game ourselves. While all methods are valid, the third way is clearly the best, especially if combined or reinforced by the other two. If we have not experienced the game personally, we are liable to commit severe misunderstandings, even if we study the mechanics and try our best to guess at their workings.

Obviously, games and game play should be taken seriously in game studies. However, we have to ask what else should be? Earlier we suggested that different histories of games need to be constructed based on different readings of what they are. If this approach is followed then it may not be necessary to have the game or acts of game play as the central focus and indeed, it may not be desirable at all. For example, in this book, Aphra Kerr's argument that questions the symmetry of the democratization of innovation processes within the games industry does not require her to go and play games. She is viewing games as products and considering their development trajectory in much the way that scholars in Science and Technology Studies have been doing for around 30 years. These scholars study the historical development of artefacts – famously the bicycle (Pinch and Bijker 1987) and fluorescent lighting (Bijker 1994) alongside contemporary sets of socio-technical arrangements such as Feminism and the Internet (Adam 2005). Within this community, although ethnography is deployed, no one demands that a lived experience is the gold standard – especially as it is sometimes neither possible nor desirable. Adam (2005), for example, discusses the Internet with respect to cyber stalking – she is not criticized for not doing it herself. Positions that suggest that the best way to understand something 'fully' is to do it, are tinged with positivism and as Knights (1997) observes, the demand for exhaustive and complete explanations is a deeply masculine construction; a demand that should be resisted.

Further, underlying the discourse that game studies must agree acceptable methods, is a desire to set games apart as different to other things. Whilst they clearly are different in certain respects, and obviously some of this discourse is rooted in challenging attempts to see games as 'like films', if we stay open to understanding games as different things, further insights could be achieved. Games are like other things. For example, Perron and Wolf (2009) state that different games offer different experiences and gamers will experience games in different ways. At a recent workshop, Nikunen *et al.* (2008) was discussing extreme pornography and the Internet. Early on she discussed the different categories of pornography and how they offered different experiences to different people. She then went on to give an example of a specific porn trailer that was re-appropriated via *YouTube* users to give it a viral-like quality. Essentially, the extreme porn trailer became the subject of user-generated videos that depicted the responses of various people to it – disbelief, horror and disgust being common responses. So the point here, is that we have a particular set of sociotechnical arrangements that come in different forms and offer different experiences to different users – just like games. If we accept that games can be different things and be parts of different sets of arrangements, then as Perron and Wolf's (2009) discussion of the interdisciplinary challenge for games in their appendix suggests, we would do well to consider the methodological norms associated with these, rather than trying to re-fight old battles. Specifically then, as related to our concern in this book, there are a plethora of accepted methods associated with those studying the Internet; these should be considered before deciding whether different rules are needed for studying online games.

Technology and the non-human

A further consideration for Perron and Wolf is the need to understand technology and its development to better understand why games are the way they are. Here they refer to such things as algorithms, processing speed, storage capabilities and access speeds, with reference to their effects on game design and experiences of play. Notably, they discuss how specific hardware configurations are required in order for a game to be accurately represented in its original form. From this they argue that a 'technological context' is necessary for understanding and researching games. Moreover, they argue that although technological features may not be relevant to certain analyses, researchers will not be able to determine this without understanding them in the first place. There is much to be gained from unpacking what we would see as sociotechnical arrangements of gaming, rather than merely technology. Paul and Philpott, in this book for example, do an excellent job of interweaving the roles of servers and players in the rise and downfall of a *WoW* guild, and similarly, Lin and Sun, enrol a consideration of servers to consider the dynamics of 'unofficial' MMORPG gaming. Yet there is more we can add to this discourse.

Whilst that which gets labelled as technology is important, we have to be careful not to privilege this in much the same way as we have discussed in relation to methodology. Specifically, whilst it is important to unpack sets of sociotechnical arrangements, we have to be clear that it is impossible to have an understanding of the whole situation. To think otherwise resonates with the design fallacy – the presumption that to meet user needs we must build ever more extensive knowledge about the specific context and purposes of various users into technology design (Stewart and Williams 2005). Stewart and Williams argue that the problem with this thinking is that it privileges prior design, it is unrealistic and simplistic, it may not be effective in enhancing design or use and it overlooks opportunities for intervention. We know that games are not delivered as complete solutions, which are sufficiently specified *a-priori*. We know games are produced in use. This idea of ongoing development in use is well known within the body of work known as the social shaping of technology (Fleck 1994; Rohracher 2005; Stewart and Williams

2005). Therefore, we should not expect all researchers to unpack games to the level of, say, the algorithms they are operating. Yes, this might mean that certain things go unanalyzed, but there will always be things missing – we have to accept that. It is in the hands of researchers to decide, based on the task at hand and the extensive literatures we have on methodology, exactly what methodology is appropriate for their study. From this, they can then rigoursly create a plausible account of their work that others can choose to find more or less interesting.

In much the same way as we have discussed the need to consider what we mean by online gaming, it is important to consider what is meant by technology. Perron and Wolf's (2009) conceptualization is very much grounded in computing, yet we know that so-called technologies can be much more than that – ideas, houses, microwaves and methods all can be thought of in this way. Even in a gaming context, we need to think of technologies in a much broader fashion. Perron and Wolf quite rightly point to the potential roles of peripherals in mediating gaming activity, which goes some way to extending the possibilities of what to consider, but we need to go further. In this book Astrid Ensslin, considers the technology of language and how this can affect the gaming experience, Frans Mäyrä enrols photography in his discussion of *Flickr* based gaming, and in two other chapters, Douglas Brown and Aprha Kerr implicate market forces. Beyond the studies in this book, others have pointed to, the roles of different things in shaping experiences of gaming. For example, Crawford (2006) talks of the role of public houses as an intermediary of football manager games, recreational drugs, alcohol, food, acoustic guitars, mobile phones and even different homes are enrolled in Fletcher and Light's *SingStar* ethnographies (2008; 2008; 2009) and costumes are brought to bear in cosplay as discussed by Newman (2008). Such examples, suggest that we should not only think about what might be seen as technologies, they encourage us to think about other things that are non-human, but that contribute to the gaming experience.

Interactivity and convergence

Perron and Wolf (2009) set out the trope of interactivity in game studies, acknowledging its problematic but pervasive nature. They suggest a comprehensive theory of interactivity is required to consider how the interaction of a game is designed and the games options and choices are structured. However, they also ask if there are universal statements and claims about interactivity that will hold up in the light of future innovations? It seems to us that no one theory of interactivity is going to help us understand the gaming experience, online or otherwise. In this book alone for example, Keith Massie, arguably approaches the subject of interactivity with respect to the gender and race identity work possible in *EverQuest*, Fern Delamere considers the potentials for the generation of social capital by disabled people in *Second Life* and Anders Drachen considers interactions in the form of communication amongst gamers playing different forms of games. Interactivity can, and often does, mean different things to different people, and can refer to human-to-human interaction, human to computer interaction, or much more beyond.

Rather than trying to invent new theories of interactivity, game studies may well be better served by considering the wealth of theory that surrounds the appropriation of different sets of arrangements, each of course having its strengths and weaknesses. For example, Aphra Kerr, in this book, draws on science and technology studies (STS) to understand the potentials for the engagement of players in games innovation and it would seem that this body of work has a lot to offer game studies more generally when considering discourses of interactivity. Contemporary STS researchers reject technologically, and socially, deterministic accounts of the construction and appropriation of technologies and focus on the mutually constitutive nature, and negotiable boundary between, society and technology [for overviews see, (Pinch and Bijker 1987; Bijker and Law 1994; Williams and Edge 1996; Mackenzie and Wajcman 1999; Sørensen 2002)]. From this perspective, technology applications do not have predictable, universal outcomes. Instead, technologies are conceptualized as being shaped as they are designed and used depending upon who is, and is not, involved along the way. Therefore, whilst they may change situations, the technologies themselves may be subject to change; resulting in intended and unintended consequences for that deemed the technology, and that deemed the social arrangements it is meshed with.

A further strand of course, is concerned with media audiences. Some writers on video games (Kline *et al.* 2003; Jenkins 2006a; 2006b) have highlighted the similarities between gamers and other media audiences, and indicated the potential benefits of taking an approach, which locates video gamers within a history of media audiences (of one kind or another). However, within the video game literature, little effort has gone into systematically and fully setting out the useful arguments and developments that audience literature has gone through, and how these both mirror and help inform, many contemporary game studies debates. This, means that many critics of a media studies approach to video games, have quite easily, and often flippantly, been able to reject the idea that other media audiences are active in their consumption, and hence comparable in any way to video gamers. For instance, as (Frasca 2003: 227) forcefully argues: 'no matter how badly literary theorists remind us of the active role of the reader, that train will hit Anna Karenina and Oedipus will kill his father and sleep with his mother'. Of course, all texts, such as books and films, do provide certain structural parameters, which most readers, most of the time, agree upon and adhere to. But so do video games. For most (if not virtually all) readers who finish reading Tolstoy's *Anna Karenina*, the principle character of the book *will* die. However, similarly, a gamer reaching the end of *Super Mario Bros.* will rescue the Princess. There is no option to change your mind here or perform another action; the end sequence, as in most video games, is as set and as rigid as any book (Crawford 2011).

A final example of work we might draw upon to engage in the analysis of interactivity is human-computer interaction studies (HCI). This field offers a wealth of theory that raises issues around user interfaces. This interface comprises software and hardware – the former thus might include how characters and objects are displayed on the screen, the later how peripherals mediate the user experience. Games of all kinds have these kind of issues associated with them and thus there is no reason not to take these as a starting point at least for developing nuanced understandings of interactivity, within the limitations of a HCI focus, as related to this area – even broadly conceived. For example, *Scrabble* as a boardgame has a structure to how characters and objects are displayed – the board acting as a screen is one such example, it also has peripherals that mediate the player experience, the bag in which the game's tiles are taken from for instance. The *Dance Party* set of games for the Nintendo Wii similarly deploy a screen comprising characters and objects coupled with peripherals such as the Wii-mote and a dance mat.

At the heart of this debate again seems to be the extent to which games are different from other things and therefore the importance that is placed on theorizing games and game studies differently. A significant problem here, for those wishing to differentiate games is media convergence. Games in a very general sense undoubtedly display differences to things we might want to compare them with. However, gaming is increasingly converging with other things – one only has to look at the Sony PS3 as a console, music player, video player and access point to the Internet. This is something Perron and Wolf (2009) are also keen to point out. More specifically, for online gaming such convergence perhaps is further problematising our ability to differentiate gaming and games from other activities and things. As Griffiths and Light (2008: 456) conclude with respect to *Habbo Hotel*:

> Habbo originally started out as a game but has since evolved to include social networking functionality. Moreover, both classes of these technologies have evolved to incorporate features of the other. Thus, social networking includes gaming and gaming includes social networking.

Frans Mäyrä's work in this book amply illustrates this 'problem' too. Mäyrä discusses contextual play, whereby playful behaviours, which are rooted in or emerge from social relations and exchanges of information, are used to maintain and expand such networks of relationships. He argues for the convergence of playful activity in and amongst other activity – crucially among online sites that are not defined primarily as an MMORPG. Thus, he demonstrates how game play is augmented by other activities afforded by the sites in question – such as, status updates in *Facebook* are possible whilst playing *Mafia Wars*. Such 'problems' of convergence are not just restricted to the uptake of so-called Web 2.0-based social media. From a different perspective, Aphra Kerr, points to the potential for the implicit gathering of data on user or player behaviour that is now available that can be easily collected and analysed. Taking this further, not only might we think about this as data collected via the 'in game' experience, one only has to look at fan wikis, forums and *Facebook* groups to see the potentials brought about by convergence here – user generated or otherwise. The *SingStar* VIP lounge hosted by the *SingStar* development team on *Facebook* is a very good example of this – it is a key mechanism by which the developers collect data about users; in addition to anything they might collect via the data generated by players in game. If we consider convergence from a different angle, a visit to humplex.com reveals links to various flash games that arguably converge with ideas of pornography. To make the gaming lineage clear, a series of masturbation games have been made which centre on characters from *Metal Gear Solid* (Solid Snake), *Resident Evil 5* (Chris Redfield) and *Final Fantasy X* (Wakka). Not only does this example make us consider interactivity and appropriation, convergence is enrolled in terms of identity. Solid Snake is rendered solider and sex object, the gamer can be seen as many things other than gamer and by many other people depending upon how they choose and are able to interact with Solid Snake and where.

In the light of such wide ranging issues associated with interactivity and convergence then further issues for the study of online games, if not games more generally, are raised. The examples above highlight the need for more comparative studies of different gaming situations. For online gaming, we also argue there is a need for multisided studies, given, at least in the context of the discussion here, the multiple potentials for different experiences of interactivity and convergence. In this case, Celia Pearce's work makes an important contribution as a story of coming together, migration, divergence and multisitedness.

Conclusion

This book demonstrates the value of considering online gaming on its own terms, as long as a nuanced contextualized approach is taken and they are not treated as a new class of cultural artefact that can only be understood in terms of play. It is important to consider what is meant by being online; because of course the advancement of Internet access through broadband, in the developed world at least, has facilitated, for some, increased access to a greater number of online games. Moreover, the recent increase in membership of Web 2.0-based sites has facilitated the creation of markets for companies such as Zynga and titles such as *Mafia Wars* as well as the transfer of older games to this platform, such as *Bejeweled.*

The online element is also particularly, and increasingly, important if we position game studies as concerned with the gaming experience generally and not just that occurring when we are engaged with games themselves. Historically, we know that many games have had online fan sites and forums. However, we argue that these have, in the main, been populated by what might be seen at the so-called stereotypical gamer – a particular kind of nerd as it were. Associated with this were, and continue to be, particular masculine and socially exclusive discourses, where credit and status is given for how much you know about a game, rather than how much you enjoy playing it. Yet at the same time we are witnessing the increased participation of new and returning gamers to the gaming arena through games such as *SingStar*, *Lips*, *Guitar Hero*, *Dance Dance Revolution*, *Wii Sports* and *Bejeweled.* All of these games have fan sites on *Facebook* for example, and through this, because of the general trend towards identity authenticity in the site, it is possible to gain insights into the diversity of people engaged in these games and also a diversity of discourses that permeate these sites. In many ways these resonate with extant games forums elsewhere on the web – there are, for instance, discourses of competition, arguments with games developers and invites to battle online. However, these sites seem different given there explicit and inextricable ties into other aspects of people's lives. This is an area definitely deserving of further attention. If we are to say that on *Facebook*, often (but not always) people are who they say they are, then is it okay that your manager can see that at 13.50 you just bit someone in *Vampire Wars*? A press release by MyJobGroup (2010) in the UK indicates it might not be, since such activity is costing the British economy, apparently, 14 billion pounds a year.

It is important to pay greater attention to the intersection of games and the gaming experience with other things. This is important beyond any of the points we made earlier in the chapter, regarding convergence, defining games and discourses of interactivity. The *Vampire Wars* case engages with notions of strict boundaries usually constructed between work and play. Moreover, they remind us to think about the context for the emergence of particular online gaming situations. To provide a different example, *Dr Who* has been running as a television programme since 1963 and at the time of writing in August 2010 its maker, the BBC, has just launched the new adventure games which run as part of its 360 degree programming strategy. Thus, it is not always the game that comes first, a particular experience may come first, which leads to the generation of gaming activity – and this point takes us back to the need to historically contextualize gaming in different ways.

Our closing point is that online gaming makes it even clearer that the study of games is political. Investigating such processes is important given that the sharing and generation of knowledge has a political dimension. For example, even though we know that there are different kinds of knowledge, and that it is provisional and situated in nature, (Blackler 1995; Fleck 1997; Marshall and Brady 2001; Sutton 2001; Mannheim 2004) somehow the ability to 'level up' in an MMORPG is often given more legitimacy than the knowledge required to gain a top score on *Hello Kitty Roller Rescue*. We thus think that greater research into what knowledge is constructed, in relation to online games and games more generally, is required. Specifically, as early science and technology studies projects attended to the process of making science, game studies might do well to consider how and why particular forms of knowledge are made.

We feel privileged to include scholars in this book who have engaged with discourses of social inclusion around age, gender, race, ethnicity and disability. However, we argue that this work is still in the minority and even where it does exist, social markers, such as, gender, race and disability, are often treated in an instrumental fashion, as variables. As, for example, Jenson and De Castell (2008: 24) point out in their paper on gender and gaming, it is necessary: "to un-learn stereotypical assumptions and challenge covertly stereotyped concepts (such as "competition") that have thus far driven gender research".

It is therefore felt that further knowledge on social inclusion and gaming experiences are needed to add greater depth to game studies. Additionally, we would also like to see further studies around queer gamers and sexuality and gaming – again a very underrepresented area with few people such as Sunden (2009) working in that area. Moreover, in the UK for example, it has been reported that 30 percent of the population still do not have, or desire, access to the Internet (Dutton *et al.* 2009). Further, those on lower incomes and with a basic level of education were cited as the least likely to use the Internet. We doubt the United Kingdom is in a unique position here and we need to ask questions about class (or the equivalent classification processes) and gaming patterns. Further, research needs to expand beyond a solitary focus on Europe and North America. Though we know of, and studies frequently refer to, gold farmers in China and Vietnam, it is difficult to find studies of, say, gaming in Ghana.

It is perhaps from this point then that the politics of gaming needs to be brought to the fore. Though discourses of addiction and aggression should be challenged, this does not mean that gaming should be ubiquitously celebrated. Also, we have to be careful that we do not position gaming as absolutely, universally and necessarily central to everyone's everyday life. Video games matter, but not to everyone, nor should they. We cannot, for instance, immediately read any potential absence of online gaming in Ghana as digital exclusion. Of course those who can and want to, can get much from playing digital games, but it is not the only way to, for example, to make money, play, learn and socialize.

References

Aarseth, E. (2003) Playing Research: Methodological Approaches to Game Analysis. *Proceedings of the Digital Arts and Culture Conference*. Melbourne, Australia.

Adam, A. (2005) *Gender, Ethics and Information Technology*, Basingstoke, Palgrave Macmillan.

Avedon, E. M. and B. Sutton-Smith (1971) *The Study of Games*, New York, John Wiley.

Bijker, W. E. (1994) The social construction of fluorescent lighting, or how and artefact was invented in its diffusion Stage. In W. E. Bijker and J. Law. (eds) *Shaping Technology/Building Society: Studies in Sociotechnical Change*. Cambridge, MIT Press: 75–104.

Bijker, W. E. and J. Law (1994) *Shaping Technology/Building Society: Studies in Sociotechnical Change*, Cambridge, MIT Press.

Blackler, F. (1995) Knowledge, knowledge work and organizations: An overview and interpretation. *Organization Studies*, 16(6): 1021–1046.

Bowker, G. C. and S. L. Star (1999) *Sorting Things Out: Classification and Its Consequences*, San Diego, MIT Press.

Cook, D. T. (2001) Exchange value as pedagogy in children's leisure: moral panics in children's culture at century's end. *Leisure Sciences*, 23(2): 81–98.

Crawford, G. (2006) The cult of champ man: the culture and pleasures of championship manager/football manager games. *Information Communication and Society*, 9(4): 496–514.

Crawford, G. and J. Rutter (2007) Playing the game: Performance in digital game audiences. In J. Gray, C. Sandvoss and C. L. Harrington. (eds) *Fandom: Identities and Communities in a Mediated World*. New York, New York University Press: 271–81.

Crawford, G. (2011) *Video Gamers*, London: Routledge.

Dutton, W. H., E. J. Helsper and M. M. Gerber (2009) *The Internet in Britain 2009*. Oxford, Oxford Internet Institute, University of Oxford.

Fine, G. A. (1987) *With the Boys*, Chicago, University of Chicago Press.

Fleck, J. (1994) Learning by trying: the implementation of configurational technology. *Research Policy*, 23(6): 637–52.

Fleck, J. (1997) Contingent knowledge and technology development. *Technology Analysis and Strategic Management*, 9(4): 383–97.

Fletcher, G. and B. Light (2008) Tech's, Drugs and Rock and Roll: Technological Complicity in Domestication of Gaming. *Paper Presented at 4S-EASST 2008*. Rotterdam, the Netherlands.

Fletcher, G., B. Light and E. Ferneley (2008) Access All Areas? The Evolution of Singstart from the Ps2 to Ps3 Platform. *Paper Presented at IR 9.0, the Conference of the Association of Internet Researchers*. Copenhagen, Denmark.

Frasca, G. (2003) Simulation versus narrative: Introduction to ludology. In M. J. P. Wolf and B. Perron. (eds) *The Video Game Theory Reader*. London, Routledge: 221–35.

Griffiths, M. and B. Light (2008) Social networking and digital gaming media convergence: classification and its consequences for appropriation *Information Systems Frontiers*, 10(4): 447–459.

Jenkins, H. (2006a) *Convergence Culture: Where Old and New Media Collide*, New York, New York University Press.

Jenkins, H. (2006b) *Fans, Bloggers and Gamers: Exploring Participatory Culture*, New York, New York University Press.

Jenson, J. and S. De Castell (2008) Theorizing gender and digital gameplay: Oversights, accidents and surprises. *Eludamos. Journal for Computer Game Culture*, 2(1): 15–25.

Juul, J. (2010) *A Casual Revolution: Reinventing Video Games*, Cambridge, The MIT Press.

Kline, S., N. Dyver-Witherford and G. De Peuter (2003) *Digital Play: The Interaction of Technology, Culture, and Marketing*, London, McGill-Queen's University Press.

Knights, D. (1997) Organisation theory in the age of deconstruction: dualism, gender and postmodernism revisited. *Organisation Studies*, 18(1): 1–19.

Light, B. and G. Fletcher (2009) Playing to Sing? Ethnographies of Music as Gaming. *Paper Presented at The British Sociological Association Annual Conference*, Cardiff, UK.

Mackenzie, D. and J. Wajcman. (eds) (1999) *The Social Shaping of Technology*. Maidenhead, Open University Press.

Mannheim, K. (2004) The sociology of knowledge and ideology. In C. Lemert. (ed.) *Social Theory: The Multicultural and Classic Readings*. Oxford, Westview Press: 213–16.

Marshall, N. and T. Brady (2001) Knowledge management and the politics of knowledge: Illustrations from complex products and systems. *European Journal of Information Systems*, 10(2): 99–112.

MyJobGroup.co.uk. (2010) Social Media Costing UK Economy up to £14billion in Lost Work Time. Retrieved 4 August, 2010, from http: //myjobgroup.co.uk/media-centre/press-releases/social-media-costing-uk-economy-up-to.shtml.

Newman, J. (2008) *Playing with Videogames*, London, Routledge.

Nikunen, K., S. Paasonen and L. Saarenmaa (2008) *Pornification: Sex and Sexuality in Media Culture*, Oxford: Berg.

Perron, B. and M. J. P. Wolf (eds) (2009) *The Video Game Theory Reader 2*. London, Routledge.

Pinch, T. and W. E. Bijker (1987) The social construction of facts and artifacts: Or how the sociology of science and the sociology of technology might benefit each other. In W. E. Bijker, T. P. Hughes and T. Pinch. (eds) *The Social Construction of Technological Systems*. London, The MIT Press: 17–50.

Rohracher, H. (2005) From passive consumers to active participants: The diverse roles of users in innovation processes. In H. Rohracher. (ed.) *User Involvement in Innovation Processes: Strategies and Limitations Form a Socio-Technical Perspective*. Wien, Profil: 9–35.

Sørensen, K. H. (2002) Social shaping on the move? On the policy relevance of the social shaping of technology perspective. In K. H. Sørensen and R. Williams. (eds) *Shaping Technology, Guiding Policy: Concepts, Spaces and Tools*. Cheltenham, Edward Elgar: 19–35.

Stewart, J. and R. Williams (2005) The wrong trousers? Beyond the design fallacy: Social learning and the user. In H. Rohracher. (ed.) *User Involvement in Innovation Processes: Strategies and Limitations from a Socio-Technical Perspective*. Wien, Profil: 39–71.

Sunden, J. (2009) Play as Transgression: An Ethnographic Approach to Queer Game Cultures. *Proceedings of DiGRA: Breaking New Ground: Innovation in Games, Play, Practice and Theory*, Uxbridge.

Sutton, D. C. (2001) What is knowledge and can it be managed? *European Journal of Information Systems*, 10(2): 80–88.

Williams, R. and D. Edge (1996) The social shaping of technology. In W. H. Dutton. (ed.) *Information and Communication Technologies: Visions and Realities*. Oxford, Oxford University Press: 53–67.

Index

Taylor & Francis
eBooks
FOR LIBRARIES
ORDER YOUR FREE 30 DAY INSTITUTIONAL TRIAL TODAY!
Over 23,000 eBook titles in the Humanities, Social Sciences, STM and Law from some of the world's leading imprints.
Choose from a range of subject packages or create your own!
Benefits for you
Free MARC records
COUNTER-compliant usage statistics
Flexible purchase and pricing options
Benefits for your user
Off-site, anytime access via Athens or referring URL
Print or copy pages or chapters
Full content search
Bookmark, highlight and annotate text
Access to thousands of pages of quality research at the click of a button
For more information, pricing enquiries or to order a free trial, contact your local online sales team.
UK and Rest of World: online.sales@tandf.co.uk
US, Canada and Latin America: e-reference@taylorandfrancis.com
www.ebooksubscriptions.com
ALPSP Award for BEST eBOOK PUBLISHER 2009 Finalist
Taylor & Francis eBooks
Taylor & Francis Group
A flexible and dynamic resource for teaching, learning and research.